MICROELECTRONICS EDUCATION

# MICROELECTRONICS EDUCATION

Proceedings of the 3rd European Workshop on Microelectronics Education

Hosted by ST University (ST Microelectronics), France, May 18&19, 2000

Edited by

B. COURTOIS
*CMP/TIMA, Grenoble, France*

N. GUILLEMOT
*CIME/INPG, Grenoble, France*

G. KAMARINOS
*LPCS/CNRS, Grenoble, France*

and

G. STÉHELIN
*ST University (ST Microelectronics), Fuveau, France*

SPRINGER-SCIENCE+BUSINESS MEDIA, B.V.

A C.I.P. Catalogue record for this book is available from the Library of Congress.

ISBN 978-90-481-5518-7 ISBN 978-94-015-9506-3 (eBook)
DOI 10.1007/978-94-015-9506-3

*Printed on acid-free paper*

# Contents

## WOMEN IN ELECTRICAL ENGINEERING



## INTERNATIONAL PERSPECTIVES



## MICROSYSTEMS

## POSTER SESSIONS

### MULTIMEDIA INNOVATIONS

# Preface

This is the third edition of the European Workshop on Microelectronics Education (EWME).

A steady-state regime has now been reached. An international community of university teachers is constituted; they exchange their experience and their pedagogical tools. They discuss the best ways to transfer the rapidly changing techniques to their students, and to introduce them to the new physical and mathematical concepts and models for the innovative techniques, devices, circuits and design methods.

The number of abstracts submitted to EWME 2000 (about one hundred) enabled the scientific committee to proceed to a clear selection. EWME is a European meeting. Indeed, authors from 20 different European countries contribute to this volume. Nevertheless, the participation of authors from Brazil, Canada, China, New Zealand, and USA, shows that the workshop gradually attains an international dimension.

The $20^{th}$ century can be characterized as the "century of electron". The electron, as an elementary particle, was discovered by J.J. Thomson in 1897, and was rapidly used to transfer energy and information. Thanks to electron, universe and micro-cosmos could be explored. Electron became the omnipotent and omnipresent, almost immaterial, angel of our World. This was made possible thanks to electronics and, for the last 30 years, to microelectronics. Microelectronics not only modified and even radically transformed the industrial and the every-day landscapes, but it also led to the so-called "information revolution" with which begins the $21^{st}$ century.

It is impossible to imagine the "planetary village" and the global "new economy" without the possibilities of information networks and the multimedia. EWME 2000 gives an important place to these tools. They are now considered as essential for an efficient though friendly teaching. Are they the substrates or the sources of an "educational revolution"?

Every one of us knows that the integration of new tools in an existent environment strongly depends on the educational, social, and cultural characteristics of the users (students). It is dangerous to accept to be driven by the tool instead of adapting it to the users conditions and to their real needs. Evidently, only teachers can rapidly adapt their address to the ever changing mind of a class-room.

Microelectronics is in the heart of Sciences of Information Technologies and Communications (STIC). Information has to be transformed by education in knowledge; information is a tool for any kind of power; information is, also, a product for the global market. It is therefore "natural" that information science and technologies are a new field of antagonisms and various struggles. The session concerning the role of women in electrical engineering points out a genuine economic and social problem: the effective parity of men and women in an advanced, efficient, and modern society.

EWME 2000 is held in Aix-en-Provence, an ancient town founded by Romans in 123 BC. Astounding monuments of the Roman period remind visitors of the Mediterranean roots of our civilization. A civilization which, up to now, has exploited nature for the development of techniques, but was also able to find, admire, and represent (by the Arts) the beauty of this nature. The various representations of the "Montagne Sainte-Victoire", mountains in the Arc Valley near Aix-en-Provence, painted by the great Paul Cézanne, are definitely of an exquisite beauty. Are we able to transmit to our students the beauty of the representation (simulation) of an electronic wave function or the beauty of a superb "patchwork" of an integrated circuit?

B. Courtois N. Guillemot G. Kamarinos G. Stéhelin

# ORAL PRESENTATIONS

# DESIGN TEACHING

# IN THE FUTURE

# ENHANCING SYSTEM-LEVEL EDUCATION WITH REUSABLE DESIGNS

D. W. BOULDIN
*Electrical & Computer Engineering*
*University of Tennessee*
*Knoxville, TN 37996-2100 U.S.A.*
*dbouldin@utk.edu*

## Abstract

Now that silicon manufacturing has enabled the cost-effective production of single-chip implementations for what previously were considered full systems, designers are increasingly reusing significant portions of previous designs. Thus, today's integrated circuits require that designers create and integrate reusable building blocks or intellectual property (IP). It is, therefore, imperative that universities provide the appropriate courses so that more students can become system-level designers. Curricula need to be updated at several points to include team projects. Universities need to collaborate with one another to share course materials, intellectual property blocks and ideas. Several possibilities for collaboration are presented and discussed.

## 1. Introduction

Multi-million transistor integrated circuits can now be produced cost-effectively [1]. Thus, designers are faced with the challenge of creating and verifying the content of these chips as quickly as possible in order to reduce the time-to-market. It has been estimated that a one-month delay in bringing a product to market can result in a loss of ten percent of the potential revenue [2]. Hence, not all of the transistors on these chips can be customized but instead must be ported from previous designs. These reusable cores or intellectual property (IP) blocks include CPUs (like ARM, MIPS and Sparc), MPEG decompression engines, PCI bus controllers, specialized DSPs, etc. Combining several complex cores using gates and standard cells is much more manageable and quicker than designing millions of transistors one at a time.

Creating and integrating reusable blocks is a system-level design style that must be adopted and implemented rigorously in our universities. Curriculum needs to be updated at several points to include team projects. However, totally revolutionizing curricula is extremely difficult with all of the pressures from other



*B. Courtois et al. (eds.), Microelectronics Education,* 5–8.

groups to include their own items of interest. However, it is relatively straightforward to adapt or upgrade existing courses to provide for formal training and experiences in system-level design.

These may include:

*Advanced graduate-level electives* can focus on open-ended design problems, use the latest CAD tools and the most advanced IP to provide students with experiences in creating and integrating IP.

*Senior undergraduate capstone sequences* can focus on more constrained problems which involve a mixture of hardware and software, thereby giving students experience integrating existing IP with a limited amount of time devoted to creating new IP.

*Freshman introductory design courses* can focus on highly supported and bounded projects which provide motivation to attract students into the more advanced curricula.

Universities need to collaborate with one another to share course materials, intellectual property blocks and ideas. Existing and potential collaborative efforts are discussed in this paper.

## 2. Requirements for Reusable IP

The myth that characterises today's IP is that these components are blocks that have well defined contents and interfaces. However, they are often fuzzy and hence appear more like patches in a quilt which must be stitched together. The components cannot be assembly blindly and rapidly but rather must be carefully pieced together to form a working system.

Therefore, design for reuse does not come free. Rather it involves much more in-depth documentation and characterisation than for a design that is not intended to be reused. Based on the experiences of software engineers [3], it is estimated that preparing a component for reuse will require about 50% additional effort. Once this has been done, the designer who is reusing the component may naively think that his design time for that component will be reduced to zero. But alas, he must take care to understand fully how the component works and how it should be integrated with other components. Again from the experiences of software engineers [3], the second design generally requires about 30% of that required to produce the component originally. Thus, the reuse is not for free but does make a significant (70% reduction) impact on the next design.

The information required to document soft IP consists of far more than just the source code. Also needed are: (1) functional description, (2) application intent, (3) interface specifications, (4) authors and owners, (5) packaging information, (6) input stimuli and output responses (test bench), (7) tools and versions used, (8) FPGA or ASIC foundry used for fabrication, (9) size, delay and power measurements, and (10) testability features including BIST, JTAG and SCAN.

## 3. Candidate Projects

In our advanced graduate electives at the University of Tennessee [4], the initial course in a two-semester sequence provides the students with the opportunity to learn how to synthesize small pieces of HDL source code into FPGAs. In the second semester, larger projects are assigned that require a team of generally four students to implement. The application requirements are first presented in narrative form and the team members must partition the design into manageable modules. Each module is the responsibility of an individual to capture in VHDL, synthesize and simulate as well as verify in an FPGA. Once each student believes his design is "known good", the team then integrates the components into a single-chip ASIC. Obviously, any deficiencies not already corrected by individual designers must be dealt with during this integration or design with reuse phase. It is not unusual for a student to recognize that the quality of his component or his documentation is substandard and hence some redesign or additional documentation is performed until the full system works properly.

Projects following the model just described are intended to provide individual students with the experience of designing *for* reuse and the team of designers with the experience of design *with* reuse. Example projects completed or underway include: Wavelet Image Compression, Huffman Encoding, LZ Data Compression, Discrete Cosine Transform, Fast Fourier Transform, CORDIC 2-D Vector Rotation, Automatic Target Recognition, Constant False Alarm Rate, Data Encryption, and Boolean Satisfiability.

## 4. Potential Collaboration

The development of large team projects like those described in the previous section are marked by the partitioning of the full design into several manageable modules. Presently in our classes, all of the modules are being designed "live" by students in the same class on the same team. However, a tougher standard for design for reuse and design with reuse would involve having one (or more) of the components be taken from a previous course. Thus, the system integrators would not have the benefit of being able to talk with the original author to gain more information or to ask for portions to be redesigned. Moreover, the IP component could potentially be obtained from a student in a course taken at another university. The component could have been designed the year before or even "live" using the Internet for collaboration.

All of this discussion brings me to one of the central points of this paper: namely, universities should collaborate on the creation of portable, soft IP that can be reused at other universities. One possibility is for a complete multi-component project to be shared so that the next team of designers can utilize one or more of the pre-existing blocks. If one of the team members is unsuccessful in doing his module, the instructor could then reveal the previously developed solution. This approach provides the instructor with a "known good project" and its solution manual. It also potentially permits a team to tackle a more complex project by adding to its members the "virtual" designers who have preceded them. In essence, this is the approach that commercial companies are seeking to employ. They want to be able to create a system which would normally require a team of 20 designers but have perhaps only 5 on hand. The efforts of the missing designers are replaced by the reusable components.

All of the examples given so far have been for developing soft IP using a HDL for implementation using FPGAs or single-chip ASICs. These are appropriate for advanced graduate electives which are targeted at developing system-on-a-chip designers. For senior capstone courses, projects must generally involve integrating existing hardware and software components with only a limited amount of time available for creating new components. Again sharing of previously developed modules could be performed by reusing those from a prior semester or another university. Similarly, sharing of resources for freshman introductory courses could also be undertaken.

## 5. Conclusions

Sharing of IP modules that have been designed for reuse is a viable means of enhancing system-level education and several possibilities for collaboration have been proposed in this paper.

## 6. References


[1] *International Technology Roadmap for Semiconductors*, http://www.itrs.net.ntrs/publntrs.nsf.

[2] Smith, M., ASICs, Addison-Wesley-Longman (1997).

[3] Design For Reuse Workshop, http://aemp.eeel.nist.gov/reuse.

[4] *Bouldin, D.*, Graduate Courses Using Synthesis, University of Tennessee, http://microsys6.engr.utk.edu/ece/bouldin_courses.

# DEMANDS ON MICROELECTRONICS EDUCATION AND RESEARCH IN THE POST-PC ERA

**H.De Man**
*K.U.Leuven/IMEC, Belgium*
*deman@imec.be*

## Introduction

Before the year 2005 we will reach the 0.1 micron deep-sub-submicron age. This leaves us barely enough time to educate today's freshman into the Ph.D. technical leader designing the ultra-deep-submicron systems of tomorrow. By that time, a single chip will contain 100 Million logic transistors embedded in 10-100 Megabytes of distributed DRAM and flash memory.

In principle, GHz radio circuitry can be combined with digital signal processing clocked at 1 GHz to provide wireless or wired broadband access to the global information highway. This opens the road to a knowledge based society were flexible silicon platforms are offered virtually free to the consumer by companies that make their money out of myriads of new services. Every multimedia terminal for the individual citizen as well as all his home appliances will have their own IP address. They will have a wireless plug and play IP connection controlled by intelligent agent software. This is the world of embedded connectivity where every citizen will be surrounded by hundreds of invisible platforms that augment his or her conciousness, health, security and pleasure.
This ongoing process is rapidly leading to a complete consumerization of the electronics world.
Embedded connectivity provides the huge volume of deep-submicron chips needed to amortize the cost of the future wafer fabs.
Hence, neither the PC chips nor the DRAM's will be the drivers of future chip technology. Instead, rather cheap, remote run-time reconfigurable, and extremely energy efficient terminals on a single chip or within a single package will be the volume products. This will cause a substantial shift in the deep-submicron semiconductor business model.
In terms of complexity, such chips are the equivalent of a few million lines of RT code to specify their operation. A very substantial part of that specification will be embedded software anyway and provisions must be made to accept real-time agents to adapt the system at run-time to communication content and channel characteristics. All high level design will be at the concurrent object oriented programming level. Hardware design therefore amounts to programming in as far as the digital part is concerned.
However, since productivity increase of a software person is only 4.8%/year, we will run into a severe shortage of design talent while, at the same time, facing very challenging paradigm shifts at the system level.

Semiconductor technology may thus outpace our ability to design the products that must fill billion dollar fab lines.



*B. Courtois et al. (eds.), Microelectronics Education,* 9–14.

Hence the need to invest in training and education of System-on-a-Chip (SoC) engineers is grossly underestimated.

It will be discussed how all of the above is based on some major paradigm shifts in future systems. Coping with them will require a radically new organization of research in system design at our engineering schools. It is our duty to attract young people to this profession instead of delivering them to law schools to waste their time in battling over IP legal issues.
One can indeed wonder what type of engineers we will need to design these marvels of the information technology of the first decennium of the next millenium.

**Paradigm Shifts**

The main paradigm shifts in design, already occurring today, can be caught in a few one-liners such as:

- Chip evolves from component to encapsulation of complete system expertise.
- Design primitive evolves from transistor to complex IP.
- Product evolves from PC to globally connected consumer ICT terminal.
- High quality images and icons will be the "units" of communication.
- From compute bound to storage bound operation.
- Specifications evolve from VHDL to concurrent Object Oriented Modeling libraries.
- Digital signal processing definitely dominates analog signal processing.
- From nsec obsession to nWatt obsession.
- From fixed hardware to flexible embedded software.
- From hardwired to run-time reconfigurable logic and interconnect.
- From monoprocessor to heterogeneous multiprocessor architectures.
- From System on a Chip to System on a Package (SoC vs. SoP).
- From a box dominated to a service dominated electronics industry.
- Software costs outpace silicon cost by 10 to 1.
- Software test will dominate hardware test.
- Key problem will be design for reliability and dependability for real-time distributed software.

In the middle of all this exciting turmoil, SoC or SoP design engineers must be able to translate domain specific system or even service expertise into IP based architectures while meeting the power, time to product and intuitive usability constraints of the typical ICT product. The world will belong to the system architects.

**Are we ready to educate SoC design engineers?**

Unfortunately, today's schools are not prepared yet to deliver this kind of engineer. They either deliver the classical VLSI IP designer or the telecom or computer specialist. Few pay attention to the link of both. On the other hand CS departments focus more on Bill Gates stuff than on embedded system design techniques. A new breed of engineers must be created and the way we create them has to change.

The fundamental problem of classical engineering schools is indeed that they are based on *individually acting* professors and students. Research domains are *horizontally structured*, meaning that there is little attention to how they interact vertically to provide the *design flow* for *product creation*. Furthermore, the complexity of global systems exceeds the capacity and domain of a single professor who is presently awarded for being extremely smart in an isolated domain in engineering science. Few engineering professors have experience in system design and have not participated in industrial system design projects.

Is the creation of schools in system design possible at all? This author believes the answer is positive, based on a new form of system research organization called *design center*[1].

## Concept of System Design Center

Recently, a number of such centers have emerged such as ALBA[2], YAMACRAW[3], Berkeley Wireless Research Center[4], Gigascale Silicon Research Center[5], Eindhoven Embedded Systems Institute[6], Intelligent Home project IMEC[7] , Human Computer Interface Institute CMU[8] etc...

All these centers are characterized by long term, *vertically structured*, system design projects resulting from a coordinated effort of *multi-disciplinary research teams with a sufficient critical mass* having an intense interaction with the systems industry. Characteristic is also that concepts are proven by working *demonstrators and publicly available software* rather than by "theorem proving" in the traditional citation index style. The latter occur rather as a by-product, which tends to be more useful as it refers to the solution of real- instead of a smart but fictitious problems. In essence, these projects result from combining the best modes of operation of the system industry and the university. This leads to the following definition:

*A design center is a vertically structured multi-disciplinary oriented research organization located on or near a major university campus where faculty, Ph.D. students and MSc students from associated universities, perform team research on design methods and tools for the design of future ICT systems.*

Its mission is to produce the technical leaders of tomorrow by immersing faculty and students alike in daring system level projects defined by academic and industrial visionary leadership.

The goal is, following[9], that: "*Ph.D. students must become true leaders of tomorrow rather than super master degree holders of yesterday*".

More in particular, design centers are characterized by the following:

1. Their research, at the basis of teaching, leads to the *demonstration* of the feasibility of a global system concept at the leading-edge of technology. Research is *theme driven* and 5..10 years ahead of current practice. Besides producing SoC engineers, the research goal is the formalization of design methodologies and the encapsulation of them in reusable software. Direct transfer of the technology *in it self* is not the goal. Actual transfer takes place through *people trained for technical leadership*, ready to create new products either in new start-up companies or in the affiliated major system houses or CAE companies (See Fig. 1).

2. The ambitious research goal can only be achieved by *faculty team building*. Faculty commit to share and build knowledge to reach a common goal each *within and at the fringes of disciplines*, where "concepts and ideas rub and spark against each other to create new approaches and new solutions" [Dan R. Olsen, ref 8]. Notice that this may not be consistent with the actual "science" dominated publication policy but it sure

helps to create the right people. It also requires a deeply respected visionary academic to take the lead of such an institute (the *project champion*).

3. Design centers also must have a strong *Industrial Affiliate Program,* which makes industry an active partner. This keeps research goals realistic, generates long term system requirements by visionary dialogues and entitles the industry to prototype background and foreground research results and a direct contact with most important product: *the well trained SoC engineering student.* A practical way for such interaction is to have industrial residents to participate in project research. In fact industrial residents can play a team leader role for the Ph.D. researchers operating in the design institutes.

Notice that such a design institute is a reflection of the modus operandi of the industry in which the engineering students will work after graduation. It provides the framework for global thinking by induction from their team professors (as opposed to the old-fashioned isolated genius).

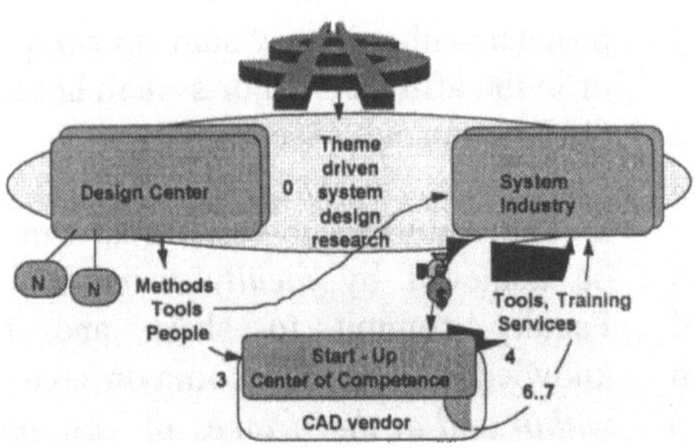


*Fig.1 : The structure, interaction and products of a design center. The numbers indicate time in years between concept and result.*

Just as virtual companies emerge, virtual engineering schools can be conceived. Networked sharing of educational IP of the best schools will contribute to a better education, a better use of education funding and a faster progress in technology. In conclusion one can state that a design institute will :

- Create professors understanding the global system issues;
- Make them able to teach it and put horizontal disciplines in a vertical context;
- Will attract more students as subjects become nearer to reality;
- Will prepare students better for the ICT industry.

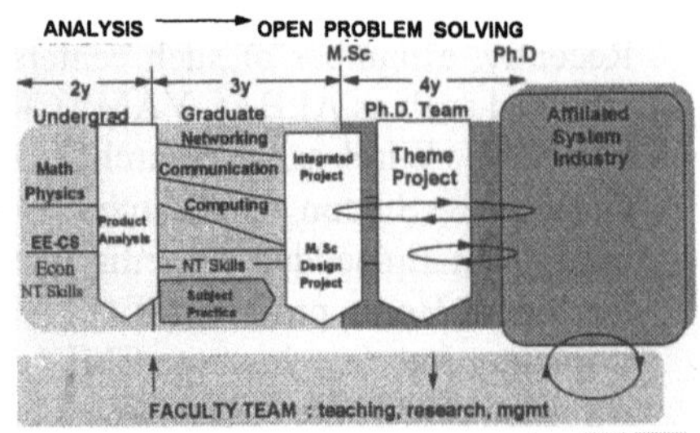


*Fig 2: Tentative 5 year curriculum of a design centric school for the ICT oriented SoC engineer.*

## Education program for SoC engineer

Fig.2 shows a tentative MSc and Ph.D. program embedded in a design institute for ICT-SoC training. In view of the broad education, a 5 year MSc program is a prerequisite.

It has the following characteristics :

1. Regular courses are based on *team teaching* and are restricted to basic concepts. Team teaching focuses on engineering aspects and context. As the white arrows indicate, at least 30% of the time budget is spent on project work of which the majority (vertical arrows) is devoted to inter-disciplinary projects.

2. The first two years are analytical in nature and should provide the scientific basis to engineering. "Basic science" should however be *thought by engineers for engineers*. Also the basics of EE, CS are thought. Basic science is linked to engineering by devoting 25% of the time on *product analysis* seminars. These are hands-on seminars whereby basic science professors *relate theory to practice* by analysing aspects of a popular product (CD-player, Modem, DVD, GSM...). Global design aspects, economics and marketing issues of such product are explained by engineers from industry.

3. In view of the important role of software in future information systems, the master program should provide a balance between EE and CS subjects whereby CS should be EE oriented i.e. pay less attention to "infinite resource programming" but more to :

- Embedded software: compilation and real time aspects, interface synthesis, parallel processing
- Software impact on power dissipation, memory architecture etc.
- Computational models (data-flow, control flow, concurrency)
- OO programming as a means for system modelling, knowledge reuse and on the fly tool development.
- Use of networking as an engineering tool
- Use of commercial and university CAE to focus on architectural aspects.

This means less computer "science" but more software/hardware engineering.

4. During the last three years of the MSc program the EE=CS subjects give gradually way to the ICT subjects. In the last year special attention is paid on how to implement IT systems in hw/sw architectures as that is the final goal of the education.

5. The emphasis of the master program is on the ability of problem solving both individually and in team. Hence at least 25% of the time budget has to be spent on inter-disciplinary design projects (vertically shaded arrows). In these projects students get involved in more and more complex design problems from specification to implementation involving several EE and CS disciplines. A *management structure* involving both faculty and students "runs" these projects. In this way concurrent engineering principles, team work and communication skills are introduced in practice.
The master thesis is devoted on a more in depth design task (including tool development if appropriate). A master thesis can also be made by a team of 2..3 students supervised by a Ph.D. student working in the design center.

6. The Ph.D. program is devoted to research within the theme project of the design institute. Ph.D. students focus on aspects of large system design problems, whereby *emphasis is placed on combining engineering disciplines*. The Ph.D. program should be oriented to create enhanced design productivity by demonstrating new design methodologies. Such methodologies should be documented by reusable software available over the network. This *form of publication* may turn out to be a lot more useful than the actual *overproduction* of academic papers of which the impact can actually not be easily tested.
Methods, tools and libraries shared by researchers will be naturally selected. Their acceptance rate can be a new and,

from an engineering viewpoint, more interesting "citation index".

## Conclusions.

Deep-submicron technology combined with the tremendous growth of global networking leads to SoC and requires a rethinking of computing and communication architectures and of new multi-disciplinary design methodologies. This creates the need for *SoC engineers* able to bridge the gap between system level concepts and their implementation in heterogeneous silicon architectures. The concept of *design center* is introduced where academic staff and students team-up around visionary information technology design projects. Building actual software-hardware prototype demonstrators is undertaken as a means to create new methodologies that give rise to new CAE tools, libraries, teaching material and training courses available over the network.

A design center is the ideal environment for educating the SoC engineer. In such schools emphasis is on broad education in the basics of computing, communications and components. Design practice (including non-technical skills) are learned by team-based integrated projects in product design

## References


1 H. De Man : "Education for the Deep-Submicron Age: Business As Usual?", Proceedings of the 34th Design Automation Conference, Anaheim, Ca. June 1997, pp. 307-312

2 http://www.london.press.net/issues/8/58/pagefour.htm, http://www.sli-scotland.org.uk/

3 http://www.gcatt.gatech.edu/yamacraw/about.html

4 http://bwrc.eecs.berkeley.edu/

5 http://www-cad.eecs.berkeley.edu/GSRC/

6 http://www.eesi.tue.nl/english/eesihome.html

7 http://www.imec.be/desics/Welcome.html

8 http://www.cs.cmu.edu/afs/cs.cmu.edu/usr/hcii/www/index.html

9 B.J. Sheu, Y.D.Lee, W.C.Young : "Higher Education for the Information Age", IEEE Circuits and Devices Magazine, Vol. 14, No 3, May 1998, pp. 23-28

# ON THE DESIGN OF MICROELECTRONIC DESIGN PROJECTS TO FULFILL GIVEN LEARNING OBJECTIVES

EINAR J. AAS
*Norwegian University of Science and Technology*
*Department of Physical Electronics, N-7034 Trondheim, Norway*

## 1. Introduction

The Norwegian University of Science and Technology recently redesigned its engineering education programmes. This engineering education (EE) redesign started as early as 1993, and was prompted by the realisation that emphasis should be shifted from teaching to learning. An extensive external Gallup survey including 751 practising engineers and 100 industrial managers provided a solid background for possible enhancements of the education programmes. The main conclusion was that the basic science and engineering competence of the graduated engineers was solid. However, skills within non-technical areas such as economics, environment and management were deemed unsatisfactory. In particular, improvements within written and oral communication and negotiation skills were asked for in the new education system. The final document, *The Will to Improve* [1, pp. 350-355], summarises the new objectives of our EE. The framework for the improved curricula is also discussed in [2]. The overall objective as stated by the Curriculum Development Committee, is:

*The overall objective should be a degree programme that gives students knowledge, skills and attitudes leading to relevant professional competence.*

One important element of the new EE programmes is more systematic use of PBL (Project Based Learning). Clearly, PBL propels the students towards their own learning. I am convinced that *electronic engineering education* (EEE) lends itself well to PBL because electronic design projects normally are integrated in EEE. The challenge is to *design* the design project such that stated *learning objectives* may be fulfilled.

In addition to this study, we have gained considerable experience in earlier student design projects since 1991. Earlier results are reported in [3-6]. In [3] and [4], we introduced a general metric for design quality, and presented statistics on more than 70 student design experiments in terms of design quality and design efficiency. Design error types and different design verification strategies were revealed, including the efficiencies of each strategy. In [5] we focused on the impact of design quality on Return on investment. [5] introduced peer evaluation of designs.

Design and verification strategies were of particular interest because we stress the importance of acquiring design skills in our curriculum. Several other engineering education institutions have also devised PBL schemes to develop design skills in EE, e.g. Kollenburg [7].



*B. Courtois et al. (eds.), Microelectronics Education, 15–18.*

## 2. Learning objectives for microelectronics design

One significant goal of our EEE is to educate students in design competence. Generally, we define competence thus: *competence is the fusion of knowledge, skills and attitudes.*

Specifically, we want our Circuits and Systems (CAS) students to develop their own design competence, including specification, synthesis, analysis and verification, and realization. Skilled use of commercially available CAD tools shall be embedded in the design competence. We will only deal with digital design in this paper. The verification shall be done with focus on high *design quality* $Q$, defined as:

$$Q = \text{Prob [the design object satisfies its specification]} \qquad (1)$$

In addition to emphasis on design quality, the realization shall be done with focus on high *design efficiency*. This efficiency is defined with respect to predefined target parameters, e.g., cost, chip area, performance, power and fault coverage.

Finally, we want our students to be aware of industrial and commercial aspects of design and manufacture, e.g., cost, time-to-market and market window. We introduce the concept of "satisficing design" coined by H.A. Simon [8] i.e., finding a solution that is good enough without reference to an "optimal" solution.

## 3. The design of design projects

In addition to the design competence defined above, we also want to develop skills such as team collaboration, communication, peer design evaluation and critique, reflection on own result, presentation and "marketing". A design project must be *designed* in such a way that all the skills listed above may be developed during the work. We communicate a list of tasks, "problems" and questions related to these learning objectives to the students at the beginning of the project. Thus, they may use their communication skills to question or challenge our understanding of their learning process. Some typical tasks related to this academic year's design of an arithmetic unit are given in Table 1.

TABLE 1. Tasks of an arithmetic unit design project

| Task | Intent or skills to be exercised |
|---|---|
| Design area-minimal unit | VHDL code style exercise; area optimized. |
| Design max frequency unit | VHDL code style exercise; time optimized. |
| Document designs above in DOC_1 | Written communication skill. |
| Swap DOC_1 with peer group; evaluate design of peers, grade design. | Evaluation practise, and constructive critique skills. |
| Update DOC_1 according to comments. | Reflect and improve. |
| Model in VHDL, simulate and document in DOC_2. Swap with peer group again. | Modeling and verification skills; awareness of design quality. |
| Synthesize with commercial tool | Synthesis tool usage skill. |
| Read SELFTEST document, and use this arithmetic methodology in own circuit. | Learn arithmetic based self-test principles, and apply to given design. |
| Embed self-test circuitry, and fault simulate to desired fault coverage | Understand self-test advantages, and relate to fault coverage metrics. |
| Oral presentation | Oral communication skill |

The design problem must be small enough to be tackled within the time given, and large enough to solve and verify nontrivial design tasks. Teams of two students are formed, milestones and deliverables are defined, and peer evaluation is done at two milestones. This academic year's problem was to design an arithmetic unit computing the expression **Y = A•X + B** with given data formats and proper rounding.

## 4. Peer evaluation experiences

The main goal of peer evaluations is to train the students in judging other professional's work. We emphasize concrete and constructive comments, including exposure of design flaws, oversights, ambiguities, contradictions or incomplete descriptions. In addition, the students shall give marks on specific aspects of their peers' reports. Two web-based evaluation sheets are used to collect the evaluation. The first evaluates problem definition and architecture exploration. The second evaluation deals with the VHDL code, and the comprehensiveness of verification. Table 2 presents the tasks to be evaluated by peers.

TABLE 2. Web-based peer evaluation tasks.

| Evaluation no. 1 | Evaluation no. 2 |
|---|---|
| Readability of DOC_1, i.e. structure and degree of completeness? | Readability of DOC_2, i.e. structure and degree of completeness? |
| Chosen architectures; are they well suited for min area and min delay, respectively? | Is DOC_1 updated based upon comments from Evaluation no. 1? |
| Is truncation properly handled? | Comprehensiveness of simulation? Are you convinced that the design will indeed work? |
| What did you learn from your peers? | May 100% functional verification be attained by the solution selected? |
| Comments to this evaluation? | Comments to this evaluation? |

Most comments offered were concrete, with suggestions on improvements, revelation of design flaws, and demands for more thorough documentation. "*Incomplete documentation*" was frequently remarked. Some comments to the question "*What did you learn from your peers?*" are listed in Table 3.

TABLE 3. "What did you learn from your peer?"

| | |
|---|---|
| Use of accumulated adder at end of Braun array. | Use of ELM adder for time optimal design. |
| Use of Manchester carry chain adder. | Importance of clear documentation for reader. |
| Rounding technique. | Peers had different approach, but ours is better. |
| Learnt new swift multiplier. | Use of Wallace tree in multiplier. |
| Treating **B** as partial product simplifies solution. | *NO!* (4 of 17 groups learned nothing from peers..) |

Another interesting aspect of these projects is the opportunity to observe the variation of *design efficiency*, measured as the inverse of a cost metric, between the different groups. The efficiency metrics chosen so far are silicon area or performance, while power might be another metric. An example is given in the next section.

## 5. Examples of student design projects

We offer a course called *"COMPUTER AIDED DESIGN AND TEST OF (DIGITAL) ELECTRONICS"*. This two-term course covers theoretical and practical aspects of design and test, from specification to manufacturing. An integral part of the course is the *design project* mentioned. As an example, the variation of design efficiency in terms of chip area for the academic year 1998/1999 is shown in Fig. 1.

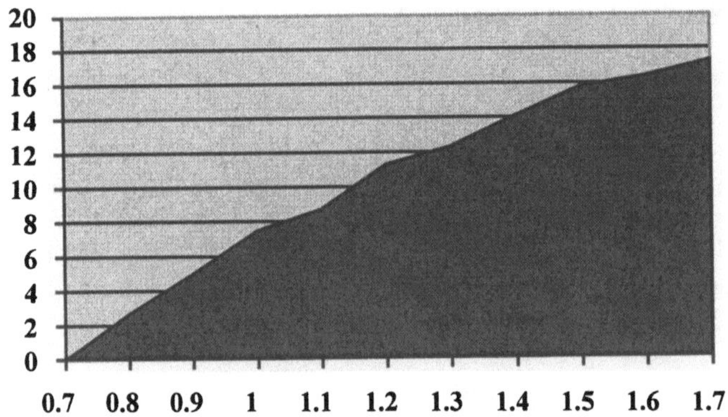


*Figure 1.* Cumulated no. of design groups versus chip area ($mm^2$)

(This year's design has not been synthesized at the time of writing, and thus results on area and time can not be reported.) The design capture was done by VHDL description, with subsequent automatic gate-level synthesis. In this project, only cost in terms of chip area was defined, and design efficiency can thus be measured by area alone. Notice that it is the skills of the designer that determine the "optimality" of the resulting design. This strengthens our belief that knowledge about *how* the tools work, i.e., both understanding the underlying theory and nurturing skillful craftsmanship, will contribute significantly to the successful designer's "tool box".

## 6. References

1. E. J. Aas, "The National System of Engineering Education in Norway", pp. 321-355 in "The National Systems of Higher Engineering Education in Europe", ISBN 88-7741-870-2, Pisa, 1995.
2. S. Nordal and E. J. Aas, "Improving the curricula for engineering education at the Norwegian University of Science and Technology", Proc. of Educating the Engineer for Lifelong Learning, SEFI Annual Conference '96, Vienna, 11-13 Sept. 1996, pp. 313-318.
3. E. J.Aas, T. Steen and K. A. Klingsheim, "Quantifying Design Quality through Design Experiments", IEEE Design & Test of Computers, pp. 27-38, Spring 1994.
4. E. J. Aas et al., "Student Design Experiments - Lessons Learned in Design and Verification", Proc. 4. EUROCHIP Workshop on VLSI Design Training, pp. 124-129, Toledo Sept. 1993.
5. E. J. Aas and I. Sundsbø, "Harnessing the Human Factor for Quality Designs", IEEE Circuits and Devices Magazine , pp. 24-28, May 1995.
6. E. J. Aas and J. Typpö, "Project Based Learning Objectives and Experiences in Electronic Education: A memory Controller Design Experiment", Proc. 6th IEEE Int'l Conf. on Electronics, Circuits, & Systems (ICECS'99), pp. 1611-1614, Cyprus, Sept. 5-8, 1999.
7. P. A. M. Kollenburg, and H. Veenstra, "Successful Integration of Industry and Education with Concurrent Engineering Projects", Proc. 25th Annual SEFI Conf., Helsinki, Sept. 1998.
8. H.A. Simon, "The Sciences of the Artificial", MIT Press, Cambridge, Mass., 1969.

# System-on-Chip Design and its Transfer to Microelectronics Education

FRANÇOIS CHARLOT
*Application Manager,*
*ARM*

*Francois.Charlot@arm.com*

**Key words:** system-on-chip, ASIC design, embedded processor, standards, intellectual property, IP, ARM, AMBA

**Abstract:** System-on-Chip (SoC) design is an industry buzzword that has yet to make its way into microelectronics education. Beyond the hype, SoC design is a complex process involving decisions that have far-reaching implications. This paper highlights some of the new issues SoC design has raised. It shows the solutions ARM offers as industry standards, and how we transfer them to the microelectronics education.

## 1. INTRODUCTION: WHY SYSTEMS-ON-CHIPS?

There is relentless pressure in the industry to increase functionality, decrease cost and hit the market in ever-shorter times. The progress in semiconductor manufacturing technology allows the integration of more features in a single chip for the same cost. This means systems made of several electronic components can now be made of a single chip, hence the term system-on-chip. The SoC is cheaper, smaller, more reliable, and requires less power than a multiple-device system.

But SoC design is not easy, and designing a discrete system is quicker. A too long development may miss a product's market window. Too high costs alter the possible savings. A bad specification or a poor execution may lead to multiple SoC revisions.



*B. Courtois et al. (eds.), Microelectronics Education,* 19–22.

At the risk of stating the obvious, making the best of systems on chips requires to produce the right device at the right time. The complexity of a SoC requires new tools and methods to achieve this goal.

## 2. PUTTING TOGETHER A SYSTEM-ON-CHIP

The first golden rule of successful SoC integration is to know what to integrate. This does not just mean the system hardware (for example, how fast it goes, how much memory fits on the board, how many peripheral functions can be added). It means a detailed understanding of the hardware/software and analog/digital partitioning, physical constraints such as voltages, power, electromagnetic compatibility...

On the other hand, the system architect must understand what is realistically possible in a SoC design. This includes choosing the processor and intellectual property (IP) blocks, devising how to validate and verify the device and how to design for volume manufacture and test. Successful SoC design requires thinking from system to silicon.

Embedding a processor in a SoC presents significant differences with board-level design. Raw processor speed is not a good enough indication the SoC will meet its performance targets. The SoC architecture must also provide sufficient memory bandwidth and low system latency. The processor must be delivered in a form that fits the SoC design flow and manufacture process. Tools must be available so the software development can start before the SoC is fabricated. Finally, good support, a secure supply, and a complete roadmap ensure longer-term benefits.

A SoC typically integrates many dedicated peripherals. Designing them again in each SoC project being not economically practical, SoC designers now consider reusing previous designs. But reusing is not just about a copy-and-paste of an in-house design or acquiring any piece of IP. Reusable IP must be developed in this aim from the beginning, using a proven and well laid-out method. Its documentation must address a variety of audiences such as SoC architects, silicon designers, or software programmers. Its deliverables must support all industry-standard tools. It must be comprehensive and include test and production support.

Peripherals not yet available as reusable IP must be developed from scratch. Integration capabilities allow implementing such blocks as software running on a small slave processor along with simple inputs and outputs. This is economically practical considering a 50-MIPS ARM7TDMI™ processor takes less than 0.5 $mm^2$ in a 0.18 μm process. Integrating multiple processors on the same chip requires multi-processor simulation models and debugging tools as well as a high-performance on-chip bus architecture.

## 3. BENEFITS OF STANDARDS

A complete processor roadmap, a rich set of peripherals, a high-performance on-chip bus architecture, and comprehensive design and development tools are some of the ingredients of SoC design. Some semiconductor companies can offer them today, but these are most often proprietary solutions. The benefits of industry-wide standards are to allow multiple sources, giving access to a broader span of technologies at the best price. They foster innovation by establishing new grounds for research and product development. They cut costs by sharing them among the partners. They shorten schedules by bringing well-established frameworks.

ARM is a British company that has been the leader in SoC solutions since 1990. It plays a key role in educating the industry in SoC design methods and reusability. It is an active member of many standard committees such as IEEE, VSIA, or Bluetooth. ARM has earned many industry awards in recognition of its contribution to the electronics industry.

ARM® processors are licensable IP blocks that power 58% of the appliances using a 32-bit embedded processor[1]. Almost 40 semiconductor companies have licensed this technology with the largest vendors such as Intel, Texas Instruments, ST Microelectronics, or NEC. This success made ARM processors an industry *de facto* standard with the broadest tool support available from third parties, such as operating systems, development and debug environments, application software…

ARM has also defined the Advanced Microcontroller Bus Architecture (AMBA™), a free open standard of on-chip bus[2]. AMBA allows interconnecting multiple processors in a SoC along with private or shared resources such as memory or peripherals. AMBA was designed for simplicity, reliability, testability, high performance, and low power consumption. Since it is a common infrastructure to all ARM processors, it has set the ground to a strong, industry-wide, reuse framework.

Being an open and free standard, most of ARM's semiconductor partners have adopted AMBA for their own reuse frameworks. Even non-ARM processors use this bus, such as Analog Devices' 219x digital signal processors. Many independent IP suppliers have also endorsed the AMBA standard to distribute their products.

## 4. TRANSFER TO EDUCATION

Transferring industry technology to the education is more than selling products at low prices, although this is a must-do. Educational institutions should have access to a complete technology range, and to semiconductor manufacturing facilities at an affordable cost.

Circuits Multi-Projets (CMP) has an agreement with ARM allowing it to provide academic institutions with a complete solution for developing ARM processors-based SoC devices. This agreement covers all ARM development tools: ARM Developer Suite™ for code generation and debugging, Multi-ICE™ multi-processor JTAG target interface, evaluation boards, Integrator™ boards for FPGA-based SoC prototyping, and Prospector™ boards for software development and demonstration. Beyond tools, CMP provides SoC design tools (synthesis, simulation...) that support ARM processors, and offers semiconductor manufacturing services through its semiconductor house partners in leading-edge processes such as 0.18 µm.[3]

ARM also works directly with universities, including Manchester University (UK), University of California at Berkeley (USA), Université Catholique de Louvain (Belgium) and Aachen University (Germany), to promote ARM-related research. ARM also supports broadly innovation in electronics, such as through the Young Electronic Designer Awards (YEDA) sponsorship.[4]

## 5. CONCLUSION

System-on-chip design is a vast domain with many branches such as reuse. Thanks to its economical implications, it gets a lot of focus from research and industry and progresses quickly. We have shown here how an industry leader makes its state-of-the-art technology truly available to academic institutions. This helps students understanding and applying SoC concepts before they join the industry, and add more ties with research institutions to help shaping the technologies and products of the future.

## REFERENCES

1. Allison A., The New Computer Industry, January 2000.
2. ARM, AMBA Specification Rev2.0, May 1999, http://www.arm.com/Pro+Peripherals/AMBA/
3. http://cmp.imag.fr/
4. http://www.yeda.org.uk/

# CONTINUING EDUCATION

# AND TRAINING

# BEST PRACTICES FROM 100 CORPORATE UNIVERSITIES WORLD-WIDE

J. MEISTER
*Corporate University Xchange*
*381 Park Avenue South*
*Suite 713*
*NEW YORK, NY 10016*
*USA*
*Phone: +1 212 213 2828*
*Fax: +1 212 213 8621*
*email: jcm@corpu.com*

Abstract not available at time of printing



*B. Courtois et al. (eds.), Microelectronics Education,* 25.

# Quality Labelling for Electronic Training Courses

*Study by CNFM in the framework of ESPRIT European Project: EUROPRACTICE - Training and Best Practice Service*

Christophe ALFANDARI / René CASTAGNE
*CNFM - E.S.I.E.E / CNFM - EUROFORM - Université Paris-Sud*

Key words: Training, Quality, Labelling, Objectives, Microelectronics

Abstract: The necessity of a study on a Quality Label arise within the framework of EUROPRACTICE - TBPS, in order to increase the added value of European-level electronic training course brokerage. Quality Labelling (QL) concept, difficulties and limitation comes from the fact that training services addresses both the trainers and the trainees. The methodology employed was first to study the present time impact of quality assessments (such as ISO) through a survey, then to develop criteria and procedure for a TBPS QL service and finally to do field test.

## Survey about the Quality in Training

The objective of the training is certainly the most important point that Quality Labelling has to refer on. On the other hand, a difference should be done between the objective of the training action as expected by the company (why the training is set-up within the company) and the objective of the training course itself (what the participant should be able to do at the end of the training session), mainly because a course provider can only expect to measure the second one. So a clear definition of the objectives is absolutely necessary from which the structure of the training session should be established.

The survey addressed three kinds of training actor's : the participant to the training courses (the attendees), the course customers (the buyer : people who decide to send somebody to be trained) and the course provider. The preliminary survey was conduced with the following questions :

1- What does quality in training means for you ?

2- Could you quote an organisation delivering a quality label for training ?

3- Have you got a plan to improve the quality of your training?

We obtained 21 answers from the course providers, out of which 10 came from academics and 11 from business organisations. The analyse confirmed that the quality in training is strongly related to the description of the objective. That point has to be connected to the selection of participants according to the objective of the



*B. Courtois et al. (eds.), Microelectronics Education,* 27–30.

session. The second point raised by the course providers is the quality of the training material. The lecturers selection is also an important factor for the quality: course providers insists on their teaching capabilities and the technical level of expertise. However the industrial background of a lecturer can also be a critical factor when practical illustration (example, hand on, etc.) has to be done. Feed back from customers is also declared a significant point. However only one answer declared that it should be done a significant time after the training session (cold feedback).

The professionalism of the organisation to welcome participants and to insure a professional level of facilities is a very classical aspect. Because it is not specific to training, that part is covered by a lot of works on quality and it is the heart of ISO regulation for training.

From the second question, only 43% of course providers quote one or more name of organisations able to deliver a quality label for training activities These organisations are generally national ones and do not operate on an European wide level. On the other hand, 16 (61%) of the questioned course providers declare themselves interested by a quality labelling service (only 5 were not interested) and 71% of them had already a plan to improve their own course quality: They are sensitive to quality label because of the marketing argument it represents.

The opinion of Course Buyers is of prime importance. For that survey, we choose 5 large companies and 5 SMEs in the field of electrical engineering in 5 European countries (France, UK, Germany, Italy, Spain).

The Course Buyers think at a huge majority **(77%)** that there is a need for a quality label in training business. Only 33% were able to quote some (very few) organisations that offers quality label on the market, and none of them refers spontaneously to ISO 9000 . We then asked them specifically about the use of ISO 9000 regulation. Only 30% used it but they judge it insufficient in terms of quality insurance (50%) and criterion (61%), may be because it is too much related to the classical services, which training is not: in its definition as well as its organisation training is a co-production between the Course Vendors and the Course Customers.

Finally a specific survey was sent to all the 397 Course Customers who had registered through TBPS and we obtained 44 replies **(11%)**.

**80%** of Course Customers are unable to quote a quality label and only **54%** of them give the name of a training organisation that symbolises quality in training sessions. **EPFL** is the most often quoted. On the other hand these replies came essentially from academics, more or less in accordance with the academic representation in TBPS customer's population. This has to be taken into account in the analysis of the result. Because time spending is not as critical as in companies academic students seem less concerned by quality. For instance they can declare satisfied even if only a part of the course is within there field of interest.

Then the conclusions of this 3 steps survey were going in the same direction and it was quite encouraging for the continuation of the Quality Labelling work package. The main problem is in meeting expectations, which can be related to objective description and course content. The search for quality should mainly focus on the following aspects : clear specification of the objectives, trainers capabilities from educational, industrial expertise and experience point of view, quality of the training material .

Criteria and procedure for Quality labelling of Electronic courses

The labelling procedure covers two steps:

First step is to evaluate, before the course takes place, the course provider organisation from 6 aspects : Organisational aspect, Objective with respect to the content of the course, Selection of participants with respect to the declared objectives and content, lecturers, training material, organisation of the feedback.

The second step is based on a « cold » evaluation procedure

Step1: Before the course take place:

The evaluation is based on information's given by the Course Provider organisation when filling a "Course Provider Training Information Form". The criteria used are the following:

- The objectives: Are they clearly stated, possibly chosen within the list?
- The participant profile : Is it clearly described in accordance with the objectives?
- Duration and course content: are they compatible with the 2 previous points?

This supposes that the courses "running" for quality label are better described than usual ones within a course catalogue (which generally gives only a brief description of the course contents) A list of possible objectives and company/participant profile is included in the "questionnaire" and proposed to the course provider.

The Training material : We propose that the evaluation covers the following aspects: " Is your training material printed ? does it include lecturer slides and comments ? have you got a specific set of leaflet for practical work (based on industrial cases) ? Could your material be used after the session as a working tool ? is a bibliography included ? how often do you update the material ?"

The lecturer selection: A brief curriculum of the lecturers is asked (maximum 5 lines /lecturer) addressing the 3 items: teaching background, industrial background, technical level of expertise.

This evaluation is conducted by a steering committee. Based on the CP forms and the course material the steering committee formulates an opinion on the quality of the proposal. The steering committee could ask for attending the session in order to deliver the approval.

Step2 : After the course has taken place:

This second step is the part of the evaluation. As mentioned previously it is based on a " cold customers feedback " runt between 2 and 4 weeks after the end of the training session.

At the end of the session the Course Provider has to communicate the names and addresses of the participants. Then QL secretarial runs the survey.

Questions are asked to attendees, within a "Cold Evaluation Form", along the four following items:

- course location and venue, accommodation, classroom and standard equipment, technical equipment (if any), administrative relation (easiness, professionalism, kindness)
- Objective/Content and participants profile: Were the objectives clear to you? Do you feel that the course was properly described? Do you feel that the course suited to you? Do you think that you have met the objective of the course as described? Do you believe that you got useful and essential knowledge from that training ?
- Training material: how was the quality of the printed training material? Does the material include lecturer slide and comment? Have you got a specific set of leaflet

for practical work (industrial cases)? Do you think that you will use the material after the session as a working tools? Is it up-to-date? How is the bibliography?

- Lecturers: Do you feel that lecturers skills were well fitted to the course objectives and content? How was their teaching or instructional qualification? How was their technical level of expertise? How was their industrial qualification?

For each items the attendees are ask to quote out of 10 and different weights can be attributed to the items.

Test of the procedure.

The procedure was first tested in an experimental run on 4 training sessions delivered by 3 different Course Providers who accepted to play fully there role from step 1 to step 2. A detailed report was communicated to the Course Provider and doubtfull cases (2 out of 4) were discussed according to the information collected during the Step 1. In spite of this reduced number, the test seems representative of the difficulties to be encountered. The main difficulty comes in establishing the comment relative to a given criterion. The number of questions to cover that criterion should be as reduced as possible. For a great part the procedure lays on a strong correlation to establish between the Course Provider Information form and the Course Customer evaluation form.

Conclusion

One of the drawbacks of such a procedure is the amount of administrative work to be done for each course, although this is mainly true at the beginning. The question then arises of the perennially/obsolescence of the label. A Quality Labelling service must be delivered on a professional basis and this necessitates a strong secretarial support.

The mission of the CNFM relatively to the question of QL within TBPS, was to prepare and run a survey and propose a procedure. Since April 99, CNFM has transferred the job to TFI, an organisation more appropriate to run the label from a professional the point of view. Between April and September 99 an additional 22 courses from 13 courses providers entered the QL process with TBPS, 8 passed QL, 8 were cancelled, 2 were at step1 and 4 at step 2.

Nevertheless there were still questions on the matter, such as :

How long the label duration?

Is it good to limit the quality labeling to the "pedagogical" objective ? Is it possible to extend the label to the industrial objective of the training action ?

Is it economically and technologically realistic to go for it ?

Acknowlegments:

This work has been performed under the European Commission initiative EUROPRACTICE - Training and Best Practice Service.

We are glad to acknowledge the collaboration of:

From TWI (UK): Adrian Duncan, Dr. Nihal Sinnadurai, Dr. Wendy Hanson.

From D.T.U (DK): Prof. Ole Oleson,

From COREP (It): Prof. Pier Luiggi Civera, Dr Andrea Di Blas.

From Technoconsult (DK): Ivan Ring Nielsen

From Technology For Industry Ltd (UK): Dr J Malcolm Wilkinson, Linda Mahoney

From CNFM: Brigitte Gautherot, Monique Louis, Pierre Bière.

# System LSI Design Methodology Training in Hitachi Ltd.

**Goro Suzuki[+],Yoshio Okamura[+],Masayuki Miyoshi[+],Yasunori Kitazawa[++]**
[+]Design Technology Development Dept., System LSI Business Division, Hitachi Ltd.
[++]LSI/PB Design Center Hitachi Information Technology Ltd.
E-mail:suzukig@cm.musashi.hitachi.co.jp

## 1. Introduction

System LSI providers are in the very hard design situation. That is, we have to design chips in time-to-market using very deep sub-micron technology [1]. In order to cope with such a hard situation, top-down and IP reuse design methodology has been proposed. How to brush up all the designers' skill uniformly is a big problem. We Hitachi has begun to provide system LSI design methodology training course for all our designers. Outline of this course will be described in this paper.

## 2. Trainee, trainer, and training environments

Hitachi Ltd. consists of about ten business groups. Among them, semiconductor group is main training targets. Trainees are not only the design engineers but also design managers. We think training of such managers are very important in order to exchange the design culture smoothly from legasy design style to new methodology. That is top-down training approach, and design engineers training is bottom-up approach so to speak. We are providing two training courses: 54 hours course for the design engineers and 36 hours course for design managers. Syllabuses are almost the same, but more HDL code are written by the trainees, and more time are assigned to the writing scripts for the logic synthesis tool in the design engineers course. Training seat capacity is ten for each course, and they can use their own UNIX machines and own tool licenses.

Trainers are system LSI design experts. We are assigning the full-time trainers for these traing courses.

## 3. Training materials

Trainees design the eight-bit microprocessor during this training course. Specification of this microprocessor is as followes:

- instruction length : eight bit, fixed lengh
- addressing mode : direct address
- machine cycle : five nsec
- process technology : 0.18 $\mu$ ,CBIC
- # of instruction sets : ten
- cycle per instruction : eight
- gate counts : about one thousand gates



*B. Courtois et al. (eds.), Microelectronics Education,* 31–34.

This microprocessor has so simple architecture that it takes only two hours to understand it. It is very important to understand the architecture deeply especially in order to use the static timing analysis tool. As you know, it is necessary to understand the # of cycle of each path and false path. We discuss them based on the block diagram and time chart.
We prepare the several application programs for the verification, and prepare the selfcheck testbench. Trainees can concentrate on the study of the design methodology. We provide the handout of the OHP, tool manual, our original textbooks of the design style. Comparing with RMM [2], these books contain more detail contents based on our design style. Associated with this book, we provide VTR, so trainers can study it by himself using these materials during our course.

## 4. Training syllabus

As I mentioned, the syllabuses of the two courses are almost same. Outline of the syllabus is as follows. Trainees design the eight-bit microprocessor following the Fig.1 design flow starting with RTL level description by Verilog HDL.

(S1) RTL level description using UNIX vi
(we teach basic grammar of Verilog HDL)
(S2) RTL level dynamic simulation for the functional verification using event driven simulator with selfcheck testbench
(we teach how to describe the testbench)
(S3) RTL level quality insurance using style check program.
(mainly coding style is checked)
(S4) logic synthesis
(we teach how to make the logic synthesis script)
(S5) gate level quality insurance using our checker
(our checker checks out whether design is satisfied with synchronized circuit or not, and checks out whether design is satisfied with design for testability rule or not)
(S6) scoring the final quality using Synopsys® Open MORE assessment program [3]
(S7) multiplexor base test circuit insertion for design for testability
(S8) test pattern automatic generation for manufacturing
(S7) formal functional verification between before and after test circuit insertion
(S9) wire load model base static timing analysis
(S10) physical layout using 0.18 $\mu$ cell base IC
(at the beginning, we run the clock tree synthesis tool)
(S11) formal functional verification between before and after physical layout
(S12) extracted RC base static timing analysis

In (S2), (S9), and (S12), if the result is not so good, (S1) or coding modification and down from (S3) are repeated.

Because making the real silicon chip is time and man-power consuming process, the training completion criteria is getting adequate result (function and speed) using DA tool.

There are IP providers and IP integrators in each above business group at a same time. For example, IP developing department in one group makes CPU core, and DSP developing department in the same group uses such IP. We have to guarantee the quality of the IP. So we are teaching the design style: the correct timing insurance design methodology, power saving design methodology, and design for testability, using text book and by the design of the training microprocessor. And we are checking the quality of training design in the RTL level and gate level by checker, and finally we are scoring the design quality using Open MORE. During the microprocessor design, design management such as bug tracking, directory structure, and usage of UNIX make file command are also being taught.

As I mentioned in the introduction section, the purpose of our training is to teach the new design methodology and to brush up design skill. Therefore, we are teaching the necessity of the tool usage clearly for the improvement of the design productivity. Pitfall is that it takes too long time to teach how to use the tool. Such a state is very bad for the design methodology training. We have to emphasize the necessity and advantage of the tool usage, rather than how to use them. The reason why we use that design metodology is very important.

## 5. Future works

Training course has to keep up with the most current design methodology. So we have to brush up our training materials continuously. Actually, we synchronize the syllabus with the design environment version in our company. We have to teach lots of design engineers and managers. Therefore, how to realize high throughput of the training is very critical issue by limited resource of trainers. We are examining the new material like CD-ROM.

## 6. References

[1] Sematech, International Technology Roadmap for Semiconductors, http://www.itrs.net
[2] M.Keating and P.Bricaud, Reuse Methodology Manual, Kluwer Academic Publishers(1999)
[3] Synopsys,, MORE:Measure of Reuse Excellence, http://www.synopsys.com

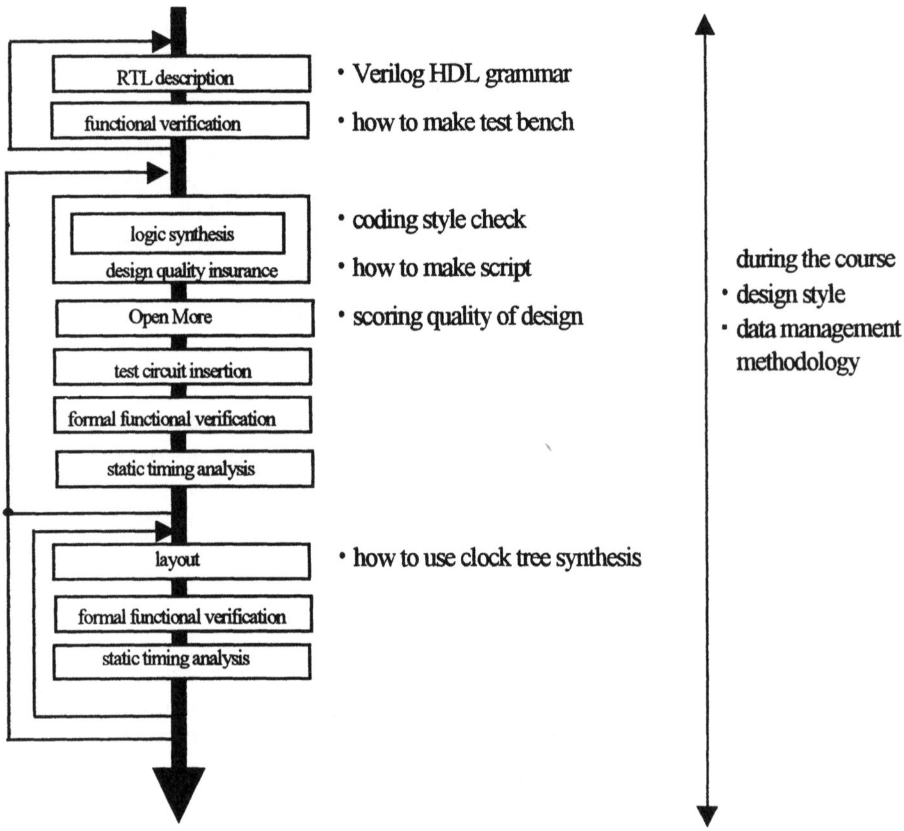


Fig.1 Training design flow and teaching items

# A UNIVERSITY NETWORK FOR THE SILICON INDUSTRY

PLF Hemment[(1)] and CM Dyson[(2)]
[(1)]*University of Surrey, Guildford, Surrey, UK*
[(2)]*National Microelectronics Institute, Edinburgh, UK*

*A new part-time modular postgraduate Masters-level programme developed by a consortium of ten universities is described. The philosophy and strategies adopted by the academic and industrial partners in designing a world class, relevant and cost effective programme for professional engineers employed in the Silicon industry are discussed. The route for exploitation is through the creation of a 'University Network for the Silicon Industry'*

## 1. Introduction

The growth in the use of electronics in everyone's daily life continues to explode. This has led to electronics becoming the fastest growing industry in the world - and it depends on a supply of semiconductor Integrated Circuits (ICs). However, the rate of growth continues to make it difficult for the semiconductor industry to match capacity with demand. It takes close to three years to bring a manufacturing plant to full capacity. By the time a shortage has arisen, it is too late to invest.

Only twelve months ago the semiconductor industry was facing temporary over-capacity and the associated low product prices. And now we are faced, once more, with lengthening lead times for products, capacity shortages and rising semiconductor prices. Plants in the UK are running at capacity. Many are expanding and taking on more staff. Filtronic have purchased the mothballed Fujitsu fab near Durham, and are ramping-up to produce GaAs circuits for mobile phones. The empty plants in Newcastle, Dunfermline and Newport are attractive facilities to bring back into production rapidly. This market dynamic results in the need for the industry to be able to staff new plants rapidly. This can only be achieved by ensuring that the industry builds a strong base of experienced engineers that is able to sustain periods of rapid growth. The situation is compounded by the rapid development of technology in the industry. It is necessary to refresh the skills of employees in this rapidly changing environment on a continuous basis so that they are in a position to introduce and operate the next generation of advanced processes.

Against this background, the semiconductor industry and academia in the UK have been collaborating for over 8 years to address the requirements for education and training. This has led to a number of initiatives, including:

- The development of a standardised skills pipeline for the industry.



*B. Courtois et al. (eds.), Microelectronics Education,* 35–38.

- A vocational education and training system to develop operators and technicians. This involves vocational education qualifications run at a network of Further Education Colleges and vocational training qualifications that are taught 'on the job'. This work is currently being extended by the colleges to provide the same material in a distance learning format.
- A part time postgraduate programme at Masters-level, which is the subject of this paper.

## 2. Justification and Development

Silicon manufacturing demands knowledge of materials, device physics, process chemistry, batch manufacturing, statistics, quality management and other disciplines. There is general agreement that this breadth of knowledge cannot be achieved through undergraduate courses. Rather the industry prefers to attract employees with undergraduate training in a wide range of relevant disciplines and for them to apply their skills in teams to address the industries specific problems. This approach brings flexibility and adaptability to problem solving. However, it does mean that there is a requirement in the industry to provide ways in which engineers can address any gaps in their education.

The new MSc programme evolved in 1997 from discussions between industry and academia to address the shortage of process and manufacturing engineers. It was recognised that there was (and is) a world-wide shortage of up-to-date specialist, postgraduate training and education courses in IC manufacture. At that time, UK industry estimated that it would require the equivalent of an additional 60 MSc graduates each year to maintain its rate of growth. Yet, ironically, during the late 1990's, a number of UK University courses had been discontinued or postponed due to the dearth of suitable applicants.

The solution adopted was for industry and academia to work together to develop a new, dedicated set of Masters-level postgraduate modules, based on the expertise in silicon processing and technology already existing in a consortium of ten UK Universities[1]. Each of the partner universities has justified its inclusion through its academic strength in a specific topic area. The active participation of industry ensures that the technical content of the course is fully up to date and relevant. It also provides a steady flow of applicants.

The title of the programme is "Advanced Silicon Processing and Manufacturing Technologies". The Engineering and Physical Sciences Research Council (EPSRC) provided £435,000 funding, over 5 years, for the development of the programme under its Integrated Graduate Development Scheme (IGDS). Industry has supported the programme by contracting to fund a minimum number of delegates each year.

---

1 University of Edinburgh
University of Glasgow
Heriot-Watt University
University of Liverpool
University of Newcastle
UMIST
University of Southampton
University of Surrey (Co-ordinator)
University of Wales, Cardiff
University of Wales Swansea

The programme is designed for graduate engineers employed in the silicon manufacturing sector. It provides specialist training for staff wishing to extend their technical knowledge and a "fast track" for career development of future senior engineers and engineering managers.

It is very flexible. Delegates can chose to study a single module or take 4 modules to obtain a Certificate, 8 modules for a Diploma, or 8 modules and a in-company project to obtain an MSc. They may join the programme at any time and, if they so wish, defer registration for a qualification until after taking a few modules.

The large number of University partners created a number of logistical and organisational problems, which have been resolved in imaginative, novel ways:

*Ownership:* the programme is jointly owned by the partner universities.

*Quality assurance:* the partners have established a single Validation Panel to ensure that the highest academic standards are maintained across all partners. It avoids each University having to validate all other University's courses before awarding a degree. This is a unique concept in the UK and has proven to be highly successful.

*Registration and Awards:* delegates register with one of the partner universities and the academic awarded is in the name of that university. They may transfer their registration between Universities.

*Common procedures:* UK universities are autonomous institutions, each with its own statutes, regulations and procedures. The consortium has endorsed a Memorandum of Agreement documenting common procedures for the smooth running of the programme. In a few situations, such as appeals and discipline, the individual partner's regulations take precedence.

The difficulties in the semiconductor sector during 1998/99 temporarily reduced demand for the course and delayed the start until the current upturn in the industry. The first intake of delegates was accepted during Spring 2000.

## 3. Organisation and Management

The management structure of the programme is shown in the following diagram:

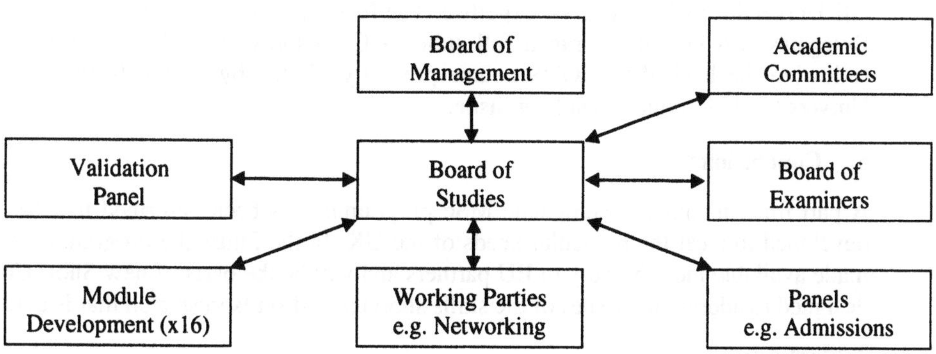


Figure 1: Management structure of the Programme

## 4. Technical Content and Delivery

The programme presently offers a choice of 16 taught modules. Each module has a duration of about two months involving distance learning plus an intensive one-week residential week held at the partner university that is delivering the module. An MSc delegate will be absent from his/her place of work for no more than 12 weeks during a typical three year period of study. Each unit of study carries a fee of £1200 so an MSc is costed at £10,800. Current module titles and the University convenors are:

| | | | |
|---|---|---|---|
| 1. Device and Process Integration for Silicon ICs | University of Edinburgh | 9. Device and Circuit Design for VLSI | University of Cardiff |
| 2. Measurement Techniques and Failure Analysis | UMIST | 10. Ion Implantation | University of Surrey |
| 3. Power Devices and Processes | University of Swansea | 11. Lithography | University of Edinburgh |
| 4. Statistics for Semi-conductor Manufacturing | Heriot-Watt University | 12. Etching: Physics and Chemistry | University of Glasgow |
| 5. Layer Deposition and Diffusion | University of Southampton | 13. Test, Assembly and Packaging | University of Cardiff |
| 6. Interconnect and Metallisation | University of Newcastle | 14. Oxidation and Isolation Technology | University of Liverpool |
| 7. Optimisation of Processes | University of Edinburgh | 15. TCAD and Compact Modelling | University of Newcastle |
| 8. Production Management | Heriot-Watt University | 16. Device Operation and Process Architectures | University of Glasgow |

The technical content of each module has been developed in close collaboration with industrial advisors. When appropriate industrial experts act as lecturers.

## 5. Future Plans

This programme provides the nucleus of a 'University Network for the Silicon Industry'. The courses and material developed will be used to provide less intensive short courses and as the basis for courses in Further Education colleges and schools in the UK.

Work is also progressing with a view to establishing a parallel Masters-level programme in IC design and embedded software. Again, modules will be provided by UK Universities with specialist expertise. The focus on distance learning will continue. Delegates will be able to run a full suite of CAD tools remotely over ISDN lines, following the lead of the IGDS course presently offered by Bolton Institute and the University of Northumbria at Newcastle.

## 6. Conclusions

A part time, modular postgraduate MSc programme has been described that has been developed to meet the particular needs of the UK. In the future the programme will be made available more broadly to EU partners and will be the basis of new Short Courses designed to address the issues of the skills shortage and skills mismatch that is evident.

**Acknowledgements**

The authors would like to acknowledge the work of David MacNicol at the NMI, Sandy Peace at the University of Surrey and all the contributors to the programme from academia and industry. Details of the programme can be obtained from s.peace@surrey.ac.uk.

# MULTIMEDIA

# EDUCATIONAL MULTIMEDIA AND MICROELECTRONICS : SHOULD WE THINK FIRST ABOUT THE CONDITIONS OF USE ?

H. Choplin*, J.-F. Rouet**, D. Degrugillier***, M. Milon****, R. Pacalet*****, J. Provost*****, A. Galisson*
** Centre de Ressources en Innovation Pédagogique et Technologies, (CRIPT), Ecole Nationale Supérieure des Télécommunications (ENST) 46, rue Barrault, 75013 Paris*
*** Laboratoire de Langage et Cognition (LACO-CNRS), Université de Poitiers*
**** CRIPT, Ecole Nationale Supérieure des Télécommunications de Bretagne*
***** CRIPT, Institut Nationale des Télécommunications d'Evry*
****** ENST*

## 1. A research-development context

Since 1997, CRIPT (Centre de Ressources en Innovation Pédagogique et Technologie) at the GET (Groupe des Ecoles des Télécommunications which include notably INT Evry, ENST Bretagne and ENST Paris) has coordinated the development of educational multimedia in microelectronics : « L'électronique en questions », which initiates an educational hypermedia collection dedicated to scientific and technical higher education. This product has been developed by several scientific and technical high schools and universities (GET, ENIC Lille, ENSP Strasbourg, EPFL Lausanne, IUT Cachan). It is now available and is aimed at students who are starting their electronics program. Its development involved important contributions in psychological evaluation provided largely by and with the Language and Cognition Laboratory LACO-CNRS at the University of Poitiers. This paper presents the results of this evaluation and especially the concrete recommendations it seems to imply.

## 2. Evaluation : aims, hypotheses and procedures

This evaluation concerned two multimedia sequences (each one involving about ten minutes of « linear reading ») dedicated respectively to adders (digital electronics) and transistors (analog electronics). Our aim was to evaluate the impact of two parameters on the learning process :

- multimedia design and especially the *importance of images and animations* with regard to the text (spoken or written) ;
- the *task* given to the student before using of the sequences.

Our hypotheses were :



*B. Courtois et al. (eds.), Microelectronics Education,* 41–44.

- images and animations can enable students to solve engineering problems better than text ;
- students use different learning strategies when the purpose of the task is to solve a problem and when it is to memorize.

Two designs for each sequence were developed. The first design (called « image ») is founded on a design which favors images and animations for four reasons : images and animations include rich information ; they are systematically provided before the text which is associated with them ; their value is spatially increased on the screen (they are often located on the center of the screen) ; they raise the level of interactivity because the user can control the unfolding of the sequence. The second design (called « text ») implements contrary design options (animations have disappeared and images are only decorative illustrations ; they are also systematically presented after the text ; spatially devalorized ; associated with « poor » interactivity).
26 students were recruited. We gave them a task which was either to memorize the information or to solve an engineering problem. After having studied one design (« image » or « text ») of a sequence for 20 minutes, each student had :

- on the one hand to answer questions concerning factual knowledge (for example : « *What is the link between the output current and the input voltage in a bipolar transistor ?* ») and
- on the other hand to solve an engineering problem corresponding to the sequence studied.

These questions and problems rely of course on the concepts included in the two sequences (we do not include the whole psychological methodology in this paper : pre- and post-test scores, etc.).

## 3. Evaluation results

The evaluation brings to the fore two major results which are independent of the content (transistors or adders) :

- surprisingly, the « image » design does not seem to imply the best (or lower) performance for solving the problems (it has no effect on the acquiring of factual knowledge, either) ; to be precise, the computer recordings of the students' activity show that *they did not really treat the images and animations* ;
- the task is very significant : the students who had the task of « solving a problem » performed better on problems than those who had the task « of memorizing ». But the latter answered the factual questions better.

## 4. Discussion

The effect of the task shows that the learner's way of studying is not the same when he is in a problem solving context or in a memorizing one. In particular, a memorization task seems to « prevent » the student from finding a complete and justified solution to the problem. This effect is independent of the type of design (« image » or « text »). This suggests that the « wealth » of the medium cannot make up for an unfavourable

learning context (task). Finally, it shows that these students have a good capacity for self-regulation. It allows them to adapt their way of studying to the demands of the task. This is not always the case with students (Butler & Winne, 1995).
The absence of psychological effects linked to the images and animations is more difficult to understand. First, it does not show that images and animations are systematically useless – even if several recent research projects (Wright, Milroy & Lickorish, 1999) suggest that the effect of animation could be disturbed by many factors : ergonomical (too poor or overloaded animations, too slow or too fast rhythm) and/or pedagogical (the animations considered as a « show » may lead the learner to a certain passivity). In fact, the evaluation and literature's results lead us to think that the use of images and animations should not be systematic but should be subordinated to four main parameters :

- the *task* given to the student before using the multimedia ;
- *the student's attitude to knowledge.* It is very possible that the images and animations included in our « image » designs concerned points which were not difficult and abstract enough *for the students we had solicited* : were their contribution sufficiently significant with regard to the (spoken and written) text offered ? (Levin, 1987) ;
- *the student's attitude to media.* « Our » students seem to consider images and animations in general as not really fundamental for learning (for example they are considered to be subordinate to mathematical equations) (Choplin, 99). These two last points (attitude to knowledge and to media) suggest that the images and the animations included in our « image » designs may have more effect on other types of students.
- the *study time.* It is very possible that assimilating images and animations takes a long time, especially as our « image » designs (with images and animations appearing before text) aimed to stimulate the student's questioning. But is it really possible to puzzle about electronics in 20 minutes ? In other words, the students may have benefitted more from the « image » design if we had allowed more time (longer than 20 minutes) to study the sequences or if the students had studied in a temporal context which was closer to that of « self regulated learning ».

## 5. Recommendations

In general, these results confirm that pedagogical innovation is not essentially determined by technological conditions and that it poses complex problems. In fact, our results lead to design educational multimedia with dependance on non-technological requirements : psychology of the student, the task given to him (especially by the teacher), the study time. Therefore, *before designing and developing pedagogical multimedia, it seems to be essential to define the pedagogical context* (the characteristics of the student, the way the teacher asks to the student to use the multimedia, the study time) in which the multimedia will be used.
From a more concrete point a view, it seems to be possible to require educational multimedia designers :

to precisely define the goals of the reading of the multimedia. For example, from a pedagogical point of view focused on the learner's activity, to define sequences through questioning (for example : « *How can you design a high performance adder with logical gates ?* », rather than the statement « *The adder's design* ») which could lead the student to solve a problem (and not to memorize information) ;

to choose images and animations directly connected with this goal (in our example : it is no use developing animations describing the way the logical gate operates if the aim is to answer to the question : « *How can you design a high performance adder with logical gates ?* ») ;

to concentrate the development of these images and animations – whose time requirements are always strong – on points which are difficult and abstract for the students at whom the multimedia is aimed. From this point of view, it is not certain that the animated diagram of a circuit is systematically useful for learning. Il is probably important to define images and animations whose function is not only to *represent* or to *organize* the text but also to *interpret* it (then the graphics describe unfamiliar, difficults or abstracts concepts) or, even better, to *transform* the text (then they transform it in order to have a direct impact on the learner : they target the crucial information to be learnt for example by recoding it in a more concrete and memorizable way) (Levin, 1987) ;

to increase the value of images and animations by adding a sound track. The comments do not necessarly have to be explanatory : substantial or dense information seems to be treated with more difficulty when it is spoken than when it is written. They can help the student to understand the importance of the images and animations and to treat them (the evaluation also led the students to complete a questionnaire concerning the balance and the role of the media : images and animations, texts, sounds). Thus, they are not directly connected with the content but are rather «meta » sounds (for example : « *Be careful, we are now going to study a crucial or/and very difficult point concerning the design of adders* !»). These comments seem to be required by the students because they seem to be able – at least partly – to make up for the teacher's absence.

We intend to develop these recommendations in future developments of our educational hypermedia collection « en questions ».

## 6. References


Butler, D.L. & Winne, P.H. (1995). *Feedback and self-regulated learning : A theoretical synthesis.* Review of Educational Research, 65, 245-281.

Choplin H. (1999). *Les élèves, Prisméo et les hypermédias pédagogiques : éléments d'analyse* in Choplin H. (éd)., Prisméo, une expérience de cédérom pédagogique (96-98), Rapport final : bilan et perspectives, CRIPT, GET , 1999.

Levin J.R., Anglin G.J., Carney R.N. *On empirically validating functions of picture in prose* in H.A. Hougton, D.M. Willows (eds), The psychology of illustrations, vol. 1 : basic research, New York, Springer Verlag, 1987.

Wright, P., Milroy, R., & Lickorish, A. (1999). *Static and animated graphics in learning from interactive texts*. European Journal of Psychology of Education, 14, 203-224.

# THE USE OF STREAMING MULTI-MEDIA IN MICROELECTRONIC EDUCATION

PROFESSOR CHRIS JESSHOPE
Institute of Information Science and Technology
Massey University, New Zealand.

## 1. Low-cost Multimedia in teaching

At Massey University we have developed and exploited tools, which enable teachers to create and use multimedia in their teaching. In a previous Microelectronic Education conference it was claimed that Multimedia was very expensive to produce, costing perhaps millions of dollars[1]. In this paper we will demonstrate that this need not be the case. We will describe a 20 hour multimedia course in Advanced Computer Architecture that was developed by the author in less than 200 hours of preparation time (actually less than $10K). See (http://www-ist.massey.ac.nz/~crjessho/comp_arch). This course (and the cost estimate) was based on starting with a set of PowerPoint presentations, such as may have been used in lectures.

## 2. The AudioGraph System

The AudioGraph system [2, 4-7], is the result of a collaborative project between Surrey and Massey universities in the development of multi-media authoring tools for use by academics - rather than multimedia professionals, for the production and delivery of course-ware via the world-wide web. The goal was to place the production of the media firmly in the hands of the educators rather than the technologists.

At the outset the aims of this project were:

a) to provide simple-to-use tools with an intuitive interface capable of producing multimedia presentations for a wide range of teaching activities, so that anyone, with a basic computer literacy, could become a multi-media author;

b) to ensure that the preparation time for multimedia presentations is orders of magnitude shorter that the industry norm of 200 hours of preparation time for each hour of presentation;

c) to ensure that the footprint of the web sites produced was small enough so that users could provide a large corpus of multimedia tuition on a reasonably sized server;

d) to enable students, if necessary, to be able to stream the media content at modem connection speeds; and finally

e) to provide cross-platform delivery of the multi-media content using the Netscape browser plug-in interface.



*B. Courtois et al. (eds.), Microelectronics Education,* 45–48.

The project has been successful and all of the above goals have been met. The tool is available from the NZEdSoft Institute at Massey University and is being used by a number of Universities around the world[7]. The key to goal 1 was to use a novel multi-media authoring paradigm, which is unique in that it treats the media as a pure sequence rather than a set of timed events. An iconic interface is that is used to represent the temporal nature of the presentation provides the intuitive interface for editing.

Goals 3 and 4 are met using some aggressive compression techniques. We use GSM sound compression and PNG image compression. Both contribute to small web sites and low-bandwidth download.

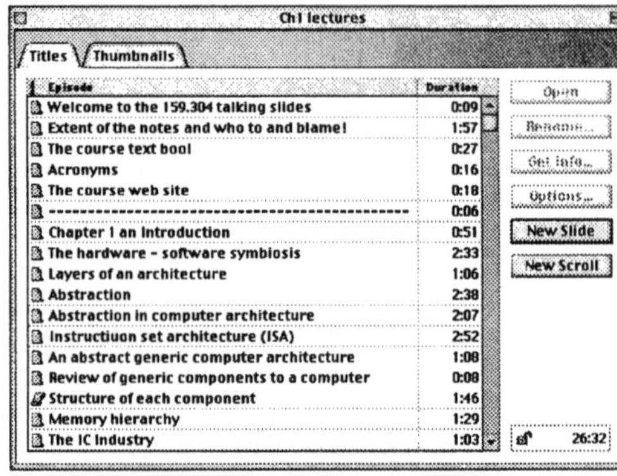


Figure 1. Opening screen showing presentation with durations

## 3. A Quick Guide to the Tools

The opening screen of the AudioGraph recorder (authoring tool) is shown in figure 1. This gives a list (or thumbnails) of the set of slides that will make up a presentation unit. One basic technique is to take a PowerPoint presentation, export it, read it in the recorder and annotate it with voice and vector graphics. The view shown is tabular and gives the approximate length of each unit as well as that for the complete presentation.

Figure 2. shows one of the slides being edited. Three windows are shown, the main display, which gives a preview of the presentation, the tools window, which selects the graphics and other media tools and finally, at the bottom of the window, the edit console that gives an iconic view of the time line of the presentation. This window also has controls for playing and editing presentations.

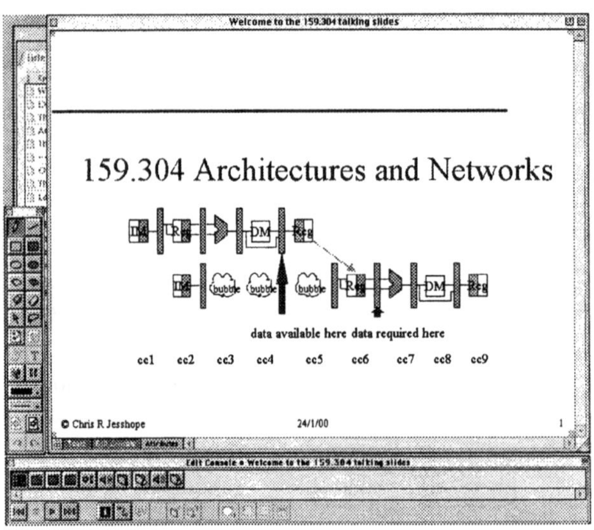


Figure 2 Editing a slide from the presentation

Once the presentation is completed a web page can be automatically created. At this stage in the production of the imedia, all sound and images are compressed into a form that can be played back using the AudioGraph web-broswer plug-in. Figure 3 shows a screen shot of the Netscape navigator playing back the presentation. Two windows are shown. One contains the same slide as in figure 2, which is being played by the plug-in.

The other shows the index window, which gives an html view of the structure of the presentation. There is a correspondence between this window and the opening window in Figure 1. If required this may be edited in a conventional html editor, to provide additional text, images or links.

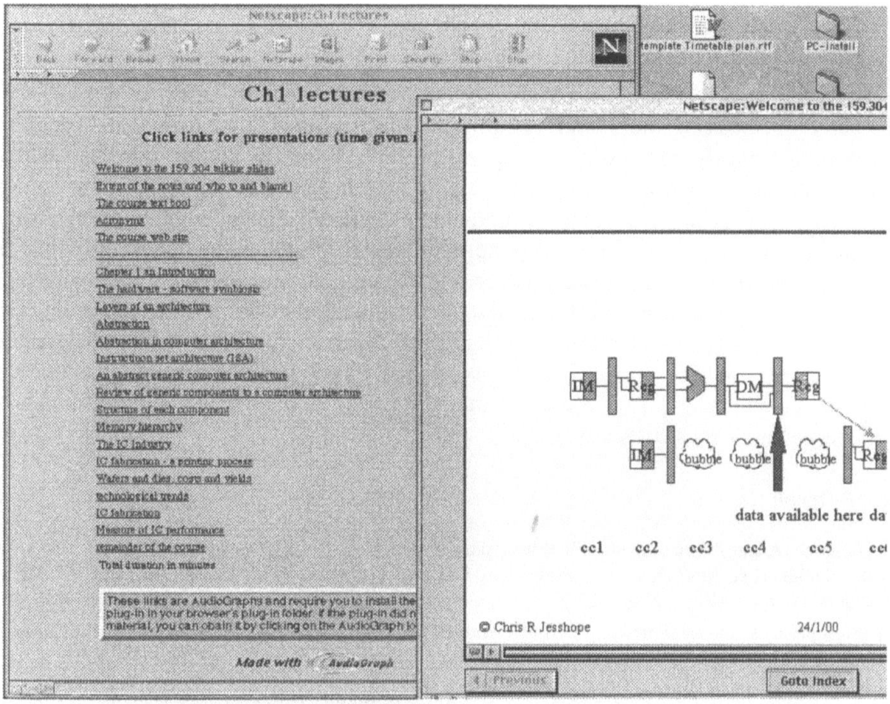


Figure 3. The presentation as a web page, showing index and one slide.

## 4. Use of the Tools at Massey University

This tool has been used to deliver 2 courses at Massey, both available on-line at (http://www-ist.massey.ac.nz/~crjessho) and described in more detail in [3]. One course is a first year digital design/general computing paper, which is delivered entirely in distance mode, where students will receive both CD and server-based web material. This mode makes use of recorded "lectures" presented in web pages and which can be taken at the students own pace. The multimedia lectures are supplemented by worked examples again using recorded multimedia and voice animated computer code. The course is assessed exactly the same as a conventionally delivered internal paper, by assignment and examination. The second example is a post-graduate course in advanced computer architecture, including the influence of microelectronics on computer design. This course is delivered internally, making use of the same techniques as in extramural teaching. Instead of formal lectures on this course, students again use CD and server-based web material, which comprises some 20 hours of recorded lectures in multimedia format. This material was based on PowerPoint presentations, many of which have been animated to better convey the concepts. The time freed up by this self paced delivery of material is used to provide tutorial support in smaller groups, where difficulties are discussed and analysed and worked examples are presented. The use of this tool has also been evaluated by independent researchers at Surrey University [8].

## 5. Conclusions

We have introduced tools and examples of their use. These tools are ideal for developing multimedia courses, especially in technically complex areas such as microelectronics and microprocesor desigm. However, the same techniques can be used in the training of software and equipment, as might be used in a microelectronic production environment. The key issue here, as with University courses, is that the target market for the courses are small and the change of the material is relatively rapid due to rapid technological developments. Under these conditions it is not feasible to use the might of professional mutimedia authoring tools, as the 200+ hours required for each hour of teaching material is too expensive. The AudioGraph tools reduce this development time by 1 to 2 orders of magnitude and make the use of multimedia for small target audiences cost economic.

## 6. References

[1] Galissson, A and Zysman, E, (1999) Write and use hypermedias for microelectronic teaching - how to take up the challenge, in Microelectronic Education (Eds. Mouthaan and Salm) ISBN0-7923-5107-X, Kluwer Academic, London, pp253-260.

[2] Jesshope, C.R. and Shafarenko, A., 1997, Web Based Teaching: a minimalist approach, *Proc. Second Australasian Conference on Computer Science Education*, ISBN: 0-89791-958-0, pp16-23

[3] Pearson, M. and Jesshope, 1988 C., Multi-campus teaching using computer networks, *Proceedings of the Third Australasian Conference on Computer Science Education* pp 106 - 111, July 1998.

[4] Jesshope, C and Slusanschi, H., 1998, Recording Formal Lectures for Publication on the Web, *Proceedings of the DEANZ Conference* , Rotorua, July 1998, pp167-179

[5] Jesshope, C., Shafarenko, A and Slusanschi, H. 1998 Low-bandwidth multimedia tools for web-based lecture publishing, *IEE Engineering Science and Educational Journal*, **7 (4)**, pp148-154. also published in: *Computing and Control Engineering Journal*, **9 (4)**, pp156-162 and IEE Computing Forum, http://forum.iee.org.uk/forum/library/, September 1989.

[6] Jesshope, C. and Slusanschi, 1998, Distance Education Using the AudioGraph Authoring Tool, *Proc. Intl. Conf. on Multi-Media Information Systems in Practice*, pp266-278, Hong Kong, December 1998 (Springer ISBN 981-4021-53-9)

[7] Jesshope, C., 1999 Web-based Teaching - Tools and Experience, *Australian Computer Science Communications*, **21**, (1), pp27-38, Springer, ISBN 981-4021-54-7, Proc Australasian Computer Science Conference, ACSC99, Auckland.

[8] NZEdSoft, AudioGraph web site, 1/12/99 http://www.nzedsoft.com/

[9] Segal, J. (1997) An evaluation of a teaching package constructed using a web-based lecture recorder, *Assoc. for Learning Technolgy Journal*, pp32—42.

# Distributed VLSI Design Education using MOCA: the Multimedia On-line Collaboration Architecture

Wayne Burleson, Jeff Peden, Chris Leonardo
*Department of Electrical and Computer Engineering, University of Massachusetts Amherst*

Key words: VLSI design, multimedia collaboration tools, distance education,

Abstract: MOCA is a suite of Internet tools to facilitate group-work in a distance learning environment, built using standard Java utilities. MOCA allows several users to work simultaneously on multimedia documents using a multimedia whiteboard and text-based chat. The system has been targeted for an advanced VLSI chip design course with student design teams composed of undergraduates and graduates at the University of Massachusetts at Amherst as well as off-campus students at a variety of industrial sites (e.g. Intel, IBM, Compaq, AMD). Advisors include University professors and teaching assistants as well as industrial mentors.

## 1. Introduction

VLSI Design is inherently project-based and collaborative, uses large multimedia data formats, and often relies on design teams which are geographically distributed. We have developed tools to facilitate a VLSI design project course in a distance-learning environment. Many VLSI design courses are project-based and many are offered through distance education, but rarely do student and faculty teams span multiple institutions due to the logistics of group meetings. In a recent collaborative embedded system design project between UMASS, ENST/Paris and Pusan National University we learned that geographical and cultural distances presented difficult barriers to communication, cooperation and learning. We now propose that a truly integrated collaborative design experience can be facilitated by platform-independent, Internet-based tools to leverage the inherently multimedia aspects of VLSI design representations while preserving the human side of informal design reviews.



*B. Courtois et al. (eds.), Microelectronics Education,* 49–52.

## 2. MOCA Groupware Tool Suite

Development of MOCA began as an effort to create software that would facilitate an abstraction of traditional "Office Hours" meetings between students and professors in an online environment. This was originally meant to allow students taking courses via pre-recorded videotape and DVD-based lectures to have a higher level of interaction with the professor of a course [1]. Many individual tools for text-based chat and shared-whiteboards were examined (e.g Lotus Notes and Microsoft NetMeeting), however, none of these tools provided a low-cost and extendable multi-platform solution. MOCA is currently being developed to provide the foundation for multimedia communication between students, teaching assistants and professors.

The Multimedia Online Collaboration Architecture aims to implement enterprise level communications and information sharing methods tailored to the requirements of academia. MOCA uses the Java platform, which gives the advantages of being hardware-independent, highly networkable, scalable, and object oriented. MOCA is organized as pairs of client/server modules with a common database, unified under a common application framework. Currently, the modules consist of a set of collaboration tools, a surveying component, and a multimedia presentation component. Also under development are various time and workflow management tools. This paper and presentation will demonstrate the capabilities of the MOCA system in the Spring 2000 offering of a project-based microelectronics course, ECE 559/659, *VLSI Design Project,* http://www.ecs.umass.edu/ece/vspgroup/burleson/courses/559/ . Details on the MOCA project and the overall Multimedia Curriculum research project can be viewed at http://www.ecs.umass.edu/ece/dvd or in [1].

A screendump of the MOCA white board used in a VLSI design review of a Huffman Coder is shown below. VLSI simulations can be displayed and annotated in a similar manner. The upper left pane is a list of users sharing the whiteboard. The lower left pane is a list of channels available to the users (i.e. other projects). The main upper right panel is a plot of a VLSI chip layout. Annotation is added using the toolbar just below to enter text and simple sketches to otherwise transparent overlays of the drawing. The red circle is an annotation overlay added by one of the users to refer to a portion of the chip that is being discussed in the corresponding text-based chat. Finally, the lower right pane lists the various users, their corresponding drawing layers, and whether that layer is currently being displayed. This allows large group discussions of a single drawing, sub-groups of which can have side discussions by specifying only certain layers to be displayed.

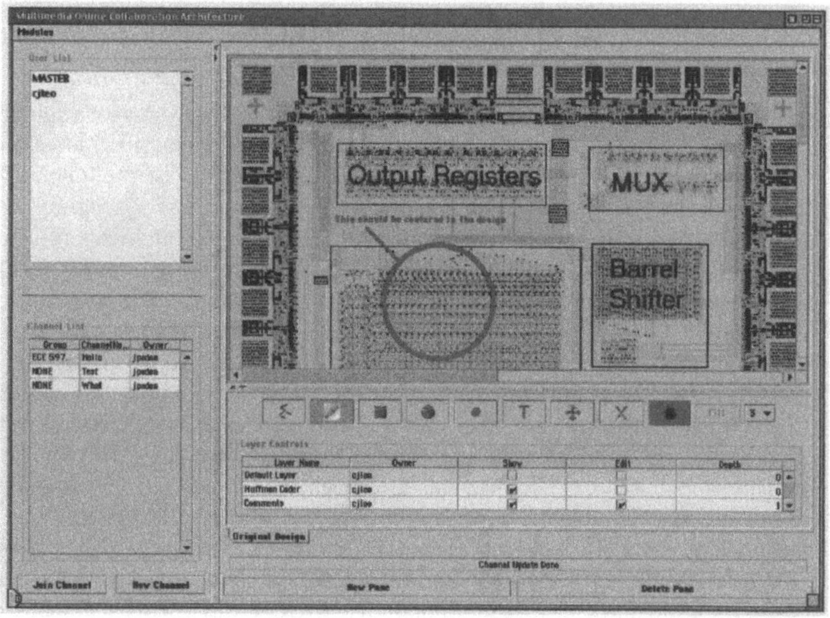


*Figure 1.* MOCA Whiteboard discussion of Huffman coder layout

## 3. Challenges in a relevant VLSI project course

In Spring semester 2000 we offered a novel version of our VLSI design project course which attempts to integrate current research at UMASS while overcoming some of the shortcomings of previous offerings of the course. Teaching an advanced VLSI course involves some difficult compromises [3]: Teaching the importance of **deep sub-micron effects** such as interconnect parasitics, power distribution and clocking while keeping the design small enough to be manageable and fabricated in the MOSIS process; Giving students **realistic constraints** on their designs, including I/O, programmability, clock-rate, power; Exposing students to **advanced research** in VLSI without giving them a skewed view of industrial VLSI design methodologies and tools; Teaching the techniques of **re-usable** cells and components, both from the standpoint of the original cell designer and the re-user; Teaching about **testability** and its importance in complex designs.

To meet the pedagogical challenges described above, we re-designed the course so that all students would be working on different aspects of a single large tiled reconfigurable architecture. The fundamentals of this architecture have been developed by Prof. R. Tessier at UMASS with VLSI implementation details developed by Prof. W. Burleson and a novel compiler by Prof. J. Babb at Princeton University. We believe that this type of inter-disciplinary and inter-institution collaboration is critical for future advanced computer architecture project which combine innovations at the architectural level, VLSI level and compiler level [3]. MOCA is used as an enabling technology to allow asynchronous and multimedia collaboration in a tool- and platform-independent method.

The overall theme of our new course is a low-power reconfigurable tiled architecture for wireless multimedia. Recent research projects [2] have argued for domain-specific heterogeneous configurable architectures as the solution to low-power wireless computing. For this course we extended this concept by looking 15 years into the future of VLSI technology [4], planning an architecture which more explicitly addresses the challenges of scalable interconnection, fault-tolerance, and compilation. The course consists of 11 groups of students, compromised of 20 ECE seniors and 13 ECE graduate students. The 3-4 student groups were organized as follows. Three groups of graduate students handled global tasks such as 1)Global interconnect, signal integrity; 2) Statically scheduled communication switch control (per tile) and global power distribution; 3) Interconnect switch circuit design (per tile) and global clock distribution. Three more groups designed tiles for general functions like FPGA, SRAM, and media execution unit. The remaining groups designed tiles for specific functions in wireless multimedia systems (e.g. motion estimation, Lempel-Ziv compression, DCT/DWT), obtaining higher performance and power efficiency through parallelism and specialized hardware. Each tile supports some amount of reconfigurability. MOCA was used for global tasks such as floorplanning, power distribution, power estimation and testing. Each global task was assigned to a graduate group however all groups provided input on the global strategies related to specific tiles. MOCA also allowed faculty, graduate students and industrial mentors to review the designs and design process.

## 4. Conclusions

The MOCA suite of tools provides a platform-independent method for multimedia collaboration. MOCA can allow much broader projects and inclusion of distance-education students and industrial mentors. Areas for future work include: 1) Full multimedia capability allowing audio and video annotation of documents and 2) Tighter integration with specific tools (e.g. schematic, layout and source code editors and simulators).

*This work has been supported by the National Science Foundation grant EIA-9812589, an NSF supplement for Research Experiences for Undergraduates and the University of Massachusetts. Many recent MOCA extensions were developed in the Fall 1999 offering of a UMASS senior/graduate-level course ECE 597M, Distributed Application Design in Java.*

## References:

[1] W. Burleson et al, "Educational Innovations in Multimedia Systems", *Frontiers in Education Conference*. Puerto Rico, 1999. http://fairway.ecn.purdue.edu/~fie/fie99/Index.html

[2] H. Zhang et al, "A 1V Heterogeneous Reconfigurable Processor IC for Baseband Wireless Applications" , ISSCC, 2000

[3] H. DeMan, "Designing Systems-on-a-Chip: Implications for Electronics Research and Education", Microelectronic Education Conference, Washington, D.C. 1999.

[4] K. Sylvester and K. Keutzer, "Getting to the Bottom of Deep Sub-Micron II: A Global Routing Paradigm", Intl. Symposium on Physical Design, 1999.

# ROAD-MAPING AND INFRASTRUCTURES

# THE MEDEA EDA ROADMAP, AND HOW IT SHOULD INFLUENCE MICROELECTRONICS EDUCATION

Joseph BOREL
*STMicroelectronics*
*850, rue Jean Monnet*
*38926 Crolles Cedex*
*France*

In the microelectronics industry there is a strong process culture because process started to be engineered in the early 70's and 80's. Device and circuit cultures are familiar to young engineers because they are the basis of circuits and products enabling European systems to exist.

CAD or Design Automation is less known because it is globally less mature and more complex to understand - being in continuous evolution. Nevertheless Design Automation is mandatory to create a product, and more and more becoming a bottleneck in the System on a Chip (SoC) industry. Therefore particular attention should be paid to this area in microelectronics education.

In a first step, we will recall what is the place of Design Automation in the supply of products (and mainly SoC). The present possibilities and future trends will be discussed. It will be shown how the field is still expanding and how engineering efforts are improving the industrial maturity of the industrial solutions.

The MEDEA EDA Roadmap will be presented. This initiative at the European level originates from the collection of users and manufacturers of SOC in the MEDEA program: the European systems companies (ALCATEL, BOSCH, SIEMENS, ...) through their needs in communication, information technologies and automotive with respect to the semiconductor companies (Infineon, Philips, ST). The EDA companies are mostly from the US, with some start-ups from Europe in strategic fields. They are providing the Design Automation tools and flows for the design of these very complex SoCs but are hardly contributing to educating electronics engineers, the development community being located in the US.



*B. Courtois et al. (eds.), Microelectronics Education,* 55–56.

Concerning advanced concepts in EDA solutions for SOC Europe is recognized as a leader and most of the larger US EDA companies have strong links and development programs in Europe. Nevertheless this is not creating the basis of a strategic industry (as EDA is today) and a dedicated effort is needed in education to reach the proper understanding and maturity level.

We will analyse the complexity of handling education in the EDA field (where a large spectrum of knowledge is needed), and the specific situation of Europe in this area (wide research knowledge but very limited business activities) will be described. The potential impact of the MEDEA EDA Roadmap will be analysed and described as a European forum for education. In this frame, engineers of various cultures can meet and grow in their perception of the future of SoC and potentially earlier bridge the gap between Technology, Design Automation and the end-user definition of needs.

# The MOSIS Service
## *IC Prototyping and Educational Program*

César A. Piña
*University of Southern California,*
*Marina del Rey, CA, USA*

***Abstract:***
*MOSIS is a low-cost prototyping and small volume production service for VLSI circuit development with a worldwide customer base. Since 1981, the service has fabricated more than 40,000 integrated circuit designs for use by commercial firms, government agencies and universities and it has served as the model for similar operations throughout the world. The present status of the MOSIS Service, current technology offerings as well as the status of the VLSI Educational Program will be discussed.*

## INTRODUCTION

The MOSIS Service was established in 1981 by the Information Sciences Institute of the University of Southern California in response to a request from the Defense Advanced Research Projects Agency (DARPA) of the U.S. Department of Defense. DARPA needed to provide a cost-effective method of access to advanced IC fabrication technologies for their research community. This effort was expanded to include educational activities in 1985 through the signing of a Memorandum of Understanding between DARPA and the National Science Foundation (NSF) to include the fabrication of student projects from University classes in VLSI design. Both DARPA and NSF provided funding for fabrication of these educational designs from 1985 through 1995. The funding situation changed significantly in 1994 as a result of the shrinking U.S. Federal budget. At this time, DARPA withdrew from the educational program, and NSF announced its planned withdrawal by 2000. Additional sources of funding from private industry together with MOSIS contributions, have made it possible to continue the educational program.



*B. Courtois et al. (eds.), Microelectronics Education,* 57–60.

**CURRENT STATUS**

Since 1994, the Service has been self-sustaining, 98% of the current income resulting from purchase orders placed by industry, universities, and research laboratories. This is in marked contrast to 1991, when 85% of the income received was the result of direct funding by Government Agencies such as DARPA and NSF.

The cost of fabricating prototype quantities is kept low by aggregating multiple designs onto one mask set. This allows customers to share overhead costs associated with mask making, wafer fabrication and assembly. The prototyping costs are controlled by the type of the phototooling used to generate the wafers. For example: full wafer lithography for 2μ feature size has an effective payload of over 6,000 sqmm. By comparison, 5x reticles for sub-micron processes have a maximum effective payload of 400 sqmm. Fortunately, the reduction in feature sizes (with its associated photomask cost increase) for the most advanced technologies has also been accompanied by an increase in the payload of the photo mask reticles (from 100 sqmm for 1.5μ in the late 80's to over 400 sqmm for 0.18μ today). This is a 4X increase in effective payload area and has made it possible to continue to produce economical multiproject wafer runs.

Interfaces have been established to CMOS processes having feature sizes ranging from 2μ to 0.18μ. A number of processes support NPN transistors, second poly and linear capacitors, as well as integrated MEM devices. A variety of CAD tools and libraries, both foundry-specific and vendor-independent, can be used to prepare designs for inclusion in the multiproject runs. Electrical test data and SPICE parameters for every wafer lot processed is provided to the users.

MOSIS has a complete, well-equipped measurement laboratory with full time staff. Equipment in the laboratory includes: an Electroglas 4080X wafer prober with hot chuck, a Karl Suss PA-200 analytical wafer prober; plus a number of smaller engineering and reliability probe stations. For electrical measurements: an IMS ATS-1 200 MHz, 208-pin digital tester; Keithley S350VX, S400UX, and S900NT Semiconductor parametric test systems; Agilent 85107 network analyzer and 4156B and 4145B semiconductor parameter analyzers; Agilent 4194A impedance/gain-phase analyzer and a 8970 noise figure analysis system. Dual pulse generator, LCR meters, digitizing oscilloscope, Keithley voltage and current sources, electrometer, nanovoltmeter, etc. as well as Agilent power supplies, frequency counters, etc. complete the electrical measurements suite. The optical analysis equipment includes Nikon inspection microscopes, Zeiss optical microscopes, and an Optronics DE1750D digital camera. A Sagitta NEXT-1 cross-sectioning system permits accurate and fast x-sectioning of devices when needed. For wafer-level reliability studies the S900 system is configured for TDDB, hot-electron, and SWEAT tests. A MOSIS developed electromigration test system completes this set.

**TECHNOLOGIES OFFERED**

Wafer fabrication runs in various technologies ranging from 1.5μ to 0.25μ are currently offered on a regularly scheduled basis. Commercial fabricators are used for all wafer fabrication: TSMC, Agilent Technologies, AMI, IBM, and Peregrine semiconductor. A detailed list of the technologies offered including design rules, process comparisons, prices, fabrication run schedules and other pertinent information can be obtained by accessing the **MOSIS website at www.mosis.org**. Several sets of design rules are supported and information about compatible design tools is also available at the website. Parametric test structures and yield monitors are inserted on every wafer, and each wafer received from the fabricators is tested electrically. The best wafers are then selected for shipment to customers. Lot-specific parametric results and SPICE device model parameters are extracted from the measurements on the probed wafers.

**EDUCATIONAL PROGRAM**

In the ten year period from 1990 to 1999, mosis processed a total of 12,130 student IC designs from universities in the us at no cost to the participating universities. These designs came from beginning and advanced VLSI classes totalling more than 50,000 students.

A MOSIS Advisory Panel for Education consisting of three professors from U.S. universities and members of contributing industrial concerns provide guidance and direction to the educational program as well as identifying and securing sources of funding.

Funding for project fabrication is provided through industrial contributions, in the form of cash, services, or products. The administrative expenses of the program are provided by MOSIS, as part of the operating expenses of the Service. The fabrication of the educational projects is done at no cost to the participating universities

Before 1997, the only technology offered to the educational community was the Orbit 2μ CMOS process. Other technologies could be accessed by VLSI classes only if permission was obtained from the National Science Foundation. In the last three years we have been able to increase the number of technologies offered in the Educational Program. During the period 1997 - 1999 we added the AMI 1.5μ and 0.5μ processes (Wafers donated by AMI), HP 0.5μ (Wafers donated by HP), and the TSMC 0.35μ and 0.25μ processes (Wafer space donated by MOSIS).

Participation in the program is open to Universities who conform to a simple set of requirements.

(1) Designs must be sent to MOSIS in CIF or GDSII format via the Internet.

(2) A report must be sent to MOSIS for each fabricated circuit after receipt and test of the completed design describing the testing of the circuits and the results obtained.

All projects charged to the MOSIS educational accounts must be designed by students in organised group classes. Submitted designs may not be for research projects, thesis work or individually directed study. Universities, however, may elect to purchase from MOSIS a chip in any of the offered technologies. Discount prices (from the standard commercial prices) have been established for university orders, both in the U.S. and abroad. Educational accounts can be established by sending an Enrolment Confirmation Form to MOSIS for each class that requires fabrication. Beginning classes are those with no VLSI design class prerequisite. Advanced classes require students to have completed a previous VLSI design class. The current technologies that are available for student designs and the required TinyChip sizes are 1.5μ CMOS: 2.2 mm x 2.2 mm and for 0.5 μCMOS: 1.5 mm x 1.5 mm

The MOSIS Fabrication Schedule indicates which fabrication runs will accept ONLY student designs. The number of designs allowed for a class is made after the requests from the participating universities are evaluated.

Each class is authorized to submit the approved number of TinyChips, in any of the designated educational technologies. The class instructor is free to choose which technology suits his class best as well as the project size. Since class budgets are issued in "TinyChip" equivalents, some instructors may choose to spend their budget by submitting a chip that is some integral multiple of the TinyChip size. Some classes concentrating on analog or mixed signal designs might choose 1.5 micron while others interested in shorter gate delays can elect the 0.5 micron process.

In addition to the "Educational Only" use runs, MOSIS has added a new feature to the educational program by providing access to 0.35μ and 0.25μ CMOS technology for selected projects from advanced VLSI Classes, at no cost to the participating Universities. The costs for this program are borne entirely by MOSIS. Fabrication of these projects is in the TSMC regularly scheduled fabrication runs. Professors teaching classes at accredited institutions must submit a one or two page mini-proposal with details on (1) project description (2) simulation plans (3) test plans and (4) estimated project size. These projects are fabricated on a space available basis on regularly scheduled runs. Projects may be any size up to 5x5 mm.

Beginning in the fall of 2000, MOSIS plans to provide a means for interested universities to participate in a program to add system level design to their curriculum by making use of System on Chip (SOC) techniques. Useful IP blocks will be collected and made available to participating institutions. Both hard and soft forms of reusable IP are planned; two libraries containing simple to moderately complex gates for the 0.35μ and 0.25μ TSMC processes have already been purchased and verified in silicon. Additional blocks will be made available in the fourth quarter of this year.

# Achievements and Advances at CMP

**B. COURTOIS, K. TORKI, B. CHARLOT, H. DELORI, S. EYRAUD, J.F PAILLOTIN**

**CMP, 46 avenue Félix Viallet, 38031 Grenoble Cedex – FRANCE**
**Fax: +33 476 47 38 14, e-mail: cmp@imag.fr, http://cmp.imag.fr**

***Abstract:*** CMP aims at providing Universities, Research Laboratories and Industries with the facility to have their integrated circuit projects fabricated for prototyping and low volume production. Presently, users are provided with a number of processes: CMOS down to .18 $\mu$, BiCMOS, SiGe HBT, GaAs HEMT and SOI/SOS. Micro Electro Mechanical Systems (MEMS) are provided in standard CMP runs in CMOS, BiCMOS and HEMT GaAs, using compatible front-side bulk micro-machining. MUMPS process is offered as a surface micro-machining. MCM processes, CAD software and design kits are also provided. This paper reviews the portfolio, addresses 2 main areas of development, namely deep submicron processes and MEMS, and discusses cooperation schemes.

## 1 – Introduction: the needs of CMP users

CMP cares to provide the Microelectronics community with up-to-date technologies:

- Universities need to teach the students with the processes these students will have to deal with in industry. Long ago, when commercial CAD software was not available to Universities, obsolete processes could be used but this is no longer true because commercial software is available only with up to date processes. Hence Universities have no choice but to teach using advanced processes.
- Research Laboratories usually need to have high performance technologies to validate new concepts. The quality of the research results depends mostly on the quality of the technologies. Accessing up to date technologies is a necessity.
- Industrial users also need to access state of the art because of product development time. Not using an up to date process would lead to introduce an obselete product.

In addition to workhouse advanced processes, there are niche processes that are becoming more and more standard. For instance, the HBT SiGe has been for several years considered as a "marginal" process, used only for research. Today, several foundries are offering this process, and the necessity to use it, is justified by the growth of RF application area for mobile equipment and telecommunication. Its use is expanding for some very high speed applications, like radar, microwave or RF applications. The process does not allow complex systems since the component density is low. Circuits using this process have a few tens of components including transistors, capacitors, resistors, lines and inductors. Silicon On Insulator (SOI) process has been introduced this year by CMP on its variant Silicon On Sapphire (SOS). This process is a 0.5 $\mu$ CMOS TLM/DLP process. The process is suitable for high performance analog and RF design since the transistors and the passive components are isolated from the substrate and can be isolated mutually. It is also possible to integrate very complex digital design (500 Kgates). This allows making system on chip, since the digital and analog parts can be also completely isolated on the same chip, avoiding substrate crosstalk. The breakdown of circuits per technology at CMP over the last 4 years is depicted by Table I.

| Year | CMOS | BiCMOS | GaAs HEMT | Dig GaAs | MUMPS |
|---|---|---|---|---|---|
| 1996 | 65% | 24% | 6% | 5% | - |
| 1997 | 58% | 24% | 14% | 4% | - |
| 1998 | 63% | 20% | 15% | 1% | 1% |
| 1999 | 61% | 17% | 21% | - | 1% |

**Table I:** *Breakdown of circuits at CMP*



*B. Courtois et al. (eds.), Microelectronics Education,* 61–64.

## 2 – CMP Portfolio

A brief summary of the portfolio is given here. Details can be found at http://cmp.imag.fr

### IC processes:

- .8 CMOS from AMS: DLP/DLM, 5V-3.3V.
- .6 CMOS from AMS: DLP/TLM, 5V-3.3V, high resistive poly optional.
- .35 CMOS from AMS: DLP/4LM, 5V-3V or 5V/3V with internal shifters.
- .25 CMOS from STMicroelectronics: 6LM, fully stackable contacts and vias, 2.5 V.
- .18 CMOS from STMicroelectronics: 6LM, triple well, fully stackable contacts and vias, 1.8 V.
- .8 BiCMOS from AMS: DLP/DLM, vertical NPN: Ft=12GHz, 5.5V-2.5V, high resistive poly optional.
- .8 HBT-CMOS (SiGe) from AMS: fully compatible with the .8 BiCMOS, vertical NPN with Ft greater than 35GHz, high resistive poly optional.
- .2 HEMT GaAs from PML: MMIC technology, (30 to 100 GHz), Ft=63 GHz and NF=0.9 dB at 12 GHz.
- .5 SOI/SOS from Peregrine Semiconductor: CMOS DLP/TLM, 3.3 V, based on Silicon-On-Sapphire substrate, Fmax greater than 50 GHz.

### MEMS processes:

MEMS processes compatible with microelectronics: .8 CMOS, .8 BiCMOS and .2 HEMT GaAs compatible bulk micromachining technologies.

A specific MEMS process: the MUMPs technology from CRONOS, USA: 3 layer polysilicon surface micromachining process.

### Multi Chip Modules:

- MCM L,C, D
- 3D MCM

### Design kits and Engineering kits

About 40 design kits and engineering kits are offered by CMP.

### CAD software

CMP has signed agreements with many major CAD vendors to provide users (Academia and Industry) with CAD software and IPs for the design of ICs, Systems on Chip, MEMS, MCMs.

## 3 – Deep submicron processes development

One main development axis at CMP has been recently the introduction of deep submicron (DSM) processes. For System On Chip applications it is necessary to use a very dense process. We call a process DSM when the MOS length is smaller than 0.5$\mu$. The standard processes actually adopted by industry are 0.35$\mu$, 0.25$\mu$ and 0.18$\mu$. CMP offers 0.35$\mu$ CMOS DLP/4LM from AMS (Austria Mikro Systeme), and 0.25$\mu$/0.18$\mu$ CMOS SLP/6LM from STM (STMicroelectronics).

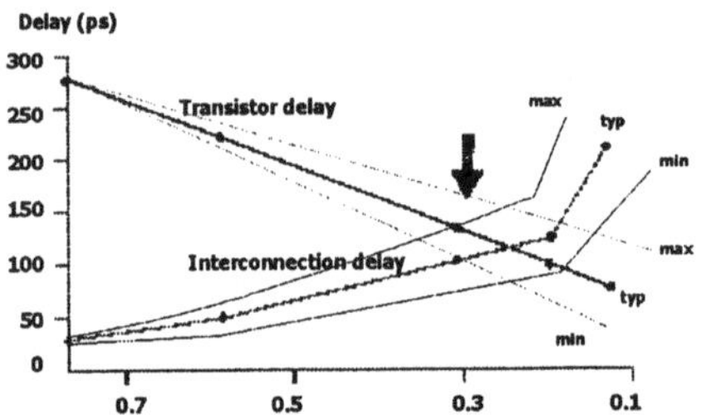


**Figure 1:** *Interconnection delay versus gate delay*

DSM processes with density up to 35 kgates/mm$^2$ (the 0.25$\mu$ CMOS from STM) allow the integration of very complex designs with several millions of transistors. Design complexity from hardware description point of view is manageable, simulation and synthesis are today accessible for such a complexity when using presently available EDA CAD tools. Place & Route and verification tools algorithms has been improved by CAD vendors to offer the power necessary to generate and verify a very complex design of several million transistors. But the only difference that allows a successful working circuit is the use of an up to date design methodology tuned for DSM technologies. It is well known that DSM processes introduce new constraints on design methodology. If traditional micronic or sub-micron methodology are used for a DSM design, the fabricated circuit will almost certainly not function as intended. Interconnection is the major actor for success or failure of a DSM design. For DSM processes, the interconnect delay has about the same value as gate delay (Fig. 1). It was expected for instance in the 1997 SIA NTRS that the interconnect delay (for Alu wires) will be approximately fifteen times the gate delay. This requires special attention to phenomena that were ignored or less investigated for micronic and sub-micron process generations. The design flow becomes more and more sophisticated, introducing new concepts: parasitics estimation during synthesis, design-planning with clock and power management, advanced place & route features

like clock-tree generation, cross-talk avoidance, power routing, accurate interconnect parasitics and timing calculation, signal integrity, and more...
The design flow becomes more complex. The consequence is that designers need to learn new paradigms, and this learning process is rather long. CMP announced the .25$\mu$ process in October 1997, the first design kits were provided to users late 1997, and the first runs started in 1998. There are a few countries who are pioneers in using such DSM processes like Denmark, Switzerland, France. A very important comment is that the later a teacher or a designer starts, the later he will be able to design on DSM processes. CMP has been a pioneer to introduce DSM processes.

## 4 – MEMS processes development

4.1 - IC PROCESSES COMPATIBLE MICROMACHINING: CMOS AND GAAS
CMP started with MEMS in 1994 by extending the IC service to the domain of integrated microsystems. Since then, a micromachining technique fully compatible with microelectronics processes, CMOS and GaAs, has been successfully used to obtain the monolithic integration of microstructures and electronic circuitry. The front-side bulk micromachining allows to build free-standing devices by stacking oxides openings in order to keep uncovered some portions of the substrate surface. Next, the structures are then released through an additional post-process of anisotropic wet chemical etching (TMAH, EDP for Si $H_3PO_4$, $NH_4OH$ and citric acid for GaAs), that removes the bulk.

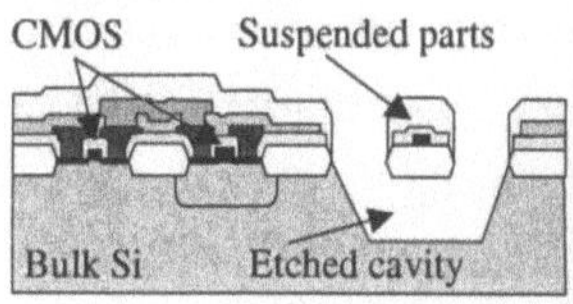


**Figure 2:** *Cross-section view of CMOS compatible front-side bulk micromachining*

**Figure 3:** *SEM photo of dies after post-process etching*

**Figure 4:** *SEM photograph of membrane shaped electrothermal converter*

Applications of such techniques are thermal sensors, such as thermopile-based devices (electro-thermal converter, IR detector, gas flow and vacuum sensors), but also mechanical devices, such as high acceleration detection, force and pressure sensors based on piezoresistive effect are available in silicon and GaAs. Finally, microwave devices like Suspended transmission lines and planar inductors present higher quality factors due to the reduction in parasitic capacitances and substrate losses. Today, the CMP service offers MEMS design using the 0.8 $\mu$ CMOS/BiCMOS from AMS foundry, and 0.2 $\mu$ GaAs HEMT MMIC process (ED02AH) from PML.

4.2 - SPECIFIC MICROMACHINING PROCESSES: MUMPS
The MUMPs (Multi-User MEMS Process) surface micromachining process is available from CRONOS through CMP since 1998.

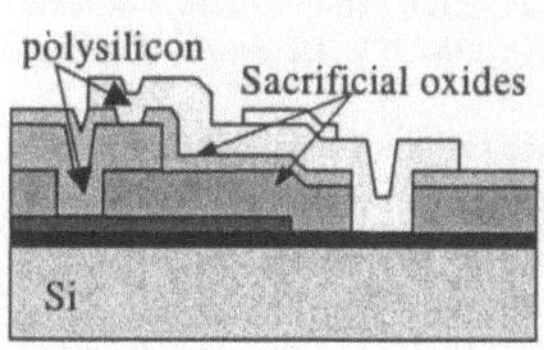


**Figure 5:** *Cross-section view of MUMPs process.*

**Figure 6:** *SEM photo of a micro hinge made of polysilicon.*

**Figure 7:** *SEM photo of a capacitive accelerometer.*

These technology offer 3 patternable structural layers of polysilicon, two sacrificial layers of deposited oxide (PSG) on a base layer of silicon nitride and a thin gold layer for optical and electrical purposes. Fixed process and design rules allows multi user runs participants to create various MEMS applications such as inertial sensors (accelerometer, gyroscope,...), electrostatic devices (micro motors, resonators, actuators,..), optical devices (Fresnel lens, blaze grating, micro-mirrors, optical bench,...) and various micromechanical applications.

## 5 – CMP: a continuous pioneer, a partner

CMP has been a pioneer in starting to provide a manufacturing infrastructure to the community in 1981. Since then, CMP has been successfully pursuing its mission and it continuously pioneered the introduction of new advanced processes like CMOS in 1983, BiCMOS in 1992, MESFET GaAs in 1992, HEMT GaAs in 1994, advanced CMOS (.5$\mu$) in 1994, MEMS in 1995, DSM CMOS (.25$\mu$) in 1997. Today, CMP is the only service offering .18$\mu$ CMOS. CMP is also a real one stop shop, offering IC processes together with MCMs, MEMS, CAD software and packaging. CMP is also a partner to qualified similar centers in the world by several ways:

- joint access to manufacturers:
  CMP/MOSIS for VSC
- global access for specified Users:
  FAPESP (Brazil) to CMP
  Swedish Universities to CMP
- Center to center agreement:
  CMP/CIC (Taïwan) on BiCMOS, HEMT GaAs
- Center to Center cooperation:
  CMP/MOSIS on BiCMOS, HEMT GaAs (CMP processes)
  MOSIS/CMP on SOS PEREGRINE (MOSIS process)

A final comment is to assess how CMP fulfills the claim to offer state of the art processes. The Table II offers a comparison of the CMP processes against industry-used processes.

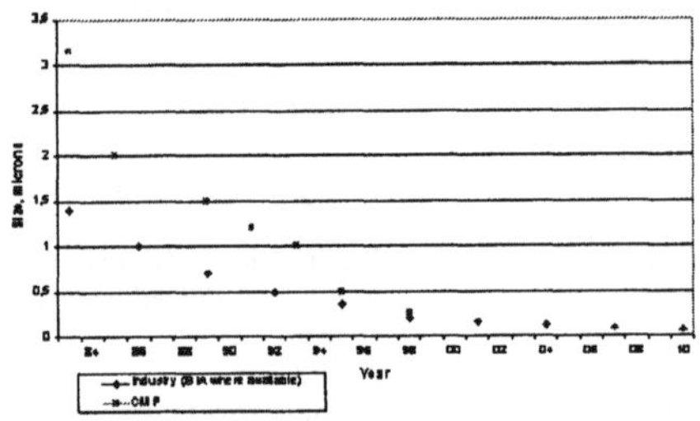


**Table II:** *Feature size, 1983-2010*

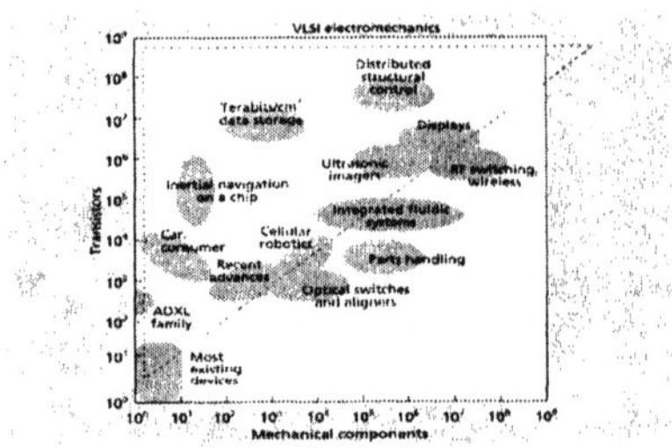


**Table III:** *MEMS development roadmap*

A challenge to CMP is to provide the MEMS community with a similar level of advancement. A roadmap like the one depicted in Table III will allow to assess the success.

## BIBLIOGRAPHY


CMP Annual Reports, CMP Information Bulletins. available upon request, send E-mail to cmp@ imag.fr - Web site: http://cmp.imag.fr

**H. DE MAN** "Education For The Deep Sub-micron Age: Business as Usual?", 34th Design Automation Conference, DAC'97, Anaheim, California, USA, 1997.

**K. TORKI, S. SAULNIER** "Accurate Timing modeling and Design-Kit Development For a Sub-micron CMOS Technology", 2nd European Workshop Microelectronics Education, EWME'98, The Netherlands, 14-15 May 1998.

**B. COURTOIS, H. DELORI, J.M KARAM, J.F PAILLOTIN, K. TORKI** "The services available from CMP", *Invited paper* at 3rd International Conference on ASIC (ASICON'98), Beijing – China, 20-23 October 1998.

**J. M. KARAM, B. COURTOIS, AND J. M. PARET**, "Collective fabrication of microsystems compatible with CMOS through the CMP service", *Materials Science and Engineering B*, vol. 35, 1995, pp. 219-223.

**J. M. KARAM, B. COURTOIS, AND H. BOUTAMINE**, "CAD tools for bridging microsystems and foundries", *IEEE Design & Test of Computers*, Apr.-June 1997, pp. 34-39.

**K. MARKUS, K. GABRIEL**, "MEMS: The Systems Function Revolution", Computer Vol. 32 N°10 – October 1999.

**B. COURTOIS**, "MPC Services available worldwide", Invited talk at IEEE Asia-Pacific Conference on Circuits and Systems (APCCAS'94), Taipei Taiwan, 5-8 December 1994.

# The EUROPRACTICE IC Service offers a new approach to deep submicron design

Dr. Carl Das[1], Dr. John McLean[2]
[1]IMEC, Kapeldreef 75, B-3001 Leuven, Belgium
[2]CLRS, RAL, Chilton, Didcot, Oxfordshire OX11 0QX, UK

**Key words:** EUROPRACTICE, CAD tools, training, prototyping, SoC, hardware-software co-design

**Abstract:** EUROPRACTICE IC Service is offering new CAD tools for hardware-software co-design and access to deep submicron technologies for System-On-Chip design.

## 1. EUROPRACTICE PROJECT

EUROPRACTICE was launched by the European Commission (DGIII) in October 1995 to help companies improve their competitive position in world markets by adopting ASIC (by EUROPRACTICE IC Service), Multi-Chip Module (MCM) or Microsystems (MEMs) solutions in the products they manufacture and to help European universities to adopt new design methodologies in their curricula and research projects. EUROPRACTICE also provides users with the training and CAD software required to design and develop their ASIC, MCM or Microsystems solution. The project has been extended for a period of 3 years under the 5$^{th}$ Framework.

## 2. EUROPRACTICE IC SERVICE

The IC Service of EUROPRACTICE brings ASIC design and manufacturing capability within the technical and financial reach of any European university, research laboratory or company that wishes to use ASICs. It offers European



*B. Courtois et al. (eds.), Microelectronics Education,* 65–68.

industry and academia a proven route to ASICs that include design support, design kits for the most popular CAD packages, low-cost prototyping and manufacture, fast turnaround, low-volume capability and a wide choice of technologies.

The EUROPRACTICE IC Service is provided by the following centres : IMEC, Leuven (B), RAL, Didcot (UK), Fraunhofer-Institut (FhG/IIS), Erlangen (D), Nordic VLSI ASA, Trondheim (N) and Delta, Hoersholm (DK).

## 2.1 CAD Tools for European Universities and Publicly Funded Research Laboratories

European industry is suffering from an acute shortage of suitable trained microelectronics design engineers. Academic Institutions require access to a wide range of IC design tools, advanced tools to train specialist microelectronics students and to undertake research projects, simpler tools are needed to introduce students to the type of tools that they are most likely to use in Small and Medium Enterprises (SMEs).

Most EDA vendors are unwilling to interact with large numbers of academic institutions. In order to overcome these problems the EUROPRACTICE IC service supports over 470 academic institutions in 32 European countries with over 25,000 licences of leading design tools for learning and research.

The normal EUROPRACTICE licences for academic institutions prohibit commercial exploitation but allow learning and non-contract research. Optional conditions that allow First-Users in companies the possibility to use academic licences, under the guidance of experienced academic staff, have been negotiated and are actively encouraged and promoted.

Publicly funded Research Laboratories are offered similar licences for their non-commercial work, subject to EDA vendor approval. Currently 48 Research Laboratories have tools via EUROPRACTICE.

## 2.2 Prototyping and small volumes

EUROPRACTICE select those foundries and technologies for which it is possible to offer prototyping as well as small volume fabrication. Also for the 0.35µ and 0.25µ CMOS technologies EUROPRACTICE offers MPW runs and small batches.

Annually, about 500 ASIC designs are prototyped on MPW runs and about 100 ASICs are turned into small volume projects.

## 2.3 A new approach to deep submicron design

In order to allow companies and universities to adopt new technologies, such as deep submicron 0.35µ and 0.25µ CMOS, a new approach was needed.

Indeed IC design has moved from ASIC design to SoC, System-On-Chip, design. For ASIC design it was sufficient to make the standard CAD tools (HDL, synthesis, layout, etc.), cell libraries and design kits and prototype fabrication available at low cost. SoC design however is more complicated and involves hardware-software co-design, integration of IP blocks and layout.

### 2.3.1 New CAD tools for European universities

Over the years of EUROPRACTICE, large numbers of European universities have adopted IC design flows from Cadence Design Systems, Mentor Graphics and Synopsys plus other tools from Avant!, Tanner Research and other leading vendors. These systems have been used extensively to design analogue, mixed-signal and complex digital systems that have been prototyped via the EUROPRACTICE MPW Service.

In order to enable and encourage this experienced academic user base to progress to the design of more complex systems such as SoC in support of their research and training programmes, EUROPRACTICE has introduced a range of tools to support hardware-software co-design. As a first step a range of tools have been introduced, all with access to the ARM7TDMI 32-bit risc processor models, to enable real hardware-software co-design activities to commence. The hardware-software co-verification tools Seamless from Mentor Graphics and Eagle from Synopsys are now available plus the leading co-design tool N2C from CoWare.

The ARM7TDMI is available at several levels of abstraction within EUROPRACTICE with compiler and development tools available directly from ARM at special academic conditions. A range of ARM7TDMI Design Simulation Models for use with the range of simulators supported via EUROPRACTICE such as Synopsys VSS/VCS, Cadence VerilogXL and Model Technology ModelSim. These are complete VHDL and Verilog models of the processor with generic timing primitives that may be altered via a timing shell to include process specific timing. Also included is "Tiny World" an example system using AMBA (ARM Microcontroller Bus Architecture) bus specification. This eases the integration of the ARM7TDMI into a users embedded system.

The EUROPRACTICE 0.35μ CMOS Alcatel Microelectronics Design Kit allows the ARM7TDMI based system to be implemented using place and route software and eventually implemented in silicon via the MPW Service.

### 2.3.2 Cell libraries and design kits for deep submicron design

Availability of cell libraries and design kits to the IC designer is one the priorities of the EUROPRACTICE IC service. For the new 0.25μ CMOS technology from UMC, one the largest pure wafer foundries in the world, no cell libraries are available directly from UMC. Therefore an agreement was made with Virtual Silicon Technology, one of the cell library vendors, in order to distribute their cell library to EUROPRACTICE customers.

### 2.3.3 Training, library support and deep submicron layout service

Whereas European universities have the possibility to purchase deep submicron layout tools for educational purposes through the EUROPRACTICE scheme, most companies do not have these layout tools in-house available. This is due to the high cost of these tools and to the high level of experience needed to take into account effects such as clock-skew, latency of interacting clocks, IR-drop on power-distribution, electro-migration, interconnect delays, etc.

Therefore EUROPRACTICE has installed a commercial deep submicron layout service allowing companies to interface on the netlist level and having their ICs layouted and prototyped through the EUROPRACTICE service.

In addition, dedicated on-the-project training sessions are organised upon request from the customers covering the use of the libraries, the design flow from front-end to back-end and all possible effects that have to be taken into account from the start of the design.

## 3. CONCLUSION

EUROPRACTICE is offering now a new approach to deep submicron design. This new approach includes full support for the ARM7TDMI processor plus integration of the CoWare N2C, Mentor Graphics Seamless and Synopsys Eagle tools into the well established design portfolio available to European universities. Distribution of industry-standard cell libraries and design kits, on-the-project training, deep submicron layout service and access to prototypes and small volumes are also available.

By making these leading tools and flows readily available to European universities, researchers are now better able to address the challenges of SoC and trainers are able to plan practical courses in microelectronics system design techniques that would otherwise be impractical. Students graduating from these institutions will be more appropriately trained to help address the growing skills shortage faced by industry as design complexities continue to increase.

## ACKNOWLEDGEMENTS

The authors wish to thank the European Commission for their support of the EUROPRACTICE project under the IST program.

# WOMEN IN ELECTRICAL

# ENGINEERING

# WOMEN IN ENGINEERING IN EUROPE

C. Dupas*, C. Hermann+ and P. Polselli*
**Institut d'Electronique Fondamentale, Bâtiment 220, Université Paris-Sud, 91405 Orsay, France - CNRS UMR 8622*
*+Laboratoire de Physique de la Matière Condensée, Ecole Polytechnique, 91128 Palaiseau, France - CNRS UMR 7643*

Although women now represent approximately half of the students in Western Europe countries, there are disciplines in which their proportion remains low. This is the case for engineering. In this paper we will first present some data on the European situation. Then we will discuss in more details the French case. Finally we will explain why it is necessary that the proportion of women in engineering should increase in Europe.

## 1. Women in engineering in Europe

Collecting gender-disaggregated data on students or women in science is often very difficult, only very few countries, like Sweden [1], are popularising, or even gathering, such statistics. Here we will rely on data which were published in the European Technology Assessment Network report « Women and Science » [1] made under the auspices of the Directorate Research of the Commission of European Communities.
Another problem is that « engineering » does not include the same fields according to the country. For example architecture is included in engineering in Germany or Switzerland, but not in France : in German technical universities in 1998-99, the women rate was 53% in Biology, 42% in Architecture, 28% in Chemistry, but 6% in Mechanical and 5% in Electrical engineering [2].

### 1.1. STUDENTS

The European comparison is reasonable at the PhD level, which corresponds to an international standard [1]. In Germany, where Engineering PhD's are defended at 33.4 years on the average, 8.3% of such degrees were attributed to women in 1997 (for all PhD disciplines the women rate was 32.1%). In Finland, women obtained 11% of the Technology doctorates in 1995 (for all disciplines : 36% of women) ; in Spain, in 1994, they got 22.9% of the Technical and Engineering PhD's (all disciplines: 37.6%), a rate very close to the French one of 22.6% of women in Physics and Engineering (all disciplines : 39.2%), defended at 28 years on the average.
In fact, in the European Union, the rate of women in Engineering or Technology is always the lowest of all disciplines, but there remains a factor 3 between the *low* rates of France and Spain and the *very low* rate of Germany !



*B. Courtois et al. (eds.), Microelectronics Education,* 71–74.

An interesting point is the evolution of the presence of women versus time : in United Kingdom there were 15.4% women postgraduates in Engineering and Technology in 1992-93, which raised to 17.0% in 1996-97, with then 23.7% of women in the new category IT and Systems sciences. Meanwhile the governmental report « The Rising Tide » had been published (see §3.b) and most of its proposals enforced. In Germany, women represented 6.4% of the total of the Engineering students in 1975, 12.3% in 1990, 18.8% in 1997. German positive actions were first voted in 1989 (§3.2).

1.2. UNIVERSITY TEACHING STAFF

Not only are women students rare in Engineering, but, like in any discipline, the way to professorship is a very « leaky pipeline » [1]. In UK in 1996/97, in Engineering and Technology, women were only 4.0% of the senior lecturers and 1.3% of the full professors ; in IT and Systems sciences they were 14.7% of the senior lecturers and 8.2% of the full professors. In Netherlands, the proportion of women full professors in Technology amounted to 2.4% in 1998.
The present very low rates of women professors can in no way be attributed to low women students populations 20 years ago : studies in Germany [1] show that from the former numbers of men and women students one would expect a female professor rate three times larger than what is now!
These low rates at professorship can reflect different situations : a gradual erosion of women percentage (Germany), or a steep decrease at the full professor level (France and Spain).

1.3. ENGINEERS

Gender-disaggregated data about women in industry are still more difficult to obtain. A thorough comparison of French and German women engineers is performed by C. Marry [2].

More generally, the reasons of such a country-dependent situation of the professorship or the engineer job are, among others, the prestige, very high in Germanic countries, the history and the social structure : good or bad image of work for middle-class wives, availability of crèches and day-cares, of deduction for child care... i.e. items which are not specific to engineering studies and jobs, but relate more generally to each culture.

## 2. Women in engineering in France

2.1. STUDENTS, HIGHER EDUCATION AND RESEARCH STAFF

The French Higher Education system is double, with universities and High Schools (« Grandes Ecoles »). In the French society, the Grandes Ecoles are mythic : many families dream of having a son entering Ecole Polytechnique (for engineering) or Ecole Nationale d'Administration (for state high administration), but few of them invest such a hope in their daughters. The Ecoles, offering privileged conditions of studies, provide an « old boys »'network for the rest of the career. The best students in science follow

this way, which lead many of them to a non-scientific business. In the majority of such elite scientific paths, the rate of women now amounts 15 to 30%, to be compared with 36% of women in scientific universities.
In Electronics, Electrical Engineering and Automatics, at the licence level (3rd year) in the university, women are only 5%, to compare to 41% for the whole scientific disciplines.
In the scientific staff of the universities in 1997, women represented 28.6% of the maîtres de conférences (lower grade) and 9.3% of the professors. In CNRS, the largest French research institution, in the Engineering sciences department in 1996, women were 21.1% of the chargés de recherche (CR, lower grade) and 7.7% of the directeurs de recherche (DR, upper grade, corresponding to university professor). This compares to the Physical and Mathematical sciences department where women were 22.6% of the CR and 14.4% of the DR. Very few women are laboratory directors in the Engineering sciences department : in 1999 they were 5 for 168 laboratories.

### 2.2. AN EXAMPLE : INSTITUT D'ELECTRONIQUE FONDAMENTALE (IEF), UNIVERSITÉ PARIS-SUD

The IEF is an important university laboratory in engineering sciences, associated with the CNRS, with 170 people (115 permanent staff and 55 PhD students), including 44 women (27%). Women are few among the permanent research staff (9% DR, 14% professors, 19 %CR , 23% lecturers). Our study relating to the thirty last years shows that on 363 theses, 66 (18%) were defended by women. It also shows a constant increase in the number of female PhD students : from 11% in the Seventies, they passed to 15 % in the Eighties to 25 % in the Nineties, which seems encouraging.

### 2.4. FRENCH WOMEN ENGINEERS

The surveys made by the National Council of French Engineers and Scientists (CNISF) include the gender categories, and the situation of French women engineers is analysed by the association of French women engineers [3]. In 1995, 22.7% of the degrees delivered by Engineering High Schools were attributed to women ; 13% of the French active engineers were women, with 30 years average age (to compare to 41 years for men). The proportion of women engineers according to the sector differs from that of their men colleagues : women are over-represented in teaching and Research and Development, under-represented in civil engineering and construction.

## 3. Some proposals for improvement

### 3.1. WHY INCREASE THE RATE OF WOMEN IN ENGINEERING ?

The too small proportion of women in Engineering and the « leaky pipeline » are a waste for the society : in France it has been shown[4] that the disciplines where women are few, like Electronics and Computer science, lead to job contracts which are permanent at approximately 93%, and better wages, whereas in those which are traditionally feminine, like Biology, only 64% of the master degrees get permanent

contracts. The non-access of women to leading positions deprives Higher Education, research and engineering of many talents.
To modify this present unsatisfactory situation, actions have been pursued at different levels.

## 3.2. GOVERNMENTAL POSITIVE ACTIONS [1]:

Since 1991, programs have been enforced by Germany, at very high costs (1996-2000 : 720 millions DM), to increase the number of women's scientific PhD's, and help women, in particular those who left research, to return and prepare the so-called « Habilitation ». In 1997 more than 10 000 women scientists received a scholarship in this framework. In UK, the report « The Rising Tide » (1994), on the situation of women in science and technology proposes concrete measures to improve the situation. Most of them have been adopted since, except the tax-deduction for child-care costs, which exists in France for example, but still not in the UK...

## 3.3. ENTERPRISES

Everybody has in mind illegal recruitment difficulties encountered by brilliant women engineering graduates... However some hopes may be founded on new management methods, imported from the U.S.A., now being introduced into a certain number of multinational companies, and favourable to women employment. Companies like Schlumberger, Motorola, have started to recruit women engineers and to appreciate their talents. In this same spirit, Deutsche Telekom is now linking « Equality of Men and Women in (the)Entreprise with the Objective of Quality Improvement ».

## 3.4. MINISTRIES OF EDUCATION AND ASSOCIATIONS

Another target is the education world. The stereotypes are long-ageing, and the picture of the engineer with greased hands remains. Many campaigns have been pursued in favour of the scientific and technical careers for women. These actions should be designed not only to the teenagers, but also to their parents, their teachers and job-counsellors. Role models are essential. This puts a heavy burden on the shoulders of the few present women engineers and scientists. Women associations are effective but their basis is insufficient.
Improving the place of women in engineering a society choice. This concerns men and women ; the task is so heavy that all parties need to collaborate !

## REFERENCES :


1. The ETAN report is available on the web site http://www.cordis.lu/improving/src/hp_women.htm.
2. C. Marry, La féminisation de la profession d'ingénieur : une comparaison France/Allemagne, Journées d'études scientifiques sur les ingénieurs et cadres, LEST, décembre 1999.
3. Les femmes ingénieurs en France, résultats issus de la 12° enquête socio-économique du Conseil National des Ingénieurs et Scientifiques Français, exploités par l'Association Française des Femmes Ingénieurs, 7 rue Lammenais, 75008 Paris, France.
4. Enquête 1999 de l'Agence Pour l'Emploi des Cadres, Bulletin de l'Union des Professeurs de Spéciales n°189, janvier 2000, p 22.

# INTERNATIONAL

# PERSPECTIVES

# ROLES AND PROSPECTS OF IDEC ACTIVITIES FOR THE PROMOTION OF INTEGRATED SYSTEMS EDUCATION IN KOREA

C.M. KYUNG
*IC Design Education Center, Korea Advanced Institute of Sci.& Tech.*
*373-1, Koosung-dong,Yousung-ku, Taejeon, Korea , 305-701*

**Abstract**

Enhancing the competence of semiconductor and electronic system industry is a matter of great concern in the information era of the 21st century. As the product technology gets complicated and time-to-market becomes short, the core problem in technological competence moves on to the education of potential engineers. Under this circumstance, the role of IDEC as a core of IC and system design education in Korea will be reviewed and its vision and prospects will be presented.

## I. Main goals and approaches of IDEC

IC Design Education Center (IDEC) was founded in 1995 at KAIST (Korea Advanced Institute of Science and Technology) with the support of MOCIE (Ministry of Commerce, Industry & Energy) and major semiconductor industries. During the last four years, IDEC has taken initiative actions to improve VLSI design education and research environment in universities. The missions of IDEC can be classified into the following categories; ①building-up and strengthening the infrastructure in the field of VLSI design education, ②producing highly qualified VLSI and system designers, and ③contributing to Korean semiconductor industries by promoting collaborations between universities and industries. Since its establishment, IDEC has elevated the level of VLSI design education in Korea through the following activities; (a) providing commercial CAD tools and computing platforms, (b) playing a chip fabrication interface role by MPW(Multi-



*B. Courtois et al. (eds.), Microelectronics Education*, 77–80.

Project Wafer), (c) offering lectures, seminars, and CAD tool training on VLSI design via remote lecture system, (d) development of educational materials such as textbooks and lecture CD-ROM's.

## II. Current Status and Achievements

As of 1999, IDEC completed its first four-year project and the activities mentioned above resulted in noticeable progress of VLSI design education. Fig. 1 shows the core functions of IDEC and its role, and the simple statistics of the achievements of IDEC during past 4 years are given in Table 1.

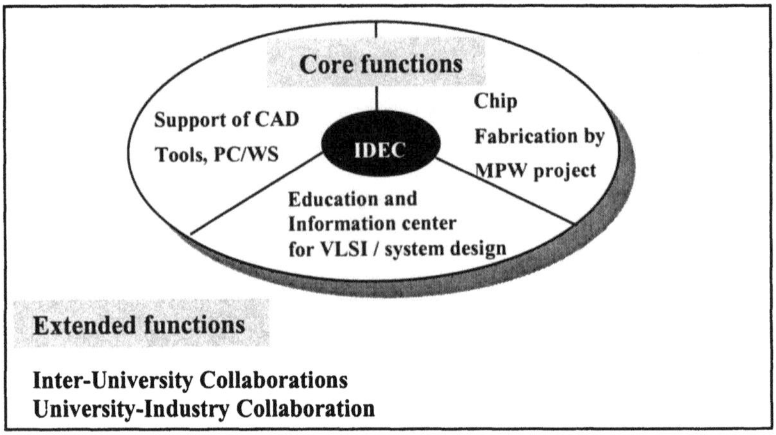


*Figure 1.* IDEC as a core of VLSI and system design education

TABLE 1. Simple statistics of the achievements of IDEC during the past

| Number | 1996 | 1997 | 1998 | 1999 |
|---|---|---|---|---|
| Number of lectures | 37 | 29 | 158 | 175 |
| Number of textbooks published | | 4 | 3 | 10 |
| Number of lecture CD-ROM's issued | | | | 30 |
| Number of chips fabricated through MPW | 65 | 113 | 183 | 143 |
| Standard Cell Library Development | | | 0.6μm, 1P3M | 0.65μm, 2P2M<br>0.25μm, 1P5M |
| Number of tool sets / supporting CAD vendors | 1,781 / 8 | 2,330 / 10 | 8,654 / 16 | 7,246 / 17 |
| Number of working groups/ Regional centers supported | 41 | 51 | 60 / 4 | 64 / 4 |
| Number of participating professors | 202 | 244 | 337 | 376 |
| Number of Technology Groups | | | 7 | 6 |
| Total operating budget (USD/Million $) | 3.7 | 3.5 | 4.9 | 5.3 |

By supplying universities with workstations, PC's, and the high-performance commercial CAD tools such as Cadence, Synopsys, Cossap, Mentor, Hspice, etc., IDEC has played a key role in improving the IC design education environment in the academia. In 1999, IDEC installed 52 workstations, 60 PC's and 7,246 full sets of CAD tools in 64 working groups in 60 universities. In addition, IDEC holds MPW(Multi-Project Wafer) chip design contest every year, which encourages creative talent and motivation of IC designers and helps exchange the information and ideas on VLSI design technology as well. As of 1999, totally 504 IC's have been successfully fabricated and 103 IC designs are under processing through MPW chip fabrication service. The technologies provided in 1999 are listed in Table 2. In February 2000, the first IDEC Conference, which is extended from MPW chip design contest, was held. IDEC Conference will emerge as a new forum for the exchange of contemporary IC design technologies and the presentation of outstanding research results. Along with these activities, IDEC opens about 175 lectures, seminars, and CAD tool training courses per year and over 20,764 registrations have been done in the overall IDEC's educational programs. Owing to the high-speed link between IDEC headquarter and four regional centers, students can take courses at the nearest site. Moreover, IDEC has developed highly-qualified educational materials such as textbooks and lecture CD-ROM's for graduate students and researchers from relevant field. 17 textbooks including 'VHDL for Digital System Design', 'Analog IC Filter Design', 'CMOS Analog Circuit Design', and 'High-Performance DRAM Design' were published by IDEC so far.

For now, over 376 professors and 2,489 students in 66 working groups from 63 associate universities are the main beneficiary of the IDEC activity. Currently, four regional centers located in Pusan National University, Chonnam National University, Kwangwoon University and Hanyang University are cooperating for above activities as shown in Fig.2.

TABLE 2. Technologies provided for MPW(Multi-Project Wafer) project in 1999

| Company | Technology |
|---|---|
| 1. Samsung Electronics | CMOS 0.65μm , SOG |
| 2. Hyundai Electronics | CMOS 0.6 μm , full-custom, 1-poly 3-metal |
| 3. Hyundai Electronics | CMOS 0.65 μm , full-custom, 2-poly 2-metal |
| 4. Anam Semiconductor | CMOS 0.25 μm , full-custom, 1-poly 5-metal |
| 5. ETRI | GaAS 0.5 μm , E/D MESFET MMIC |

For 2,3,4 processes, IDEC Standard Cell Library is available.

## III. Future Perspectives

As mentioned above, the outcome of the first four-year project is remarkable. On top of strengthening the infrastructure in the academia, the number and quality of the IC design engineers trained through IDEC are expected to significantly add values to semiconductor industry in Korea and eventually contribute to the semiconductors and systems design industries as a whole.

In closing the first-stage (4 year) project and commencing the second stage lasting for 5 years beginning 2000, IDEC will be strengthened with the following new strategies; (a) to foster highly-qualified system IC design engineers who are required in the entire field of electronics industries, encompassing telecommunication, home appliances and computer architecture, etc., (b) to pursue and maximize synergy effects between education and R&D beyond the existing basic education framework, and (c) to establish an innovative education model through collaborative research programs between industry and academia.

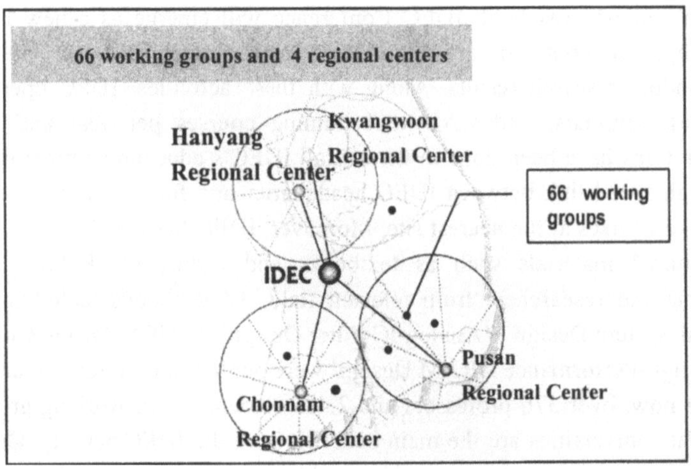


*Figure 2.* Organisation of IDEC

# THE SPECIFIC CHALLENGES OF ELECTRONICS AND IT EDUCATION IN NEW ZEALAND

R. M. HODGSON,
*Institute of Information Sciences and Technology, Massey University, Private Bag 11222, Palmerston North, New Zealand.*

## 1. Introduction

New Zealand is probably best known for its geographical isolation, clean and green image, friendly people, the quality of its agricultural products and for its sporting prowess. It is not widely known that New Zealand is also a country that has rapidly adopted and adapted emerging information and communication technologies. The concept of the knowledge-based economy has recently become central to the country's Research and Development and industrial development strategy into the $21^{st}$ Century. New Zealand has the strong and sound fundamentals to succeed in this drive. It is especially strong in infrastructure and the use of Information technology. At the same time the country faces significant challenges in its drive towards becoming a true information society, prominent among them are the challenges in tertiary engineering education.

The main aim of this paper is to describe the current status and trends in the broadly defined information technology degrees sector in New Zealand. This includes electronics and microelectronics, mechatronics and MEMS, and computer systems. Information sciences are not included in the discussion. The paper will also discuss a strategy for the achievement of the dual goal of getting more students into the area and then providing them with the best possible education within the constraints of available teaching resources.

## 2. Information Technology Usage and Education

Recently New Zealand has become, in effect, a testing ground for innovations in information technology. It is not commonly known that the country is among the world's leaders in terms of critical knowledge economy parameters, including the number of Internet hosts, public expenditure on education, enrolments in tertiary institutions and rates of ownership of personal computers. It has also recently emerged that with an investment of NZ$10.4 Billion, New Zealand is the world's leader in per capita expenditure on information and communication technology, moreover it has achieved this distinction for five of the past six years [1]. It is not surprising therefore



*B. Courtois et al. (eds.), Microelectronics Education,* 81–84.

that the International Data Corporation predicts that New Zealand will have one of the highest growth rates in electronic commerce and that by 2002 a third of all New Zealanders will be buying on-line.

Even less well known is the fact that New Zealand has a very vigorous electronic design and manufacturing industry that has been growing at a rate of about 25% compound per annum. In 1999 the electronics industry is expected to turnover in excess of NZ$1 billion, with more than 90% of this being exported. The largest section of the industry is devoted to telecommunications and line products [2]. Compared with major countries, the industry is small, but in context of the population of 3.5 million it is significant and becoming increasingly so.

There have been a number of significant and exciting information technology developments in New Zealand, among the best known and most promising of them are the development of software products such as JADE, possibly the successor to C++, the world's most popular programming language, Norton Ghost PC cloning software, a number of the world's most widely used encryption tools, Galaxy Kids multimedia educational products and the Audiograph web based teaching software. It is thought that JADE is about to become a globally significant product, particularly as it finds use in the development of electronic commerce products.

Unfortunately it is known that an openness to innovation and knowledge does not necessarily lead to economic development. In the absence of a suitable supply of graduates from the tertiary sector, New Zealand has a worrisome problem. Industry has a significantly lower than optimum proportion of technical graduates in key areas; computer science, mathematics, and engineering and one of the lowest ratios of scientists and engineers engaged in research and development when measured against those 15 top information technology countries. The proportion of science graduates in the population is also low compared with other developed countries. For these reasons educational progress in relevant high technology areas such as microelectronics, information engineering, computer science, mechatronics etc., is becoming an urgent necessity.

There are eight universities in New Zealand: Auckland, Canterbury, Lincoln, Massey, Otago, Waikato, Victoria and the recently designated Auckland University of Technology. Currently over 80,000 full-time equivalent students are enrolled each year for University study. In addition there are 25 Polytechnics and Institutes of Technology with almost 60,000 full-time equivalent students enrolling for degree and diploma level study. At first sight these are rather impressive numbers for a country of 3.5 million people, also the source of some pride. Unfortunately the proportion of University graduates in the technical subjects, especially at the masterate and doctoral level, are alarmingly low in proportion to the importance of technology to the economy and society in general (at around 4.7%).

These figures only tell part of the story. Other academic programs including mathematics and statistics provide the students with significant computer skills. Almost all commerce students now take papers in information systems and information technology and it is noted that the practical IT skills taught by the polytechnics are much sought after by industry. The current challenge is how to increase the number of students studying IT related subjects. This is a topical problem in most countries in the world. In contrast to most of them, New Zealand can only place a limited reliance on controlled immigration as means of acquiring necessary skills.

A further challenge faced by IT and electronics educators in New Zealand is related to the fact that the country does not have any full scale commercial microelectronic, IC, microsystem, or optoelectronic fabrication facilities, and it is most unlikely that such facilities will ever be developed. As a consequence the engineering tertiary institutions of the country have to find effective means of educating their graduates in the full range of modern electronic technologies but on the basis of limited local facilities.

## 3. The New Zealand Situation – A Change in Direction

Just as New Zealand has been a testing ground for innovations in information technology, it has also been the cradle of a major economic and social experiment. Over the last 15 years New Zealand has enjoyed a series of governments committed to a non interventionist approach to industrial development with a minimal range and level of incentives and support. Tariff barriers to imports have been removed and a regime of deregulation vigorously pursued. At the same time social spending has been tightly controlled while individual and corporate taxes have been reduced. Education has increasingly been treated as a private good or gain and the education sector directed to a competitive business model. While New Zealand has grown, it has done so at a slower rate than most OECD countries. At the same time its education has been in relative decline. Once ranked in the top five in most measures, the country now ranks with the recently established nation states of East Europe, including Estonia.

It has now been widely recognised that the non-interventionist approach and business models have not worked to the nation's advantage, though some benefits have accrued as a result of deregulation. In November 1999 a new coalition government was elected with clear mandate to implement sweeping policy changes. At the same time it has been noticed that the majority of smaller nations that have shown rapid technology driven economic growth, treat education as an investment and rely on planning and intervention rather than market forces to guide national development. Prominent in the model nations are Finland and Ireland. Singapore is now recognised for its planned economic achievement based on intervention, direction and national thrift.

## 4. The Need for More IT and Microelectronic Graduates

Recently Motorola has been investigating New Zealand as a potential location for its proposed third Australasian software development facility. This would initially employ 200 software and engineering graduates, the majority of positions would be for holders of postgraduate degrees. This investigation has revealed that the total (first degree) graduate output in information technology subjects will be about 1200 in 2000, to increase by 30% by 2003. If information sciences graduate numbers are included then the corresponding figures are 1700 in 2000 and 2300 in 2003. About 16% of these graduates have postgraduate degrees and about 40% have four year undergraduate degrees.

The new Auckland University of Technology and the remaining polytechnics add about 10% to the overall figures. Given the unsatisfied demands from the established electronics and software industries and the potential of major international corporations to, at a stroke, take up a significant part of the total output, the need for major education expansion in IT has become very evident.

## 5. The Microelectronics Content of Existing Programs

The three year degree programs offered by the Universities and Polytechnics include only a minimum of descriptive material on the design and engineering of integrated circuits. The professional engineering degrees in electronic engineering and information engineering typically include a paper on integrated circuit engineering in both the third and final years of study. In total this corresponds to about 25% of a year's study. Typical programs include an introduction to CMOS design (including the lambda design rules) and the simulation of devices and simple systems. The final emphasis is placed on the use of gate arrays and standard cells in system design leading to design exercises with FPGAs. The design cycle and the problem of testability are given emphasis [3]. A survey of leading institutions has led to the conclusion that MEMS has yet to appear in the taught curriculum.

## 6. Towards a National Strategy for Education in IT

It is rapidly becoming evident that there is a need for a national strategy for more and better education in IT in general and microelectronics in particular. This will need to call upon national strengths. These include a vigorous electronics sector, a very well developed IT infrastructure and a recent commitment by government to an element of selective funding. New Zealand's largest electronics company, Tait Communications, already provides specialist lecturers to its local University. The well developed IT infrastructure makes it possible for a national pooling of contributors from industry and specialist research groups in the Universities [4]. When there is a strong consensus on the content of electronics and IT degrees there is little logic in continuing to prepare and present courses and programs on a strictly local basis.

## 7. References

1. OECD, *Science, Technology and Industry Scoreboard: Benchmarking Knowledge-based Economies*, OECD, Paris, 1999.
2. Cornwall, J.B., *Electronics – A Successful New Zealand High Tech Sector*, Industrial Research Limited, Auckland, 1999.
3. Hodgson, R.M., Browne, R.F., and Demidenko, S., The Microelectronic Component of an Information and Telecommunication Degree, *Proceedings of SPIE, Education in Microelectronics and MEMS*, vol. 3894, pp.159-165,SPIE, Washington, 1999.
4. Bourne, J.R., Mayadas, A.F., and Campbell, J.O., Asynchronous Learning Networks: An Information-Technology-Based Infrastructure for Engineering Education, *Proceedings of the IEEE, Special Issue*, Vol. 88, No. 1, pp. 63-71,IEEE, New Jersey, 2000.

# SITUATION AND DEVELOPING TREND OF MICROELECTRONICS EDUCATION IN CHINA

Zhi-Gong Wang
*Institute of RF- & OE-ICs, Department of Radio Engineering*
*Southeast University, Nanjing, 210096, P. R. China*

## 1. Introduction

The human society is entering the information era. The cornerstone of the information techniques is microelectronics. Therefore, the research, the development, and the education of microelectronics are extremely important. This is true not only in developed countries but also in developing country such as in China. As the door of China to the world is further opening, the world people are interested in all aspects including the education of China. In this paper, the situation and the developing trend of microelectronics education in china will be introduced.

## 2. Survey of Microelectronics in China

The market of microelectronics products is huge in China. In the year of 2000, the estimated demand on ICs and other electronic devices is about 50 billion US dollars. However, the manufacturing capability of Chinese microelectronics industry is limited relatively: Only twenty-percent of the domestic need can be covered by national products. There are 6 main IC manufactures. The technologies are 1.5-0.35μm on 4"-6" wafers.

One of the main reasons for the backward situation is that the industry lacks high-qualified engineers for IC design. In China, there are only about 3000 IC design engineers and the number of IC design experts is very small. Because of lack of design engineers, both the number and the scale of IC design houses are small. There are less than 50 design houses and the scale of them is normally less than 50 bodies. The scale of the design teams in universities and institutes lies between 20 to 100 bodies.

## 3. Education and Research on Microelectronics in Chinese Universities

China has more than 1000 universities and institutes. Most of them have faculty or institute of electronics. Many universities have a microelectronics center for education and research. Following are four examples.



*B. Courtois et al. (eds.), Microelectronics Education*, 85–88.

**Institute of Microelectronics of Tsinghua University (IMETU) [1]**

The goal of IMETU is as a new facility for the fabrication of special ICs and device structures, an advanced research and development center in the ASIC design and semiconductor device physics and an education base of high level scientists and engineers in microelectronics field. IMETU has a $3000m^2$ clean room for 1μm-level CMOS 3" pilot line. The wafer size is 4". The research areas include novel semiconductor device, device physics and device models, CAD, design, technology, testing methods of VLSI, and MEMS. Various VLSI ASICs have been designed and fabricated. As an education center, the enrolled master and PhD-students are about 50 and 40, respectively.

**The Institute of Microelectronics of Peking University [2]**

The Institute of Microelectronics of Peking University is one of national key units on novel device/ICs and their applications, one of the important R&D bases of microelectronics and MEMS science & technology, as well as a cradle of future microelectronics talent.

The institute has established three key labs: National Key Laboratory of Micron/Nano Fabrication Technology, Emerging Technology and Novel Device/IC Laboratory of Northern Microelectronics R&D Base and Beijing High-Tech Laboratory for Software and Hardware Codesign. The total processing area is about $900m^2$, equipped with more than 80 pieces of advanced processing machines and test instruments for the manufacture of CMOS, Bipolar ICs and MEMS, etc. The Institute has an additional CAE center ($300m^2$), where many workstations and associated CAE software are installed.

In the institute, a strong working team of outstanding professors with high scholastic attainments and young scientists has formed, thereby ensuring research and teaching works in a state of prosperity. Among her over 90 staff, about 40 hold high professional titles and 30 others hold medium ones. The institute is divided into six research groups to carry out associated researches: Micro-Electro-Mechanical System (MEMS), very high speed IC, SOI (Silicon-on-Insulator) devices and circuits, ASIC design, novel device, and small dimension device physics.

Besides the research work, the institute has also concentrated its effort on talent training. Each year about 40 undergraduate students, 10-16 graduate students and 8-10 Ph.D. candidates majoring in microelectronics are enrolled, and 4-6 post-doctorals are accepted every year.

**The ASIC and Systems State Key Laboratory in Fudan University[3]**

It was established in 1992 using a world-bank loan and is supported by Chinese government. The lab's facilities include a large network connected with SUN, HP workstations and high performance Pentiums as well as some test equipment. The EDA-tools include products from Cadence, Mentor, Avavt!, Synopsys, Xilinx, Altera, HP and so on. The main research activities are concentrated on new VLSI architecture for modern information systems, ASIC design methodology including design for testability, advanced process and device characterization for design optimization. Based on microelectronics, CAD/CAE tools, circuits and systems design technology, the research area covers the common problems in the fundamental theory and actual development for

ASIC design and system integration. The training center gives training courses to persons who work in the lab or come from outside the lab. The training center is also in charge of multi-chip program. It is open to undergraduate and graduate students and scholars in micro-electronics, information system and other closely related disciplines in China and overseas.

**The Institute of RF- & OE-ICs of Southeast University [4]**

The Institute of RF- & OE-ICs of Southeast University is a new-established institute since October 1, 1997, supported by the university and the Ministry of Education of China. At moment the institute has a researching team of 10 staff and 10 post-graduate and PhD-students. The institute is equipped by many Unix-workstations and more than 30 high-performance P-II/P-III. The EDA-tools include SmartSpice®, Cadence's Opus-97®, HP's ADS® and others. The international partners include the Information Science Institute of University of South California (MOSIS Program), the Ruhr-University Bochum and in Stuttgart-University, Stuttgart, both in Germany, the Department of Electric and Information in the National Institute of Application Science in France, and so on. The main tasks are to educate talents of IC design and to make researching works supported by national and international science and technical programs and by electronic industry. The researching fields include RF, high-speed, and opto-electronic ICs and VLSI, especially for data-, mobile-, microwave-, satellite- and optic-fiber-communications. The technologies used include CMOS, silicon bipolar, GaAs, and InP.

Besides of the Institute of RF- & OE-ICs, the IC-related units in Southeast University include the National Research Center of System Engineering of ASIC and the Microelectronic Center. The total number of staff is over fifty and the students accepted are over 100.

Other universities possessing an institute on microelectronics include Zhejiang-University, Shanghai Jiaotong-University, Xi'an Jiaotong-University, the University of Electronic Science & Technique of China, Harbin Institute of Technology, and so on.

## 4. Challenge and Chance

One problem with microelectronics education in the past is that there was less chance for teachers and students to realize their design and therefore they did research more on paper and somewhat blindly.

Now, the situation is altered a lot. There are three positive aspects for microelectronics education in China:

d) The government is paying a high attention to the education, the research and industry of microelectronics. The IC design and fabrication is laid as the main task of the information industry.

d) The industry is eager to get more and more engineers in the field of microelectronics. For those who have studied microelectronics, the career is very good. Thus, more young people are attracted to this field.

c) In China, there are enough students in the subjects related to microelectronics. In Southeast University, for example, more than 1000 students are related with microelectronics. That means, there is a huge resource of manpower and brains to be exploited.

d) The easily accessible foundry technologies enable students to begin IC design in their study period.

Many universities have begun to improve their environment of microelectronics education. In Southeast University, for example, a three-level education system of IC design has been established for students beginning in the second and third year of the undergraduate up to and including post-graduate. At the second and third level, students can design chips in projects.

## 5. Measurements for Promoting the Education of Microelectronics

To promote the education of microelectronics, a great program is progressing now in the national range. A special fond for talent education of IC design will be established. The aim is that in 5 years about 20,000 high-level engineers can be supplied to the IC industry.

The successful practices in microelectronics education through the programs, such as MOSIS of the USA, CMP of France, Europractice in Europe and so on, have been introduced in China. On one hand, the international cooperation makes the Chinese university possible, to get the service of MOSIS, CMP, and other programs. In Southeast University, for example, ICs have been designed by graduate students in 0.35μm CMOS and successfully realized directly through MOSIS. On the other hand, some universities try to offer the service like as MOSIS. In Fudan University in Shanghai, for example, a service center for microelectronics education and multi-project IC implementation has been functioning since 1997 and has organized many runs of IC realization.

On February 20, 2000, the first national IC Design Base has been opened just in the city center of Shanghai. The 20 000 $m^2$ base is supported by the Ministry of Science and Technology of China and by the Science and Technology Commission of Shanghai Municipality. The aim is to supply a complete environment for the IC design functions as an incubator of IC design houses. In the base, there is a center for IC design training, a service center for IC fabrication, as MOSIS in USA, several IC testing places, and other service offices. Up to 100 design houses can enter in the building. Such a base will be opened in other great cities such as Beijing, Xi'an, and Nanjing in this and next year.

The first virtual IC design center is preparing now cooperatively among the Tsinghua University, Southeast University, Peking University, Xi'an Jiaotong-University, and Hefei University of Technology.

To promote the education of microelectronics, China executes the opening politics. Chinese universities will increase collaboration with international partners.

**References:**


1. http://dns.ime.tsinghua.edu.cn/english/index.html
2. http://ime.pku.edu.cn/INTRO_E.HTM
3. http://www.fudan.edu.cn/research/index.html#3
4. http://iroi.seu.edu.cn/english/eintroduce.htm

# MICROSYSTEMS

# COMPUTER SIMULATIONS OF MICRODEVICES AND MICROSYSTEMS: GAINING INSIGHTS BY VISUALIZING PHYSICAL CONCEPTS

G. WACHUTKA
*Physics of Electrotechnology, Munich University of Technology*
*Arcisstrasse 21, D-80290 Munich, Germany*

## 1. Introduction and Motivation

Technology development in the field of electromechanical microdevices and systems (MEMS) is a very complex interdisciplinary task which requires a profound knowledge in different technical and scientific areas such as semiconductor process technology, multi-energy domain device physics, circuit and system design etc. The microsystems industry largely demands for well-trained engineers being equipped with all this knowledge and expertise, but the education at university and other institutions is still in the process of adaptation to the industrial demands.

One complication to cope with is the following. The traditional regular lectures provide a solid theoretical basis for the individual areas of microsystems technology, but it would be necessary to combine these areas in application-oriented laboratory courses in order to create a synergetic connection of the single disciplines and to embed the theoretical knowledge of isolated areas into a wider context. It is, however, very expensive and practically nearly impossible to base such a laboratory course on real experimentation and real technological processes.

But just as well the educational goals can be reached by substituting a "virtual microsystems factory" and a "virtual instrumentation area for testing" on the computer for the real device fabrication and the real test and characterization work. The computer simulations can be performed by the use of modern software packages for process, device, and system simulation which accurately model all relevant fabrication steps and physical effects governing the device and system operation, have reasonable response times, are easy to use and allow the illustrative visualization of input quantities and results.

Moreover, computer simulations lend the trainees unique insight into all the details of a fabrication step or the functioning of an electromechanical or electronic microdevice by visualizing the distributions of various physical quantities in the device interior, which otherwise is hardly accessible to



*B. Courtois et al. (eds.), Microelectronics Education*, 91–94.

immediate experimental diagnostics. Therefore, the trainee is able to study in full detail all the involved physical effects the device function is based on and, thereby, he gains the necessary microscopic understanding. Using a technology CAD (TCAD) system for MEMS, complete simulation sequences from the layout design up to the system behavior can be performed, thereby revealing the relation between structural and technological design parameters, operating conditions, and the resulting operational behavior on device and system level. This, in turn, allows the trainee to draw further conclusions for an improved re-design.

## 2. General Aspects and Simulation Methodology

Modeling and numerical simulation of MEMS and electronic microdevices is widely recognized as necessity, even though it becomes increasingly complex, so that the effort as well as the time spent into model development, validation, and parameter calibration have to be carefully adjusted to the actual needs, as it has been proposed by the concept of "tailored modeling" [1], for example. Obviously we have to focus on two main objectives for computer simulations [2]:

- Enhanced physical understanding of fabrication processes and operational principles ("verification of concepts", "visualization of physical effects")
- Aid in making decisions during (re-)design ("analysis of design variants", "study of trade-offs").

For both objectives, the indispensable prerequiste is the availability of predictive simulation and design tools, which fit in with today's far advanced design environments used in the semiconductor industry and, in particular, are conform with the widely accepted bottom-up and top-down modeling hierarchies.

All TCAD systems for microdevices and microsystems comprise several successive descripitive levels. The order of the model levels reflects the functional sequence of steps during fabrication and test. The first level is occupied by layout and design, where device topography and process flow are defined. This data is passed to the level of process modeling, where all the fabrication steps are simulated in the sequence required for the given design. As a result we obtain the structural description of completely processed microdevices, in particular geometry and material composition. These descriptive elements are then passed to the level of device simulation. Here, after selecting the proper physical models and defining the external operating conditions and control parameters, the internal operating behavior is studied by stationary or transient mechanical structure analysis, electrostatic or magnetostatic analysis, thermal analysis, fluid and gas transport analysis, electronic transport analysis, and all possible combinations thereof (coupled-field analysis). Finally, we arrive at the system level, where the constituent elements of a microsystem are represented by compact models which have been extracted from the device level. Linking the compact models together to a system macromodel allows to study the behavior of the system as a whole. A natural and physically-based methodology for this is provided by Kirchhoffian network theory [3].

## 3. Microdevice Modeling

It is an inherent problem of MEMS modeling that most of the constituent components, by their function as sensor or actuator element, couple different physical energy and signal domains. As a consequence, the models underlying the simulation tools must be capable of accounting for a large variety of (desired or parasitic) physical coupling effects. Since our goal is to ensure the predictiveness of simulation, the use of simple behavioral models is not appropriate, because such models just reproduce the operation of known devices, but can hardly be used to inter- or extrapolate in the space of design parameters and operating conditions. Instead, one has to follow a physically rigorous approach to build device models by a thorough physical analysis of the underlying operation principles.

The most natural description of the physical operation of single micro-devices as well as their interplay as constituent elements in a microsystem is provided by continuum models and conservation laws (balance equations) which couple the relevant field quantities (mechanical, thermal, electric, magnetic,...) consistently in terms of a (typically quite complex) system of partial differential equations. Solving these equations by FEM or related methods, the primary result of a simulation run is the spatial distribution of the respective field(s), possibly as a function of time, which constitutes the basic information for the visualization of the internal behavior of a device.

However, with a view to assessing the quality of a microdevice in terms of a few characteristic parameters ("figures of merit"), the analysis often aims at integral quantities such as input-output characteristics, response functions and transients, which have to be calculated from the field quantities in a post-processing step. In a similar way, an equivalent network description can be derived from the simulated field distributions. The dynamic behavior is now described by a set of ordinary differential equations where the coefficients have been extracted from the physically-based but much more complicated device simulation. Sometimes one is led to an explicit analytical expression for the response function, from which easily scaling laws and other useful practical rules can be deduced as guidance for an improved design.

## 4. Microsystem Simulation

Considering the numerical effort required for realistic microdevices and systems, the use of continuous field models (CFM) becomes easily prohibitive. Therefore, the number of degrees of freedom has to be reduced to an amount that is still manageable and yet accurate enough for predictive simulation. A proper approach is analog network simulation based on compact models (CM) of the constituent microsystem components derived from the CFMs and a network theory which preserves the basic conservation laws expressed in the CFMs.

A systematic way to develop CMs is, among others, provided by a thermodynamic system description in terms of driving forces and resulting fluxes of the relevant physical quantities such as energy, particle numbers, mass, charge etc. [1]. Partitioning the system into blocks and lumping the exchange of flux quantities between adjacent subsystems along common interfaces into single nodes eventually yields a full system description as "Generalized Kirchhoffian Network". This is, in a certain sense, the natural extension of electric circuit theory, where the branch currents and node voltages are determined by the systematic application of the "current sum rule" at the nodes and the "voltage mesh rule" along a closed loop of the network, but now with the extension that also physical quantities other than electrical charge are allowed to flow through the network.

A generic software approach to CM is a hardware description language like VHDL-AMS [3], which constitutes a standardized model interface in analog network simulators and, furthermore, allows the description of arbitrary physical energy and signal domains in addition to the electrical quantities.

## 5. Summary

The educational success of laboratory courses in MEMS technology can strongly benefit from computer simulations. Visualizing the fabrication and operation of microdevices on the continuous field level provides the trainee insight in the device interior not readily accessible to experimental diagnostics. Combination with network simulation for full microsystem analysis reveals and illustrates the relation between design parameters and system performance.

## 6. References

[1] G. Wachutka, in *Micromechanical Sensors, Actuators, and Systems*, **DSC-40**, ASME, New York, pp. 183-197, 1992.

[2] G. Wachutka, *Sensors and Actuators* **A 41-42**, pp. 279-283, 1994.

[3] Standard VHDL Analog and Mixed-Signal Extensions, IEEE, New York, 1997.

# TEACHING MODULES FOR A CLASS IN MECHATRONICS[1]

FRED R. BEYETTE, JR. AND CARLA N. PURDY
*ECECS Department, University of Cincinnati, Cincinnati, OH 45221-0030*

Key words: MEMS, mechatronics, rapid prototyping, VHDL-AMS

**ABSTRACT.** In a previous paper ([3]) we described tools we are using to introduce mechatronic system design to students at the University of Cincinnati. Here we describe the current state of this project. Two main goals for this academic year are:

- to have students carry out a prototype process to design, fabricate, and test some simple mechatronic devices using the L-Edit tools, the MOSIS fabrication facilities, and some simple in-house post-processing;
- to develop VHDL-AMS models of devices from the mechanical, fluidic, and temperature domains so students can simulate mixed-domain systems.

Our eventual goal is to integrate our tools and models into a mechatronics design course for advanced undergraduate and beginning graduate students.

## 1. INTRODUCTION

The ECECS Department at the University of Cincinnati offers undergraduate and graduate degrees in electrical engineering, systems engineering, computer engineering, and computer science. Established research programs include digital and analog simulation and synthesis, controls, MEMS fabrication, and optoelectronics. Senior year specializations in VLSI design and in photonics are also available. The department therefore offers a rich environment for development of teaching modules to introduce mechatronic system design into the engineering curriculum. While several courses already in existence focus on design of state-of-the-art MEMS devices, the teaching material we are developing focuses instead on production of standardized devices using well-established technologies available through the MOSIS service and on training

[1] We use the terms *mechatronics* and *MEMS (microelectromechanical systems)* interchangeably to describe systems consisting of components interacting in multiple energy domains.



*B. Courtois et al. (eds.), Microelectronics Education,* 95–98.

designers who will be able to create multidomain "systems on a chip" which will soon become commonplace. Currently the materials we are developing are being used as modules in a variety of courses, including a graduate course on MEMS simulation, an advanced undergraduate course on "silicon programming", and as individual student projects; soon they will be combined into one unified course at the advanced undergraduate-beginning graduate level. The key to making this course feasible is the demonstration of a complete robust design-simulate-fabricate--test cycle which can be implemented in a classroom setting and completed in two quarters. An absolute requirement is that reliable tools with reasonable learning times be employed. We are currently instantiating a prototype for this process, with the help of funding from the U.S. Air Force Wright Labs, with final device testing planned for Spring 2000. Simultaneously, we are developing a suite of VHDL-AMS models which can be used to familiarize students with multidomain behavior and which can also be used as building blocks in classroom simulations of complex systems.

## 2. TOOLS USED; DEVICES STUDIED

***Introductory Material:*** We begin with the foundational article by Petersen ([12]). Additional material is based on chapter 5 (micromachining) and chapter 9 (scaling) of [10], while [15] summarizes current work in MEMS and problems to be solved.

***Fabrication Technology:*** We are using the fabrication technologies available through the MOSIS CMOS foundry service. In fact, our MEMS structures are fabricated directly from preprocessed (and packaged) CMOS chips using a post processing/packaging wet chemical etch step ([9]). Structural features are created on the chip surface using polysilicon, metal, silicon nitride, and $SiO_2$ layers common in a CMOS device process. These structural features are selectively “released” from the silicon substrate using a wet chemical etch process that removes silicon from regions that are photolithographically defined and limited by either the orientation of [111] crystallographic planes or by implantation of p+ etch stop regions. Details of the post processing release step can be found on the MOSIS web page [11]. To facilitate the release process, it is necessary to define two additional CAD layers (called ‘open’ and ‘pstop’) in the technology definition files used by common CAD tools like Magic or L-Edit. It should be noted that these logically defined CAD layers are created from multiple mask levels currently available in conventional CMOS fabrication processes. The ‘open’ layer creates a via that cuts through the overglass, field oxide, and gate oxide layers, leaving a base silicon surface exposed to ambient conditions. Features drawn in this CAD layer are used to open regions of silicon substrate for etching in the post processing fabrication step. The ‘pstop’ layer is implemented within the p+ diffusion mask and used to help confine the wet chemical etch process which does not etch highly doped p+ regions in the substrate. As suggested above, these modifications/additions to standard CMOS CAD and fabrication processes can be implemented without change to current CMOS fabrication procedures. Thus, standard CMOS foundry services can be used to fabricate MEMS devices that incorporate both mechanical structures and the electronic circuitry required to control the MEMS devices. Further, the ability to incorporate CMOS logic, analog circuitry and optoelectronic components [4] into the mechatronic chips provides a rich environment for the design and implementation of mixed technology information processing components. Finally, the ability to fabricate these devices using a standardized fabrication process leads to a mixed-domain device technology that builds on the low cost, high reliability, high yield and stable performance metrics that have established conventional VLSI as the technology of choice for employers of electrical and computer engineering students.

***Example Devices:*** While the limitations associated with working in a conventional CMOS process may prohibit the design and implementation of some complex mechanical structures, several important mechatronic structures can be implemented with the technology described above. The NIST MEMS library archived on the MOSIS web page includes layouts for a polysilicon heater, a thermal actuator, a micro-hot-plate, and a gas sensor based on the micro-hot-plate. Further, Marshall et al. [9] have described an IR emitter based on using the incandescent emission from structures similar to the polysilicon heater. Finally, the technology will easily allow for a variety of electrostatically actuated structures including thin-film diaphragms, cantilever beams, and torsional rotation platforms. Figure 1 shows the layout of two cantilever beam structures that are designed to be electrostatically actuated deflection mirror structures. These devices form the bases of an optical beam steering system that is capable of directing an optical beam such that it can be reflected off of a large (off chip) fixed mirror and reflected back to an on chip photodetector. The incorporation of photoreceiver circuitry, digital logic, and actuation control circuits will enable the implementation of a robust optical MEMS based information processing system. The development of standard MEMS cell libraries and the design expertise necessary to utilize them in the implementation of mixed technology information processing systems is a major objective of our educational program.

***Design Tools:*** As described in [3], we have found that the L-Edit tools ([14]) work well for design. An alternative, available in the public domain, is LASI ([2]).

***Simulation:*** Simple simulations are carried out with SPICE ([6,7,8]) with translation of electrical parameters into their mechanical equivalents. Students are also introduced to the VHDL (Very High Speed Integrated Circuit Hardware Description Language) extension VHDL-AMS (VHDL with Analog and Mixed Signal Extensions, [5]) to give an integrated simulation environment for components from both the electrical and other domains (Figure 2). A VHDL-AMS simulator is available to educators ([13]). We have also experimented with Mathematica and Ansys ([1,6,7,16]).

## 3. CONCLUSIONS

We have described a process and supporting tools to support student design, simulation, fabrication, and test of simple mechatronic systems. Use of this material in appropriate courses will provide students with the skills they will need to succeed as complex multidomain systems become commonplace.

## 4. REFERENCES

1. *Ansys Primer for Stress Analysis for Revision 4.4,* Swanson Analysis Systems, Inc., 1991.

2.Baker, http://boise.uidaho.edu/langroup/jbaker/wwwbook/winlasi/winlasi.htm.

3. F. Beyette, Jr. and C. Purdy, Tools for integrating multidomain technologies into microelectronic circuit design curricula, *MSE 99,* July 1999, 86-87.

4. Beyette, Stanko, Feld, Mitkas, Wilmsen, Geib, and Choquette, Demonstation and Performance of a CMOS/VCSEL Based Recirculating Sorter", Optical Engineering, 37(1) Jan. 1998, 312-319.

5. Design Automation Standards Committee, IEEE Computer Society, *IEEE Standard VHDL Language Reference Manual (Integrated with VHDL-AMS Changes)*, Std. 1076.1, IEEE, 1997.

6. Gibson and Purdy, Extracting behavioral data from physical descriptions for MEMS for simulation, to appear in *International Journal of Integrated Circuits and Signal Processing.*

7. Gibson, Hare, Beyette, and Purdy, Design automation of MEMS using behavioral modeling, *Ninth Great Lakes Symposium on VLSI*, Ann Arbor, Mich., March 1999.

8. Lo, Berg, Quakkelaar, Simon, Tachiki, Lee, and Pister, Parameterized layout synthesis, extraction and SPICE simulation for MEMS, *ISCAS 96*, May 1996, 481-484.

9. Marshall, Parameswaran, Zaghoul and Gaitan, "High-Level CAD Melds Micromachined Devices with Foundries", IEEE Circuits and Devices, 8(6), Nov. 1992, 10-17.

10. Madou, *Fundamentals of Microfabrication*, CRC Press, 1997.

11. See http://www.mosis.org and follow links for technical support, design tools, and NIST MEMS Library.

12. Petersen, Silicon as a mechanical material, *IEEE Proceedings* 70 (5), May 1982, 420-457.

13. SEAMS Simulator Project, Distributed Processing Laboratory, ECECS Department, University of Cincinnati, 1998.

14. Tanner Research, Inc., *Tanner Tools User's Manual*, 1996.

15. Wise, VLSI circuit challenges in microelectromechanical systems, *Proc. Ninth Great Lakes Symposium on VLSI*, March 1999.

16. Wolfram, *Mathematica.* Addison-Wesley, 1991

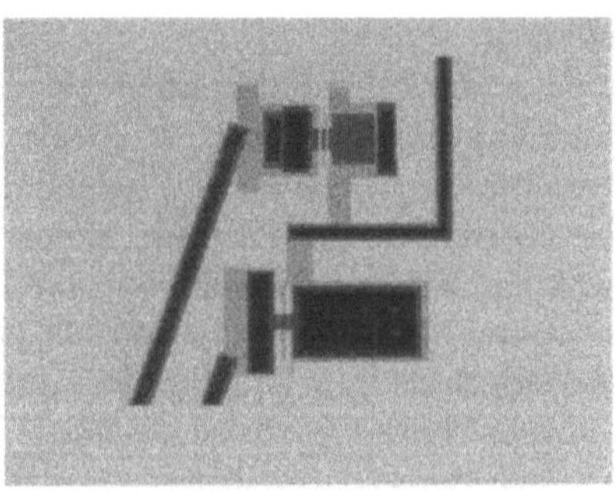

*Figure 1. Example cantilever beam system.*

```
entity mechsensor is
  port(f: in force; c: out
       capacitance);
end mechsensor;

architecture twopart of
mechsensor is
  component mechinput . . . ;
  component capacitor . . .;
  signal y: distance;
  begin
     ml: mechinput port map
          (fb=>f, x=>y);
     c1: capacitor portmap
          (xb=>y,cb=>c);
end twopart;
```

*Figure 2. Portions of VHDL-AMS sensor model.*

# MULTIMEDIA CD-ROM: "AN INTRODUCTION TO THE WORLD OF MICROSYSTEMS"

PHILIPPE FISCHER[1], MICHEL BERGER[2], MARCEL ECABERT[1]
[1] *FSRM-Fondation suisse pour la recherche en microtechnique*
*Rue Jaquet-Droz 1, CH-2007 Neuchâtel, Switzerland*
[2] *mib - génie logiciel*
*Rue du Plan 3, CH-2000 Neuchâtel, Switzerland*

## 1. Objectives

The objective of the multimedia CD-ROM is to provide a basic understanding of what microsystems are and what technologies are involved.

The FSRM launched in 1993 its course programme entitled "Training in Microsystems", which today includes 23 short courses on microsystems related topics. These courses run all over Europe and over 1700 researchers and engineers have already attended. They form the priority target group of the CD. In future, all course participants will receive the CD in advance of the course, in order to acquire a common basic knowledge on microsystems. This shall prevent the course tutor to spend too much time, up to a quarter of the total course duration, on an introduction to microsystems.

In addition, the CD is designed to be used for the promotion of microsystems in industry, schools and in the media.

Finally, it will be a valuable tool for engineering professors to use in their lectures on microsystems and microsystem technologies.

## 2. Content

The CD provides an interactive journey inside six commercially available microsystems. This approach was chosen to overcome the generally shared image, that microsystems are very nice technological challenges, but often useless and not economically viable.

The six examples were selected in order to cover a large spectrum both of application fields and fabrication technologies.



*B. Courtois et al. (eds.), Microelectronics Education,* 99–102.

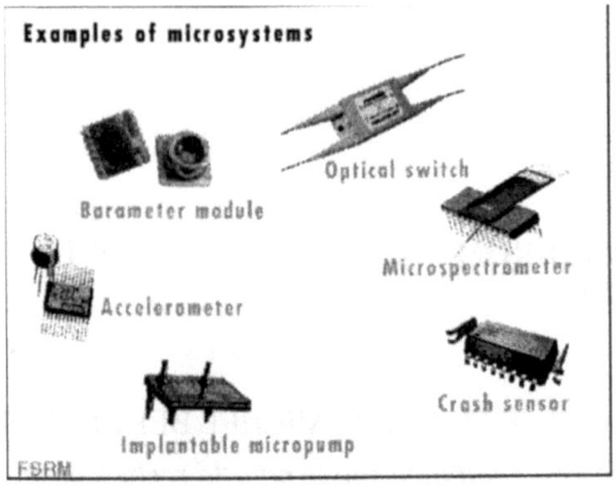


*Figure 1. Main entry screen of the multimedia CD*

Figure 1 shows the main entry screen with the six examples:

1. a barometer module for consumer applications, developed and commercialised by Intersema, Switzerland;
2. a 2 by 2 optical switch for telecommunication, developed by the University of Neuchâtel and commercialised by Sercalo Microtechnology, Switzerland;
3. a microspectrometer for medical and industrial applications, developed by Forschungszentrum Karlsruhe and commercialised by Microparts, Germany;
4. a crash sensor for automotive applications (air-bag), developed and commercialised by SensoNor, Norway;
5. an implantable micropump for medical applications, developed and commercialised by Debiotech, Switzerland;
6. an accelerometer module for industrial applications, developed and commercialised by CSEM, Switzerland;

For each of these examples, five different aspects are shown: a short description, the functionality, the fabrication, the applications and the economics.

The description of the fabrication processes used for the production of these six devices leads to a detailed description, enhanced with animations, pictures and video sequences, of many single processes used in microsystem technology. This technological aspect is also available directly by a second entry screen. Here again, the descriptions are not limited to technical aspects of how the processes work, but also show the required equipment and its cost.

## 3. Navigation Concept

The design of the navigation concept and its implementation were realised by mib-génie logiciel, Switzerland.

3.1. REPRESENTATION OF THE INFORMATION

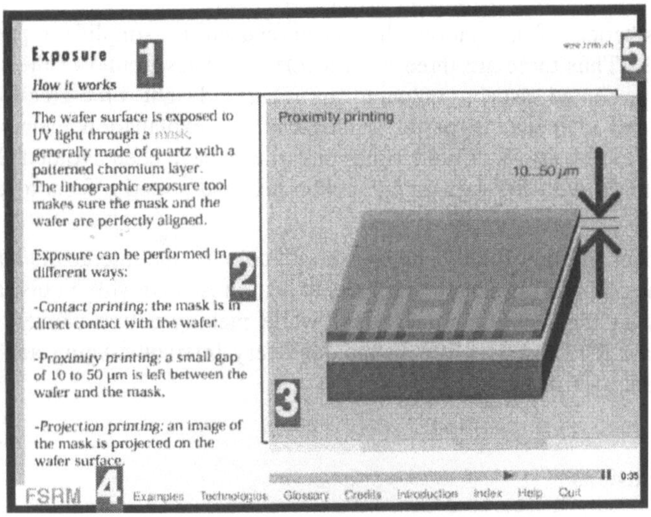


*Figure 2. Typical screen with all different zones*

The information elements are the examples and the technologies. They are represented through electronic cards in the size of the screen, as shown in figure 2. Besides a title zone (1) on the top left, the visitor finds the summary zone on the left (2), the scene on the right (3), navigation elements on the bottom (4) and additional information such as links to web sites on the top right (5). Attention has been paid to the choice of typography for good readability, to permit the use of the CD in classroom projections. The zones remain throughout the whole CD to give the visitor a perception of stability.

The information elements are further subdivided into the different aspects. The examples contain following aspects : *general description*, *how it works*, *fabrication*, *applications* and *economics*. The technologies have following aspects : *general description*, *purpose*, *how it works* and *equipment*

Colours have been assigned to each aspects for easy identification and are used as background for the scene. Thumbnails allow the access to the different aspects.

3.2. THE SCENE

Illustrations, animations and larger chunks of text are presented on the scene which is the main zone for the presentation of information as it represents more than 40% of the whole surface. Inside an aspect of an information element, the change of information (animation, video clip, etc.) is limited to the scene. The perceptive stability is thus reinforced.

Each time a movie plays on the scene, a corresponding navigational element gives the visitor the possibility to control it. A clock indicates the remaining time, thus giving immediate and constant feedback of the movie's progress.

### 3.3. ANIMATIONS AND AESTHETIC INTEGRITY

The aesthetics of animations has been chosen to simplify representations to the essentials. Thus there are three main levels : the background of the scene is, according to the presented aspect, coloured in relative bright blue-green-yellow tones, the background elements in pastel shades and the explanation elements in red. Visual elements like shadows or borderlines of surfaces have been avoided. The animations are synchronised with the explaining spoken text and sound effects to enhance the visual experience.

As the background of the scene uses bright colours, the interactors have been chosen in grey tones. This allows hypertext links to be represented without disrupting the aesthetic integrity of the whole summary text while making active elements easily spottable. Immediate feedback is given in red for every interactor being activated by a mouse movement over it.

## 4. Conclusions

The CD gives a real impression of what microsystems are today, what they are used for, how they are fabricated and what can bring them to economical success.

The CD is designed as an aesthetic appealing work, using fundamental principles of ergonomics and human-computer interaction as in [9], [8], [7], [11], [10] and exemplified for instance in [6].

## 5. References


1. Madou, M. : Fundamentals of Microfabrication, CRC Press, 1997.
2. Wechsung, R. and all: Market Analysis for Microsystems, a NEXUS Task Force Report, 1998.
3. Wilkinson, M.: Microstructures & Microsystems Technology Applications & Markets 1997-2002, TFI, 1999.
4. Sze, S.M.: Semiconductor Devices, Physics and Technology, John Wiley & Sons, 1985.
5. Popovic, G. and all: Manufacturing Processes for Micromechanical Components, FSRM course, 1999.
6. Berger, M.: Ergonomie et sécurité des tunnels à Neuchâtel, Bulletin du VSS, february 1994.
7. Daniellou, F.: L'opérateur - la vanne - l'écran, L'ergonomie des salles de contrôle, Agence nationale pour l'amélioration des conditions de travail, Montrouge, 1986
8. de Montmollin, M.: L'ergonomie, Editions de la découverte, Paris, 1986
9. Nievergelt, J. and Ventura, A.: Die Gestaltung interaktiver Programme, Teubner Verlag, Stuttgart, 1983
10. Sperandio, J.-C.: L'ergonome du travail mental, Masson, Paris, 1988
11. Valentin, A. and Lucongsang, R.: L'ergonomie des logiciels, Agence nationale pour l'amélioration des conditions de travail, Montrouge, 1987

# POSTER PRESENTATIONS

# MULTIMEDIA INNOVATIONS

# Action-Based Learning System for Teaching Digital Electronics and Test

R. Ubar and H.- D. Wuttke
*Technical University of Tallinn, Estonia: Ilmenau Technical University, Germany*

**Key words**: action-based learning, CBT, WBT, Internet, game-like environment, test

**Abstract**: The Internet and multimedia open new possibilities for learning methods. We want to present a learning method uses so-called *living pictures.* On one hand teachers can show more complex examples and immediate demonstrate the influence of changing parameters by using computer simulated living pictures in their lessons. On the other hand students can use the same simulations on their home computer, if the living pictures are available on the Internet. We demonstrate an example of a living picture, dealing with digital electronics and test.

## 1. INTRODUCTION

Traditional teaching methods either start by explaining a theory and showing some examples or giving an example to introduce a theory. However one of the both methods is chosen and fixed by the teacher independently of what is best for most of the students. The students mostly get some accompanying materials such as scripts, books etc and make some own notes. After listening to a lecture they can consult only their notes and try to solve some problems by using the new learned method as good as they remember. Mostly there is not enough time for lots of examples during a lecture and the students' notes include some errors. To correct these errors and to give some more examples the students usually replicate the subject of the lesson at home and during an exercise. How can computers, connected to the Internet, be used to make this learning process more effective? What kinds of software modules are required? How can the same software be used in several phases of the learning process? To answer these questions we'd like



*B. Courtois et al. (eds.), Microelectronics Education,* 107–110.

to present an action-based learning system realized in the project DILDIS carried out in cooperation between TU Ilmenau and TU Tallinn.

## 2. METHOD

The following figure 1 shows the four phases of the learning process supported by the education system: listening, replication, examination and practice phase. For each phase there exists a special application service adapted to the learning process, which allows different views and actions using the same interactive module. Thus the education system supports the action-based training.

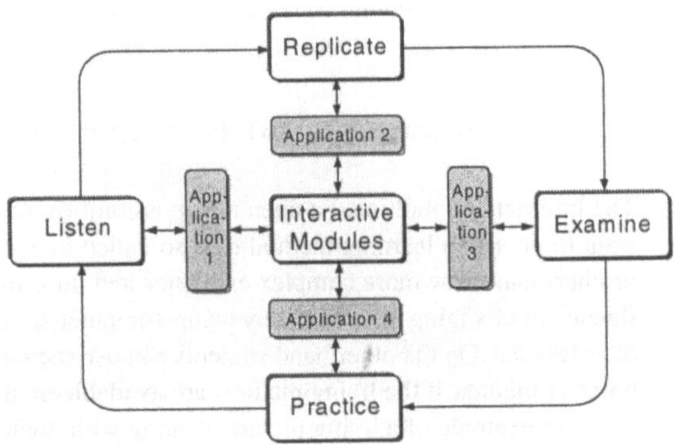


*Figure 1.* [Phases of the learning process]

Action based training via internet names a new teaching concept where the learning process is characterized by several features like computer aided learning and teaching, offering a set of tools to inspect the subject, access to multiple learning modules, a big reservoir of examples and the possibility to generate new ones, focus on correct solutions, easy action and reaction (click and watch) by using "*living pictures*", the possibility of distance learning, multilingual descriptions, individual depth and duration, learning by doing, funny and game-like context. This concept offers teachers and students the possibility of free acting in the learning process. Core of that concept are some JAVA-applets (the interactive modules) running on any browser connected to the Internet. We call these applets "***living pictures***" and explain it as follows:

Given a tricky, quite complicated situation of the learning subject in a graphical form on the computer. The graphic has to be self-explanatory and involving interaction possibilities. By using interaction possibilities the students can generate examples that are interesting enough to encourage own experiments but not too complicated for learning. They can produce input stimuli and watch the reactions. In reaction of the inputs a *simulation component* starts, executing the method that has to be taught, and presenting its results using a *visualization component*. Thus the students immediately get a correct reaction of their inputs. In that phase it is important that there is a simple interface and no assessments of the students' inputs occur. They can use it like a game: they act and the system reacts by using a yet unknown method (that one, that should be learned). In addition to the simulation component there is also an *explanation component*, describing the unknown method step by step, using the actual chosen or generated example.

Comparing this method with traditional learning methods using books and notes, taken during a lecture, the learning process becomes more effective. During the same time the number of examples that can be inspected is higher and the students loose no time in going wrong ways because of wrong notes they made. They can always be sure that they get the right results. The game-like character of the *living pictures* raises the students' curiosity and encourages them to do some own experiments and examples. The same software described above can support several phases of the learning process as written in [1]. In difference to the case of self study described above, the same modules can also be used during the replication and the examine phase and during a lecture (in the listening phase). In that case the teacher can immediately generate examples that can answer students' questions by showing changes caused by teachers inputs. The teacher can also use some prepared examples to show characteristic effects and specialties. The use case "Practice" requires the tool features of the modules, such as export/import components and connections to professional used tools.

## 3. EXAMPLE

In the following figure, a *living picture*, dealing with the fault simulation in digital circuits is represented. The *living picture* includes a graphic of a circuit with the facility to insert all possible faults and to watch how they change the circuits behavior at different input patterns or how they can be detected by test patterns.

The program is written in Java 2. It can be run via network, using standard browsers like Netscape and Internet Explorer with Java 1.2 runtime plug-in, or with Java 2 applet viewer. The work window consists of three main parts as shown in figure 2:

- vector insertion panel,
- view panel for design schematics and
- view panel for data tables and waveforms of simulated test patterns.

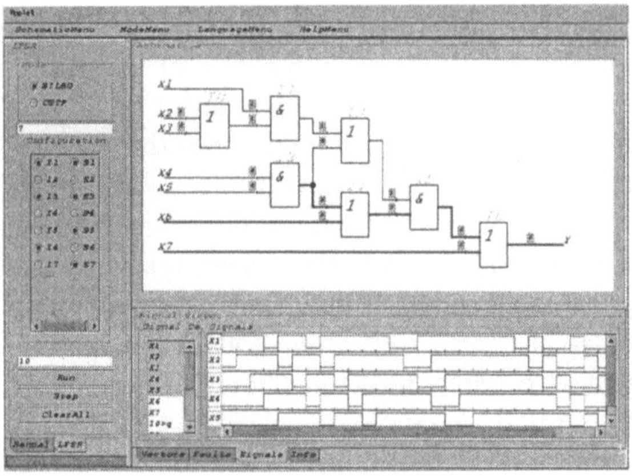

*Figure 2.* [Screenshot]

The following topics of the course "Digital Electronics and Test" are supported by this program: fault simulation, test generation, testability analysis and design of built-in self-test architectures. The key problems of these topics can be taught and learned using the program on different examples. After the theoretical investigation of these topics, a laboratory work follows with more complex designs, where the design software (Cadence, Synopsys or Xilinx), and diagnostic software Turbo-Tester [2,3] is used.

## 4. CONCLUSIONS

By the use of web-based media we achieve: presentation of course material independent of place and time, individual learning according to the students' own needs, quick cross-referencing by hyperlinked texts, new forms of communication between teachers and students (chat, joint editing), up-to-date course material.

## REFERENCES:

1. Wuttke, H.-D., Henke, K. and Peukert, R.: Internet Based Education - An Experimental Environment for Various Educational Purposes, in M. H. Hamza (ed.) Conf. on Computers a. Advanced Technology in Education, *IASTED/Acta Press No. 292*, pp. 50-54, 1999.
2. Jervan, G., Markus, A., Paomets, P., Raik, J. and Ubar, R.: Turbo Tester: A CAD System for Teaching Digital Test. In *"Microelectronics Education"*, Kluwer Academic Publishers, pp.287-290, 1998.
3. Ubar, R.: Teaching Dependability Issues in System Engineering at the Technical University of Tallinn. *Global J. of Eng. Education*, Vol.2, No 2, UICEE, Printed in Australia, pp. 215-218, 1998.

# MEDICIS : AN ENVIRONMENT FOR REMOTE FPGA CIRCUIT TESTING BY INTERNET

P. KADIONIK, P. NOUEL, P. GRESSIER, P. DUFRENE and S. LEMASSON
*ENSERB-IXL Laboratory, 351 Cours de la Libération 33405 TALENCE CEDEX, FRANCE Contact : kadionik@ixl.u-bordeaux.fr*

## 1. Introduction

During FPGA (Field Programmable Gate Array) circuit conception, it is generally necessary to perform after placement and routing a post simulation including all time delays. It permits to know global performances of the circuit and to detect critical paths after routing. All available commercial synthetizers are able to produce a standard delay file used for post simulation.
However it is more interesting to perform additional real measurements on designed FPGA circuit and to compare results with post simulation in the same design environment. A new tool answering this need has been also developed.

## 2. MEDICIS Presentation

We have developed the tool named MEDICIS (stands for "MEsure à DIstance de CIrcuitS" in French language, i.e. remote FPGA circuit testing) based on the Internet Client/Server architecture [1]. Its general architecture is given figure 1.

MEDICIS has been written in Java language as a client/server application. A Java client application is used as a Graphical User Interface to collect measurement parameters (stimuli signals, time step...) and ask a Java server application to perform real measurements by driving all devices (FPGA programming, logical analyser, pattern generator...). Communications between all client applications and the server are based on a UNIX socket mechanism that permits remote control by the Internet network.
MEDICIS allows also designers to perform their measurements on the designed FPGA circuit in real time. It can be used by several persons involved in different projects : exclusive access to the FPGA circuit for programming and testing is then used, so only one testing board is necessary.

MEDICIS uses a HP16500A logic analyser and its pattern generator board. For remote control, MEDICIS controls a RS.232 port of a workstation connected to the Ethernet



*B. Courtois et al. (eds.), Microelectronics Education,* 111–114.

local network by using the standard Sun Java Communication API (Application Programming Interface).

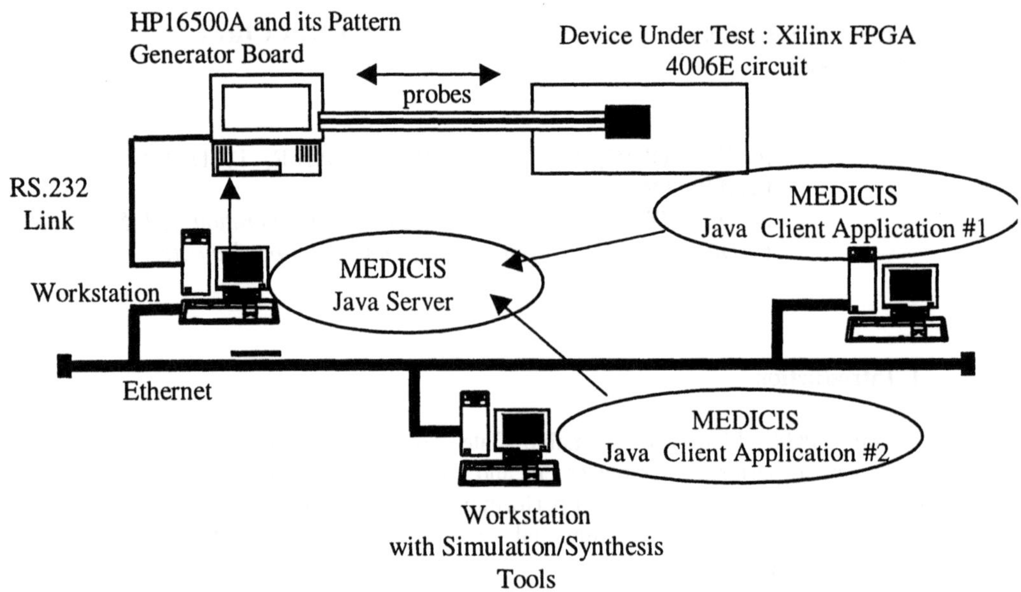


*Figure 1.* MEDICIS Environment

MEDICIS tool also implements an interface between our Mentor Graphics™ ModelSim™/simulator Leonardo™/synthesis environment and the HP16500A logic analyser.

## 2. FPGA Design Flow

In order to design a numeric circuit, a VHDL (VHSIC High level Description Language) design entry is generally used. VHDL stimuli issued from simulation are saved as a file and then used by MEDICIS to drive the FPGA inputs through the pattern generator board. FPGA outputs are then sampled and recorded by the logic analyser and saved as a file by MEDICIS. This file is compatible with the ModelSim simulation environment and can be also displayed. Figure 2 presents the general design flow from conception to real measurement : A Top/Down approach is used for FPGA synthesis. After the design entry in VHDL language, a simulation is performed with the ModelSim tool. Generic synthesis is then performed with the Leonardo tool with the pin attributs of the Xilinx FPGA target. Placement and routing is done afterwards with the Xilinx™ tool and the .BIT file is finally created. In order to use MEDICIS environment, one must have the .BIT file issued from placement/routing, the .LST file issued from the ModelSim simulation defining stimuli and a specific .PAD file defining the probe cabling of the Device under Test board. These files are used by the MEDICIS client application according to figure 3 showing the MEDICIS GUI.

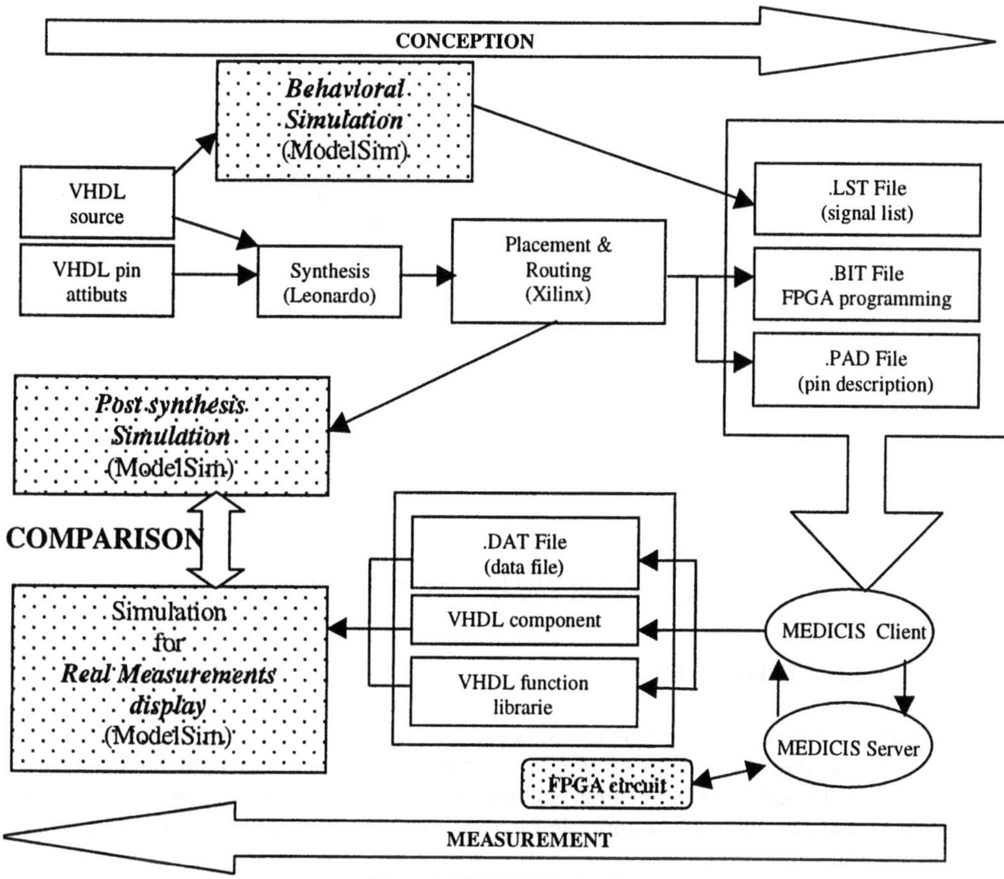


*Figure 2.* MEDICIS Design Flow

Communications are then performed by the Internet network with the MEDICIS server application in order to apply stimuli on the DUT inputs and sample outputs. A .DAT file is generated with real measurement results. In order to use it with the ModelSim simulator tool, we have created a dummy VHDL component that reads the .DAT file. After its VHDL compilation, one can visualize real measurement results with the ModelSim simulator and then compare in the same time with post synthesis simulations.

With MEDICIS, it is also possible to compare VHDL post simulation with time delays and the real measurements by using just one simulation environment. It is a useful tool for FPGA timing optimisation.

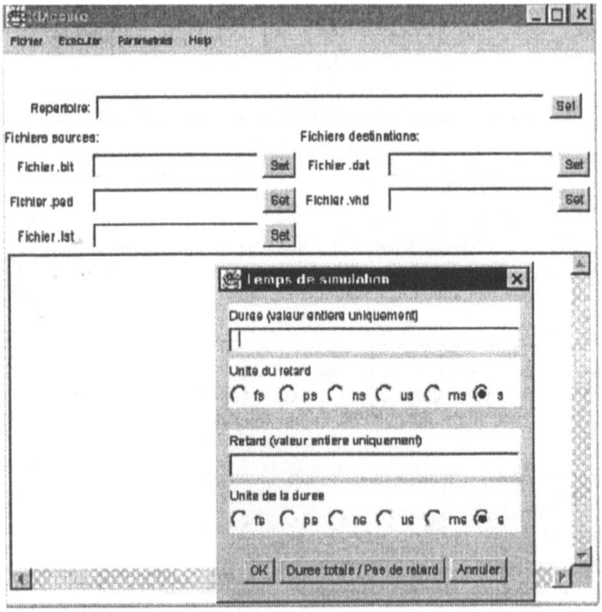


*Figure 3.* MEDICIS Graphical User Interface

## 3. MEDICIS Portability

One can note that MEDICIS is based on Java language and its object oriented programming. With help of the Java interface mechanism, MEDICIS can be easily ported to different simulation environments and logic analysers. CPLD circuits (Complex Erasable Programmable Logic Device) can be tested too. MEDICIS is really hardware independant.

## 4. Conclusion

MEDICIS is currently used at the ENSERB School of Electrical Engineering (Ecole Nationale Supérieure d'Electronique et de Radioélectricité de Bordeaux) for practical courses on VHDL language and FPGA synthesis by 120 students. The DUT FPGA circuit is a 4006E Xilinx circuit placed on a specific board.

## 5. References

Kadionik, P., Zimmer, T. and Danto, Y. (1999) RETWINE: A New Distributed Environment for Microelectronics Instrumentation Learning and Measurement, IMTC 99, the 16th IEEE Instrumentation Measurement Technology Conference, Venice, Italy, pp. 1488-1491.

# WEB based exercises on Computer Arithmetic

A. GUYOT
*TIMA laboratory*
*46 av. Félix Viallet, F-38031 Grenoble - France*
*Tel.: (33) 4 76 57 46 16 e-mail : Alain.Guyot@imag.fr*

Abstract: This paper discusses the issues of including interactive exercises in WEB based course. "http://tima-cmp.imag.fr/~guyot/Cours/Exercices" (in French)

## 1. Introduction

We want to take advantage of the fact that every student in Electrical Engineering or Computer Science has access to a computer, whether personal or provided by the university, to offer him or her stand-alone interactive exercises. In the present paper we deal with exercises associated with lectures on computer arithmetic given in the "Telecom department" of INPG.

## 2. Figures included in lecture notes

To help understand principles, printed course notes usually include many figures, which are of course stilted. Those figures fall into four categories:
1- numerical examples (tables or list of intermediate results),
2- diagrams,
3- numerical algorithms (written as computer programs),
4- architectures generated by synthesis algorithms.

## 3. What interactive exercises do offer

Let us review what animation can do for each of these figure categories:
1- Computer arithmetic algorithms simulation displays on the computer screen animated tables of intermediate and final results both in binary and for convenience of verification in decimal. Operands are entered either in decimal, hexadecimal or binary. Copy and paste are allowed. The number of bits as well as the number of steps (whenever applicable) is parameterised. For algorithms that converge towards the result by error minimisation, measures of the precision are also displayed.
2- Diagrams consist of a fixed drawing plus an animated part that the student can initiate and control by clicking the mouse.



*B. Courtois et al. (eds.), Microelectronics Education,* 115–118.

3- Numerical algorithms usually translate into a control part and an operation part. Here the student plays the role of the control part while the computer emulates the menial tasks of the operation part (binary add/sub and shift with a parameterised number of bits). Those exercises have a stand alone demo mode as well as a blind mode in which the student sees only the test bit (sign of the last intermediate result).
4- For any arithmetic operator, there are many design approaches. The synthesis exercise can draw figures similar to the ones included in the lecture notes with various combinations of design styles and constraints. Measures of complexity and delay (both in term of gate number on the critical path) are displayed. There are meant to let the student quickly explore the design space.

## 4. WEB based exercises

The exercises listed below have been developed in HTML, JavaScript and Java. JavaScript is very well suited for prototyping the exercises, but eventually they have to be translated in Java that offers more interactivity. Besides most of the exercise provides links (URL) to the authors of the theory or the synthesis algorithms.

| | |
|---|---|
| Activity in carry propagate adders | Simulation |
| Carry propagation free addition (Carry Save) | Simulation |
| Brent-Kung carry look-ahead tree generation | Synthesis |
| Serial multiplication (add/sub and shift) | Simulation |
| Generation of multiplier partial products | Synthesis |
| Generation of multiplier reduction tree | Synthesis |
| Simulation of previously generated multiplier | Simulation |
| Generation of modulo reduction (for self-checking operators) | Synthesis |
| Animation of "Robertson's diagram" for division | Diagram |
| Datapath of a sequential divider | Operation part |
| Sequential division (add/sub and shift) | Simulation |
| Datapath of a square root extractor | Operation part |
| Sequential square root extraction (add/sub and shift) | Simulation |
| Floating point add/subtract/multiply/divide | Simulation |
| Datapath of a sequential logarithm (Cordic) | Operation part |
| Sequential logarithm and exponential (add/sub and shift) | Simulation |
| Layout of the convergence domain for Sine/Cosine (Cordic) | Diagram |
| Sequential sine and cosine (add/sub and shift) | Simulation |

Most of the exercises share a similar presentation. The screen is split into two windows. The left-hand one is a form to enter the parameter while the other starts with displaying instructions. When the user hit the "generate" or "compute" button on the form, the figure is drawn in the right-hand window. One can then click on signals in this figure to display information, edit the figure or simulate the operator.

## 5. Examples

High speed adders are typically based on tree structure formalised by R.P. Brent and H.T. Kung in 1982 after early realisations by J. Sklansky in 1960.

An heuristic transformation proposed by J.P. Fishburn in 1990 and made deterministic by R. Zimmermann in 1998 allow to choice the trade-off. After setting the parameters, one can watch the adder synthesis algorithm at work.

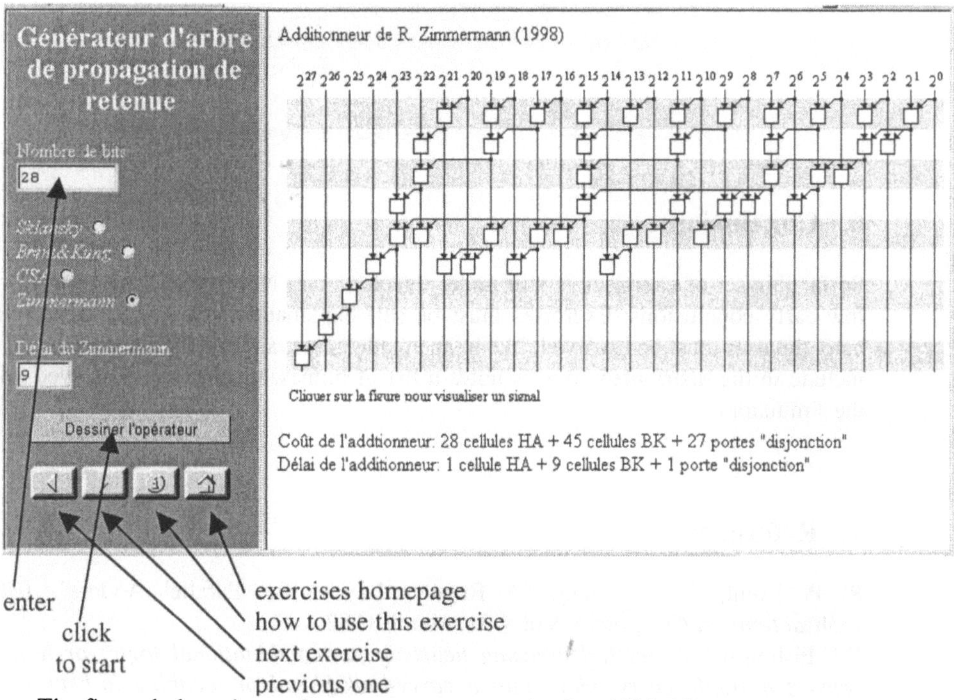


The figure below shows the square root extraction by shift/add/compare.

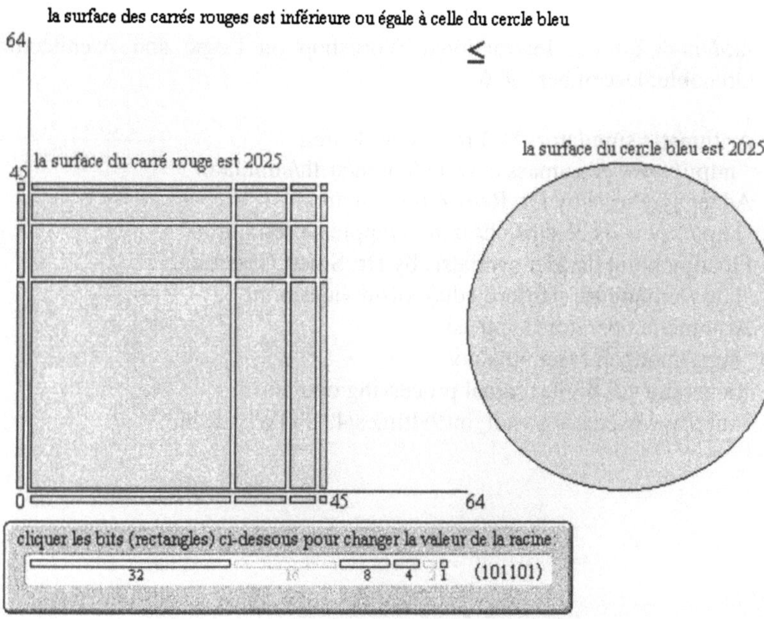

For most of the exercises, a "demonstration mode" teaches by example. The arrow is redrawn every 200 ms.

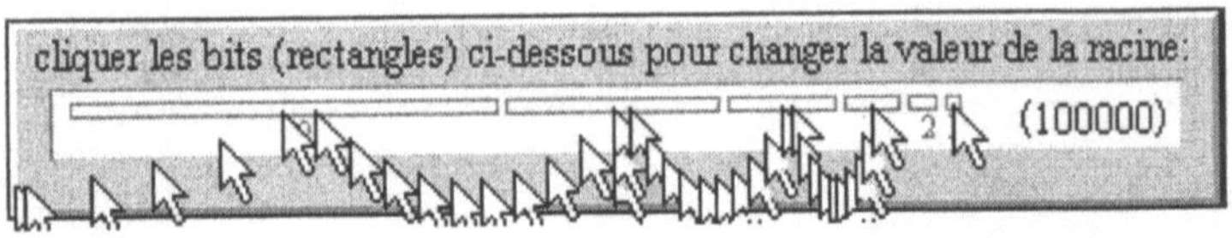


## 6. Conclusion

So far this set of exercises is still under development. Nevertheless we can draw a few early conclusions: Exercises must be self explanatory and include an on-line help, the user must be involved, like in an arcade game, and finally it seems useful to include in the distributed courses notes a list of homework that can be solved with the simulators.

## 7. References

R. P. Brent, H. T. Kung, "A Regular Layout for Parallel Adders", *IEEE Transactions on Computers*, Vol. C31, March 1982.
J.P. Fishburn " *A depth-decreasing heuristic for combinational logic; or how to convert a ripple carry adder into a carry-look-ahead or anything in-between* ", proc. 27th Design Automation Conference, 1990
R. Zimmermann, " *Non-Heuristic Optimization and Synthesis of Parallel Prefix Adders* " , proc. International Workshop on Logic and Architecture Synthesis, Grenoble, December 1996

Arithmetic simulators by Prof. Israel Koren
"http://www.ecs.umass.edu/ece/koren/arith/simulator"
Adder synthesis by Dr. Reto Zimmermann
"http://www.iis.ee.ethz.ch/~zimmi/applets/prefix.html"
Floating point divider synthesis by Dr. Stuart Oberman
"http://umunhum.stanford.edu/tools/divider.html"
Arithmetic operators synthesis
"http://modgen.fysel.ntnu.no/"
Generator for digital signal processing operators
"http://www.eda.org/vhdl_intl/vltimes/43-POWELL.html"

# DISTANCE INTERACTIVE MICROELECTRONICS TECHNOLOGY COURSE: A TOOL TO MAKE EASIER THE UNDERSTANDING AND THE LEARNING

E. CHARTIER[1], PROF. C. DIACONESCU[2], PROF. O. BONNAUD[1]
[1] *Groupe de Microélectronique et Visualisation. UPRESA-6076*
*Pôle du CNFM: Centre Commun de Microélectronique de l'Ouest*
*Campus de Beaulieu 35042 Rennes Cedex France*
*Tel: +33 (0) 2 99 28 60 71 e-mail: bonnaud@univ-rennes1.fr*
[2] *Department of Computer Science and Automatic Control, School of Engineering, Universitatea "Lucian Blaga" Sibiu, Emil Cioran Str.,4, RO-2400-SIBIU Romania*
*Tel: +40 69 41 79 28 e-mail: dinu@vectra.sibiu.ro*

## 1. Introduction

Since many years, the majority of French students as well in engineer schools as in universities have found the technology lectures mainly boring and/or difficult to understand. In order to make easier their understanding and thanks to the recent improvement of computing equipment as well as multimedia possibilities, a new tool was developed in the field of microelectronics technology.

The present course deals with basic knowledge on microelectronics technology, and can complete the lecture, which prepares students to work in a clean rooms, that are accessible in the microelectronics centres of the (French) National Committee for Education in Microelectronics. One can connect to this tool via Internet. The document is written in html format and uses Java applet possibilities for special calculations and animations.

The first part consists in several chapters about the basic steps of a technological process with some animations allowing the understanding of basic physical phenomena. The second part is devoted to the description of the main complete processes such as bipolar, NMOS or CMOS ones. The interactivity allows showing the evolution of the structure, step by step, layer after layer, giving evidence of the importance of the sequences.

## 2. Origin of this web course

This work is the result of a European support in the frame of INEIT-MUCON Socrates Intensive Programme and of a bilateral collaboration between Universities of Sibiu and Rennes 1 supported by French Foreign Office and University of Rennes 1. The base of this product is the result of the evolution of the lecture supports of microelectronics technology lectures during the last five years. On the base of a monograph devoted to students of DESS "Composants Electroniques" of University of Rennes 1 [1], a



*B. Courtois et al. (eds.), Microelectronics Education,* 119–122.

document was adapted in the frame of the setting of a francophone module at University of Sibiu presented at the European Workshop on Microelectronics Education [2] and published by the Embassy of France in Romania [3]. Following this first step, a booklet was written for accompanying a CD-ROM on Virtual Microelectronics Cleanroom [4-5]. To improve the interactive aspect, this new version is an interactive course, which can be used via Internet.

After a presentation of the goal and of the methodology adopted to develop this product, the two main parts of this course will be described thanks to several examples.

## 3. Concerned students and goal

The present course is devoted to students at bac +4 or bac +5 level of Electrical and Information Engineering formations. This supposes some prerequisites in physics of semiconductor, electronics circuits and physics of condensed matter. This product can complete the lecture, as well as the CD-ROM *Virtual microelectronics clean room* [5], which prepares students to work in clean rooms; let us note that several microelectronics centres of the (French) National Committee for Education in Microelectronics have such an equipment [6-8]. That is the case of Rennes where the Centre Commun de Microélectronique de l'Ouest has facilities to fabricate basic MOS (Metal Oxide Silicon) technology wafers which can be characterized by the students. The aim is thus to give to the students the approach of a technologist.

## 4. Product structure

One can connect to this course via Internet, the present Web address is http://gmv.spm.univ-rennes1.fr. The document is written in html format and uses Java applet possibilities for special calculations and animations. This allows to any user to minimize the transfer of data via the network and thus to save time. Each page contains on the left-hand side the contents of the course with links to each chapter. The text is presently written in French, bu new versions in other European languages are expected in the next months.

The Java applets have different functions: the first type is an animation of special steps, which allows better understanding the time dependent processes. A second type concerns basic physical processes such as the diffusion of atoms in a silicon lattice. The third applet type allows simulating a physical parameter such as the growth rate of a film. The last type is a description of a complete fabrication process with the temporal sequences of basic steps using cross-section of the devices.

## 5. Web course content

The document contains two main parts. The first part of the course consists in several chapters about the basic steps of a technological process with some animations allowing the understanding of basic physical phenomena such as diffusion of atoms in a crystal, or ingot fabrication. These animations as shown figure 1 are supposed to give a new dimensional approach; indeed, it is always difficult to describe for example the evolution of a doping profile using only static cross-section drawings of a wafer.

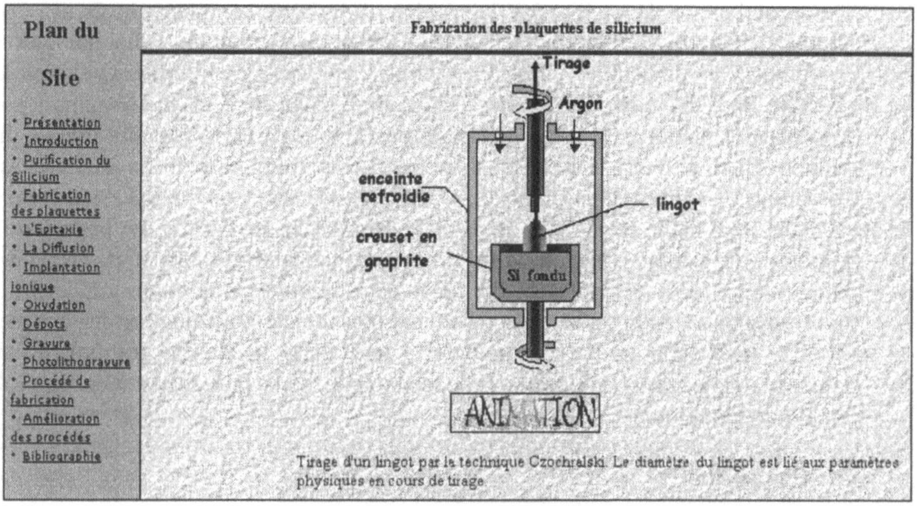


**Figure 1 : Example of the screen showing the animation possibility that corresponds to an ingot growth. Rotation of the ingot and consumption of the melt silicon are explicit.**

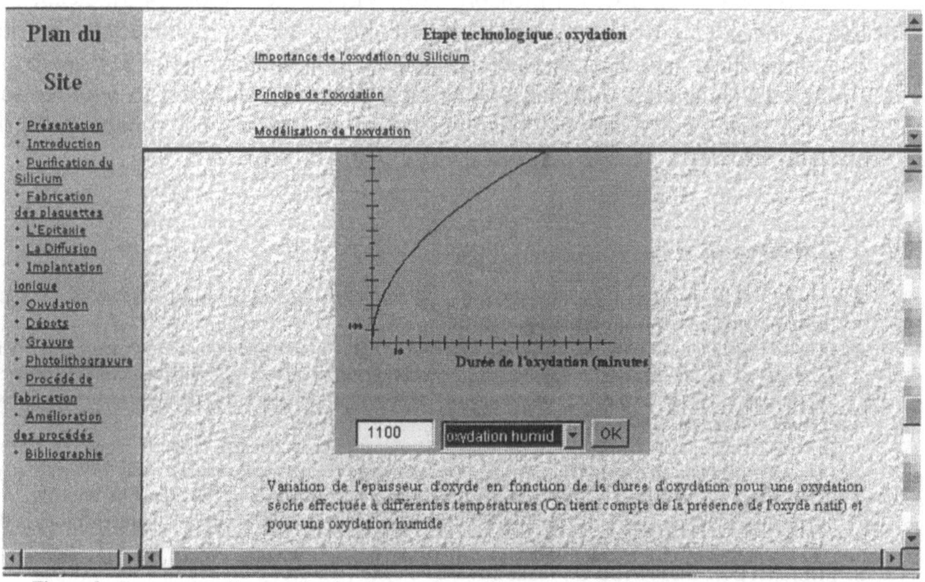


**Figure 2 : Java applet allowing the simulation of an oxide growth. The influence of parameters such as temperature and nature of the gases is well shown.**

The basic models describing the steps are also presented and applications are available. For example, it is possible to simulate the growth of an oxide using dry or wet oxidation's in function of the furnace temperature (figure 2); the used modelling is really basic but allows however an understanding of the role of the main parameters. Curves are plotted showing the variations of the thickness in function of the time with temperature as a parameter.

The second part is devoted to the description of the basic complete processes of bipolar, NMOS or CMOS (Complementary Metal Oxide Semiconductor) technologies with a progression in difficulties. The interactivity allows showing the evolution of the structure, step by step, layer after layer, giving evidence of the importance of the sequences of photolithography, self-alignment, etc.... It highlights the role of the step ranking which is really strict. The software allows going back if the student has some difficulties of understanding the role of the previous steps or the links in between. For each step, the name of the step is displayed in the upper part of the page. The cross-section of the sample is then modifies in accordance with this new step. The end of this part contains some examples of the up-to-date improvements of the processes, which involved planarisation techniques, chemical-mechanical polishing, or selective epitaxy for the VLSI submicronic technologies (Very Large Scale Integration). Finally, to demonstrate the strong evolution of the technology, some curves based on the G. Moore law [9] are presented.

## 6. Conclusion

This new tool acts as a complementation of lectures and other types of education products such as CD-ROM (Silicapolis [10], Virtual Clean-room [4-5], etc...). It gives the advantage to only need a connection to the server via Internet. It is also possible, to add at any time Java applets corresponding to new animations or new modelling in function of the evolution of the technological processes, and thus, to improve and up-date the tool. Finally, this work is the fruct of an international collaboration, that is specially exciting.

## 7. References


1. BONNAUD O., "Technologie microélectronique intégrée", Document polycopié DESS "Composants Electroniques", Université of Rennes 1, 1994.
2. DIACONESCU C., TOMA E., VESA E., BONNAUD O., "Francophone module in microelectronics: an example of an international co-operation in education between a French microelectronics center and a Romanian engineer faculty", Proceedings of 1st European Workshop on Microelectronics Education, Grenoble (France), (1996), 137-140
3. BONNAUD O. and DIACONESCU C., "Microélectronique Intégrée Monolithique et Conception Associée". *Edition du Service Culturel de l'Ambassade de France en Roumanie*. May 1996, 120 pages
4. BONNAUD O., LHERMITE H., "Virtual microelectronics clean-room for preparing students to actual technological training period", 7th EAEEIE'96 Conference, Oulu (Finland), Oral presentation, 12/14 Juin 96. Proceedings of the EAEEIE'96
5. LHERMITE H., BONNAUD O., "Salle blanche virtuelle", CD-ROM, *Edition du CNED*, (1998)
6. BONNAUD O., "The French training system for the specialists in microelectronics", Invited communication, CAS'92, Sinaîa (Roumanie), (1992).
7. BONNAUD O., "French microelectronics education organization", Proceedings of EAEEIE'92, *Towards a European Curriculum for Electrical Engineering and Information Engineering*, Ed. Univ. of Stafford (UK), (1994)
8. AUCOUTURIER J.L., BONNAUD O., DANDREL P., ROBERT M., "CNFM: the French microelectronics training network", Invited paper, Proceedings of 1st European Workshop on Microelectronics Education, Grenoble (France), (1996), 3-7
9. G.E. MOORE, Electronics Magazine, vol. 38, 1965, p.114
10. CD-ROM Silicapolis, The Microelectronics Universe, SGS-Thomson Microelectronics &CCST-Grenoble, (1998)

# TRANSTEC – online training for innovative micro-fabrication techniques

A. PRIEBE, R. BRÜCK, UNIVERSITY OF SIEGEN
M. OHLER, A. SCHMIDT, IMM MAINZ
*Hölderlinstraße 3, D-57068 Siegen*
*Phone: +49 271 740 2375*
*Fax: +49 271 740 2473*
*email: priebe@rs.uni-siegen.dehttp://www-ttec.rs.uni-siegen.de*

Key words: micro technology, multimedia course, distance learning, web-based training

Abstract: The high development speed in modern high-tech fields like micro technology imposes novel demands on life-long learning for students and professionals in these areas. Only time and location independent strategies like Internet courses can cope with this demand. Most current learning servers, however, are not sufficiently flexible to serve the demands of learners as well as course authors. This article presents the results of the TRANSTEC project, a highly flexible Internet-based learning server that offers dynamic link management for easy course maintenance, multi-thread navigation for advanced learners, sophisticated self-assessment schemes and various online and offline communication features permitting information exchange among learners and between learners and tutors.

The TRANSTEC project provides a novel concept of interactive Internet-based training entities which are combined with CD-ROM and print media. The project is related to the advent of novel technological areas such as information technology, micro optics, micro mechanics and micro fluidics. These disciplines have stimulated a world-wide development of new micro fabrication methods with a rapidly growing demand for professional education. Since micro technology is characterised by frequent changes due to improved fabrication possibilities, this education must be a continuous process. Thus, life long learning becomes increasingly important in design, production or application of micro-engineering technologies.



*B. Courtois et al. (eds.), Microelectronics Education,* 123–126.

For the training topics of the TRANSTEC course, the LIGA technique, laser-based micro technologies and micro-engineering related aspects of precision engineering are addressed in detail. The potential user groups for the course mainly span from graduate students to design engineers and executives in small and medium-sized companies.

Together with modern software components, the course is composed of different media material. The text units are mostly accompanied by graphics, pictures or drawings and further supported by life media like video/audio sequences or animations. The course also provides a problem solving environment employing design tools and a selection tool for an appropriate technology to a given technological problem. In addition, every course unit is supplemented with a self-assessment module, where the prospective learner can test the newly acquired knowledge. Single learners are supported as well as group learning participants. Different users of the same page can identify each other and internet-based chat tools allow to form a virtual class-room environment. Finally, the user can contact directly specialists of each technology.

This paper in the following will focus on the technical aspects of the realised server software. One of the great advantages against normal web servers is the dynamic generation of all navigation elements and all links between pages. Conventional web servers store links pointing to other pages in the page itself. Adding or deleting contents in a course for example requires the maintenance of many links in many pages. This is a real challenge as the author of a page has to know which other pages link to the current page. Therefore the TRANSTEC server software generates all elements needed to link pages, build chapters or complete courses and so on while requesting a page. To realise this functionality, TRANSTEC is based on the Hyperwave server [1]

The user will not recognise any difference to "normal" web servers, after logging in he will see a main window like the following:

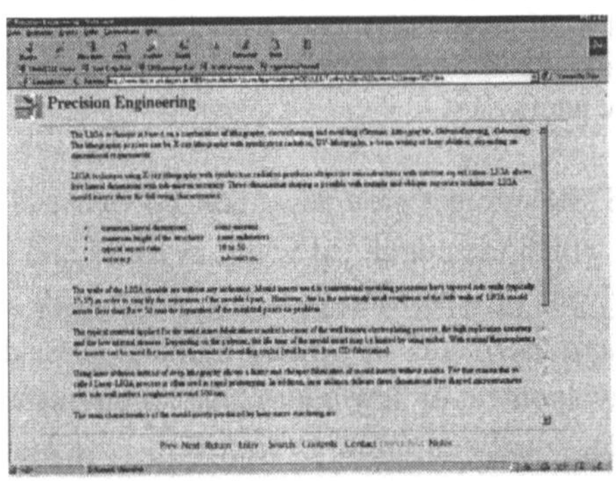


Apart from the usual Netscape elements there are three areas (i.e. frames). The top frame contains the title of the page and an (optional) icon, representing the actual learning topic. The middle frame contains the "real" content, in this example it is an introduction to precision engineering. The bottom frame holds the "navigation bar".

The term "Navigation" is a little bit misleading as there are also buttons for non-navigational functions, they will be explained in the following.

The buttons "PREV" and "NEXT" seem to be redundant as Netscape already has a "Back" and a "Forward" button. But they differ completely. Each course on the TRANSTEC server has a recommended path through it's contents. The knowledge to achieve is arranged in a logical order, the user will never be lead to pages dealing with unknown contents. So moving around with the "PREV" and "NEXT" buttons will guide the learner step by step building up his understanding like reading a book from the first to the last page.

The next two buttons "Return" and "Entry" allow the user to move more than one page with one click. Clicking on "Entry" will simply jump to the main entry page of the server. Clicking on "Return" will navigate to a page depending on the history of the user. The TRANSTEC server provides functions to keep track of the users movements. This is used to realise a very useful feature: when the user leaves the mentioned recommended path e.g. by clicking on a result of a full text search, on "normal" web servers he will get back to the page from where he jumped away only by accessing the browser's history (if available), searching the page title (if he remembers) and selecting the appropriate entry. The very length of the description shows that this is not a comfortable way. The TRANSTEC server stores those take-off pages and allows the user to return with just one click.

The mentioned search engine can be accessed with the "Search" button. It provides full text search, search in metadata like keywords and (page) titles. The range and type of documents to be searched can be reduced, e.g. one can search only for text documents changed after a specific date.

The button "Contents" opens up an additional window showing the environment of the actual page (see next figure). Already visited pages are marked with a check mark, the current page is marked with an arrow. This gives the user a very good overview of what he has already seen. The information is stored permanently on the server, not on the user's computer. This has the great advantage that from wherever the user logs in, he always gets his actual environment.

Another useful feature is implemented behind the "Contact" button. The intention is to provide a mechanism to directly contact an expert belonging to the current page. Clicking on the "Contact" button e.g. will open the user's mailer with the mail address already configured.

On some pages it will happen that some information does not fit directly into the actual context, but may be interesting. To enable authors to include such contents, the TRANSTEC server provides a mechanism which is accessible for the user via the "More Info" button. This button only is active when information is available and then opens an additional window.

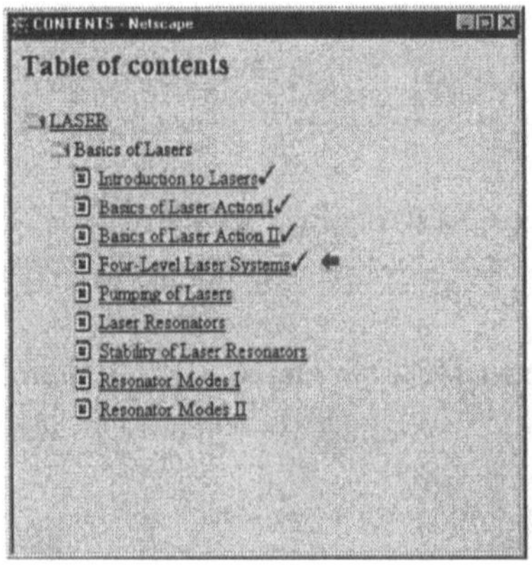


The last button is the "Notes" button. Notes are short texts written by users to annotate something to a specific page. Notes can be private, public or addressed to specific user groups (e.g. members of a company using TRANSTEC).

The implemented navigation and communication mechanisms provide easy to use and nevertheless powerful features to use the TRANSTEC server. Users with lower knowledge can step through contents in a logical and pedagogical pre-pared way to achieve better understanding, users looking for specific topics can use the contents window or the search engine to quickly access the content and if the facts not yet are sufficient users can contact experts. Besides that all users or experts can communicate via the built in functions asynchronously or synchronously.

The TRANSTEC server and courses have been realised by a consortium of seven partners from four European countries funded by the EC Educational Multimedia Taskforce under project number no. MM 1026.

Gentle WBT, http://wbt-3.iicm.edu/courses
Lotus LearningSpace: http://www.lotus.de/home.nsf/_/Produkte
Hyperwave: http://www.hyperwave.com
R.Brück, A.Priebe et al.: TRANSTEC – Ein internetbasiertes multimediales Lernsystem für innovative Hochtechnologien, page 70ff, Tagungsband GI-Jahrestagung, 1999
R.Brück, K.Hahn et al, Web based learning by engineering for MEMS technologies, page 100ff, Proceedings of SPIE Vol. 3894, 10/1999
J.Popp, Ein Internetbasiertes Lehr- und Lernsystem für die Mikrotechnik, Diplomarbeit, Universität Siegen, 03/2000

# THE ROLE OF CAL/CBT TOOLS IN MICROELECTRONICS EDUCATION

D. DONOVAL, V. DROBNÝ AND J. RACKO
*Department of Microelectronics, Slovak University of Technology in Bratislava, SK-812 19, Slovakia,* donoval@elf.stuba.sk

## 1. Introduction

The growing importance of knowledge enhanced by a very easy access to it via the Internet/WWW environment is a natural consequence of the strengthening information society in the world today. This is particularly true in microelectronics education, thus in a branch of technology which is approaching physical limits and is considered to be "brain limited". The difficult curricula and low interest of young people in technology in general is the origin of a continuing manpower shortage in microelectronics industry. Therefore the searching for new educational concepts of creating, maintaining, and increasing knowledge through research, subsequent dissemination and selection of the obtained information has increased the role and responsibility of teachers. They should select the most essential, durable information from the ever increasing knowledge pool in time, they should decide which new material should be included and which should be dropped out in a course. The new educational concepts, frequently based on multimedia technologies should attract more students to the field [1, 2].

What are the new problems which should be resolved, adopted and implemented into the microelectronics educational scenario?

- Shorter technological cycles in microelectronics industry due to the implementation of advanced technology require permanent updating of the curricula and retraining and lifelong learning. There are demands for a higher level of knowledge to cope with the "scientification" of industrial branches based on advanced technologies which should be provided for ever greater number of employees. The learners will be older and distributed over large areas. The contents, methods, and forms of education must be changed to fulfil these personal needs in new specialization and skills.
- Providing learners with the required skills, widely spread professional tools for the design, modelling, simulation and verification - supported by a user friendly interactive graphical interface - should be used in computer based training (CBT).
- Inadequate motivation of teachers for the development of new Computer Aided Learning (CAL) courseware should be overcome by an adequate number of authoring tools and/or adequate support from either government or industrial partners.
- Dissemination of standard educational tools is in contrast with the personality and individuality of teachers. On the other side the Internet/WWW environment creates the basis for a wider collaboration on national and international levels. Bringing together different approaches, culture, experience and traditions results in a new quality of the developed curricula, courses, modules, and a variety of teaching methods.



*B. Courtois et al. (eds.), Microelectronics Education,* 127–130.

- Implementation of new information and communication technologies enhances the globalization of the world. Universal education and general skills are necessary to promote the problem-less transfer and mobility of people. Some degree of standardization looks like a necessity. Evaluation, assessment, and accreditation of the implemented CAL/CBT courses is a serious problem of the microelectronics society.

## 2. CAL/CBT tools

To address the above mentioned problems and to develop new paradigms for learning, the contents, methods and forms have to be changed to fulfil the growing personal needs for a higher level of knowledge in the new specialization and skills [3].

The wider use of CAL/CBT tools which are simultaneously accessible by the Internet/WWW environment could soon create a significant domain involving three learning scenarios. They consist of a virtual, teacher-based classroom, supported learner-based self learning and group-based collaborative learning. As the complexity increases, a more tutorial style is used to present the less intuitive aspects. High degrees of animation and interaction are of crucial importance to the presentation. Animation is a particularly valuable technique when it illustrates phenomena, properties, any topics which cannot be effectively described by printed lecture notes, which is particularly true in microelectronics. The higher motivation is another benefit of the use of CAL/CBT tools.

Although the implementation of CAL/CBT multimedia based courses and modules into general educational system changes the roles of teachers and students, they were, are and will be the protagonists of the teaching/learning process. These changes have a strong impact on pedagogical, psychological, organizational aspects of the teaching/learning process. A greater degree of self discipline in self learning is required from anyone involved in the learning process. However, the direct, face-to-face contact between the student and teacher and between students themselves does not seem to diminish its importance. The established teacher-student relationship, particularly between a good teacher and a good student, can create a new quality in the learning by discovering, it can stimulate significantly research in a specific domain and subsequent implementation of the achieved results into traditional learning. It is essential for students to acquire the methodology ability how to analyze complex situations, creativity in solution finding.

Collaborative learning/training provides the learners not only with technical knowledge and skill but also with the ability to work in a team, to get experience in designing experiments and to acquire presentation skills. The industrial sector via technology transfer and mobility could even enhance and support international collaboration, which is a natural consequence of the globalization of the world. This is very important also from the financial point of view. As the government funding of universities is being reduced, a higher efficiency of the teaching/learning process is a must. By taking part in the international projects aiming to develop the new CAL/CBT tools, particularly in advanced technological fields, the expenses are shared by individual participants, while the resulting tools can be used efficiently by all partners involved.

The process of general assessment and evaluation of CAL/CBT based courses is one of the goals which should be addressed by governmental or professional bodies (e.g. IEEE in electrical engineering) in correlation with the wider use of this modern teaching/learning methods.

## 3. Example

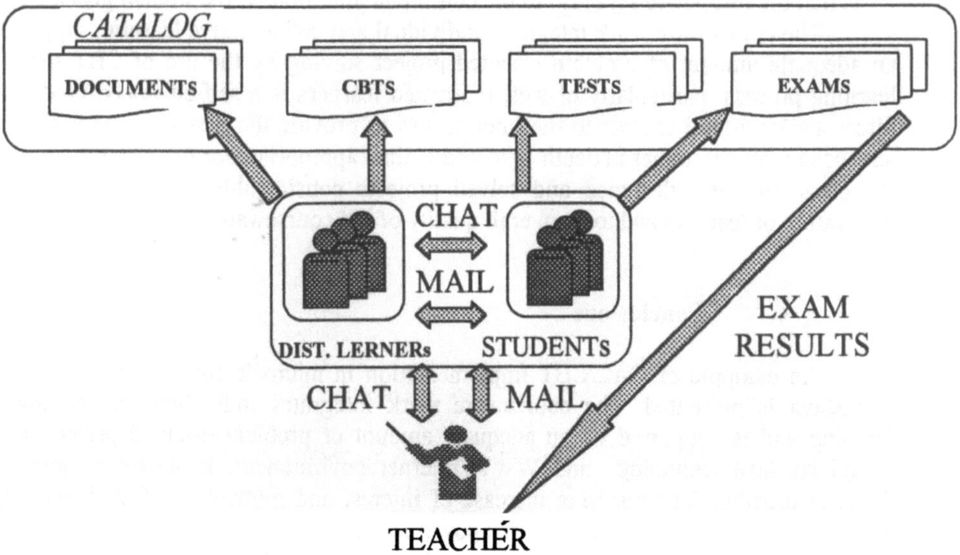


Fig. 1. A client –server application of CAL/CBT courses

Nowadays our courseware in the Physics of Semiconductor Devices is a normal client-server application (Fig. 1) [4, 5]. The client part is a JAVA 1.1.x [6] enabled browser (MS Internet Explorer 4.0x and Netscape Navigator 4.06 and higher). The server part is mostly realized in a WWW server and some server parts (e.g. chat) are written in C language. The navigation through the system is very easy and intuitive. When logging to the system, the learner enters the main part, the CATALOG. It is based on standard Windows Explorer functionality and allows the student to browse through all directories. Documents are sorted to directories so that the student can easily find what he needs.
Each directory has six basic views:

- Documents - all documents in the topical catalog are displayed like in the Windows Explorer, clicking on the name will open the selected document and the student can read/view it. Each document has an Author and a Consultant (mostly the same person), clicking on their names will open the mailer and the student can send an email to them.
- CBT's - in the topical catalog are displayed like in the Windows Explorer, clicking on the name will open it. These modules are mostly Macromedia Director files or any special JAVA programs, e.g. DEVSIM - device simulator of electrical properties.
- Chat - chatroom where all students (and consultants, of course) can communicate together. Each directory of the CATALOG has its own chatroom.
- News - similar to the chatroom but the messages are stored in a database and can be displayed again. The consultant can put any message here, he can inform that a new version of the document has been published and a description of changes can be given.
- Tests - show a list of test pools. The student can select a pool of test questions and then he/she will answer them. The questions are randomly generated from the pool. At the end of the test the student gets summary information about his/her results.

- Exams - show a list of exam pools. The process is analogous to the test, the difference is that the results are sent also to the consultant and, finally, the student gets the mark.

The courseware work integrates individual and collaborative learning supported by an adequate amount of problem oriented project solving by the use of CBT tools. The learning process, particularly of well motivated learners is a self-enhanced process. The direct and/or e-mail contact to the tutor allows to provide the best students with a more advanced material, either in depth or in width, than appropriate for the general demands in the class. Properly designed and solved projects considerably contribute to a higher motivation of learners and to an overall quality of the courseware.

## 4. Conclusions

An example of CAL/CBT implementation in microelectronics education at STU Bratislava is presented. The courseware work integrates individual and collaborative learning and is supported by an adequate amount of problem oriented project solving. Based on Java technology and WWW/Internet environment, it allows on site and/or distance learning. The resulting increase of interest and motivation of students helps to attract more students and to fill the manpower shortage in semiconductor industry.

Although the microelectronics teachers in general are naturally familiar with advanced information and communication technologies, there is no identity between the knowledge and pedagogical skills. To provide systematic teaching/ learning process using the above mentioned technologies and to prevent the learners from overfilling by information the development of more supportive and finely tuned methodologies and authoring tools is a necessity.

Due to the general importance of the education process and responsibility for the lifelong learning, distance education and professional retraining, the European Community has launched the key action "Multimedia Content and Tools" within the second theme "Creating a User Friendly Information Society" in the Fifth Framework Programme [7]. It aims to improve the functionality and usability of future information products and services to allow for a diversity of knowledge and to stimulate the creativity of teachers as well as the learners. This is particularly important for such scientific curriculum as is the microelectronics education.

*Acknowledgement* - This work was supported by grants No. 1/6098/99 and KEGA 75 of the Slovak Ministry of Education.

**References**


[1] Higher Education in the 21st Century, Vision and Action, UNESCO, Paris 1998

[2] Schmockel, H., Falter, T. and Beinvogl, W.: The view of Siemens on Training, Collaborative Projects, Continuing Ed, In: Proc. of the 2nd EWME, Ed: T.J. Mouthaan and C. Salm, Kluwer Acad. Publ. 1998, p. 3

[3] COLEMAN. J.N., et. al., Effectiveness of CAL as a Direct Replacement for Lecturing in Degree Level Electronics, IEEE Trans. on Education, Vol. 41, p. 177, 1998

[4] DONOVAL, D. et al., A Multiple Window Demonstration of the Operation of PN Diodes Using DEVSIM, Int. J. Engn. Ed.,11, p.36, 1995

[5] DONOVAL, D. and DROBNY, V., Modelling and Simulation – the Core of the Internet/ WWW Based Learning/Training, In: Proc. of the 4th Int. Conf. CALISCE 98, Ed: C. Alvegard, Goteborg 1998, p. 198

[6] Homepage for Javasoft's, http:/WWW.JavaSoft.com/HotJava/

[7] IST 1999 Workprogramme, http:/www.cordis.lu/ist

# GEOGRAPHICALLY DISTRIBUTED TEACHING LABORATORY FOR MICROELECTRONIC DESIGN AND TESTING

G. MARINOVA[(1)], V. GOULIASHKI[(2)], C. FERNANDÈS[(3)], P. DAPONTE[(4)], M. MARINOV[(4)]

[(1)] *Technical University-Sofia, galiaim@yahoo.com Ph.: 33 (0) 1 40 27 24 87*
[(2)] *IIT, BAS-Sofia, vgoul@iinf.bas.bg, Ph.: 3592 700391*
[(3)] *CNAM, cfernand@cnam.fr, Ph.: 33 (0) 1 40 27 20 81*
[(4)] *Università del Sannio, daponte@unsannio.it, marinov@mail.unisannio.it Ph.:+39 824 305817*

A Geographically Distributed Teaching Laboratory (*GeoLab*) for microelectronic design and testing is realized. The new open architecture concept of GeoLab is not problem oriented. It supports the dynamical configuration of the laboratory. A stress is given to individual training possibilities in the network. Three universities in three different countries are connected for teaching microelectronics design and testing. The approach is illustrated with distributed FPGA synthesis and test. Through GeoLab new laboratories could connect to the created network.

## 1. Introduction

The university networks established recently [1] don't offer considerable possibilities for individual training. There are few applications [2-3] of distant measurement in the networks. Another restriction for individual training in university networking in microelectronics is the license limits for distributed simulations.

The target of GeoLab is to provide an educational network for microelectronics with more accessible distance measurement laboratory and to introduce programs for distributed simulation in order to permit low budget laboratories to perform practical training and to satisfy students demand for individual training.

## 2. Implementation of GeoLab architecture

The architecture implemented makes possible different laboratories geographically distributed to cooperate and to achieve common goals. The laboratories can export on Internet the hardware and software resources they are supplied with, and render them available for inter-university projects. To achieve this, objective concepts



*B. Courtois et al. (eds.), Microelectronics Education,* 131–134.

like object oriented programming, modularity, interoperability were applied [5], and mechanisms like RMI and CORBA were adopted.

The proposed architecture is composed of four layers. The first one deals with the instruments and software available locally and makes them exportable on and configurable through Internet. The second layer has to guarantee the inter-laboratories and the inter-instruments communication. It makes possible the remote composition of the laboratory and the remote configuration of each instrument. The third layer is aimed to manage the data storage in a data base for further statistical analysis. The last layer offers an user interface of the distributed laboratory and permits to interact with the instruments and to monitor the results of the ongoing processes.

The measurement hardware in LESIM, University of Sannio, made available for GeoLab is based on the VXI standard. The respective software representation of the laboratory is illustrated on figure 1.

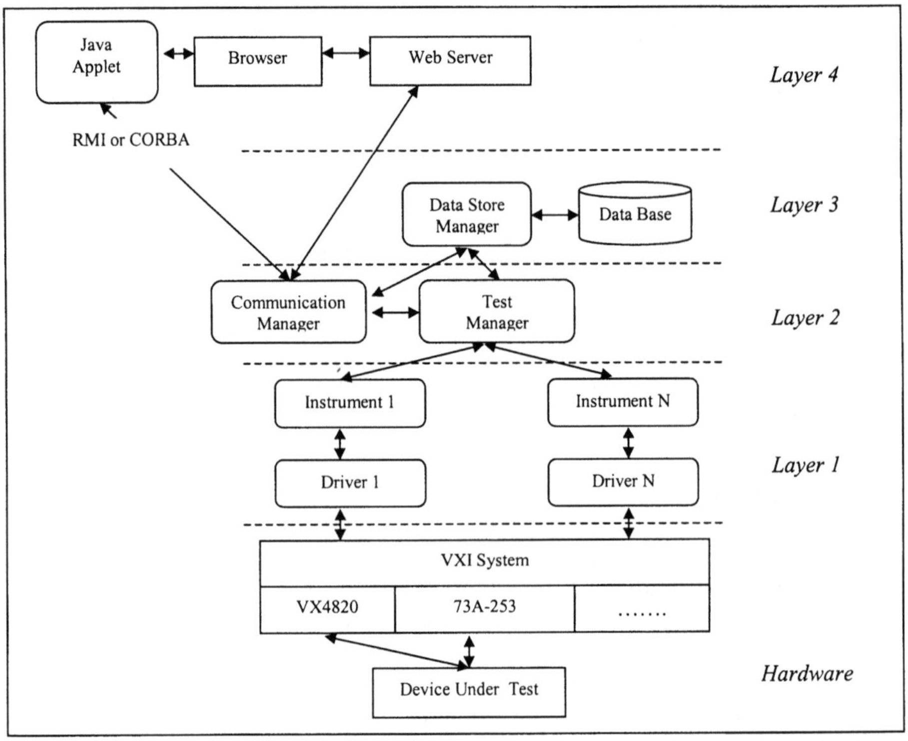


*Figure 1.* The architecture of the measurement laboratory.

## 3. Application for teaching microelectronic design and testing

GeoLab teaching materials for microelectronics are developed following the curriculum of the course "CAD in microelectronics", B6 of the Electronic department, CNAM Paris [4]. It represents also the base of the new course of

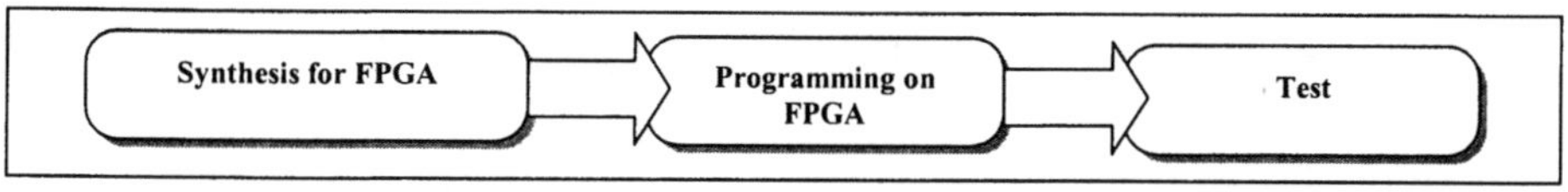


*Figure 2.* Task distribution in teaching FPGA synthesis, programming and test.

"Numerical Synthesis" in the Bachelor's degree in the Telecommunications faculty and in the French faculty in Technical University-Sofia. GeoLab teaching material leads step-by-step students through VHDL and FSM modeling, simulation, synthesis, programming, measurement and test of FPGA. The examples and exercises cover combination logic, flip-flops, latches, counters and FSM. The physical measurements in the experiences were effectuated at LESIM in the University of Sannio-Benevento, Italy. The teaching and training experiences in the examples were effectuated both from the Laboratory CAD in Microelectronics in CNAM, Paris and from the Laboratory CAD in electronics in the Technical University – Sofia.

GeoLab task distribution for these courses is shown on figure 2 and the concept of GeoLab training is shown on figure 3. The software tools applied are WARP and ISR [7,8].

An example of Security system design treated in GeoLab is illustrated on the figures 4 and 5. Figure 4 represents WARP simulation results and figure 5 represents results from distant measurement of the security system programmed on FPGA.

The Internet addresses of GeoLab are:
http://lesim1.ing.unisannio.it/geoLab.htm and http://excalibur.cnam.fr/~geomicro.html

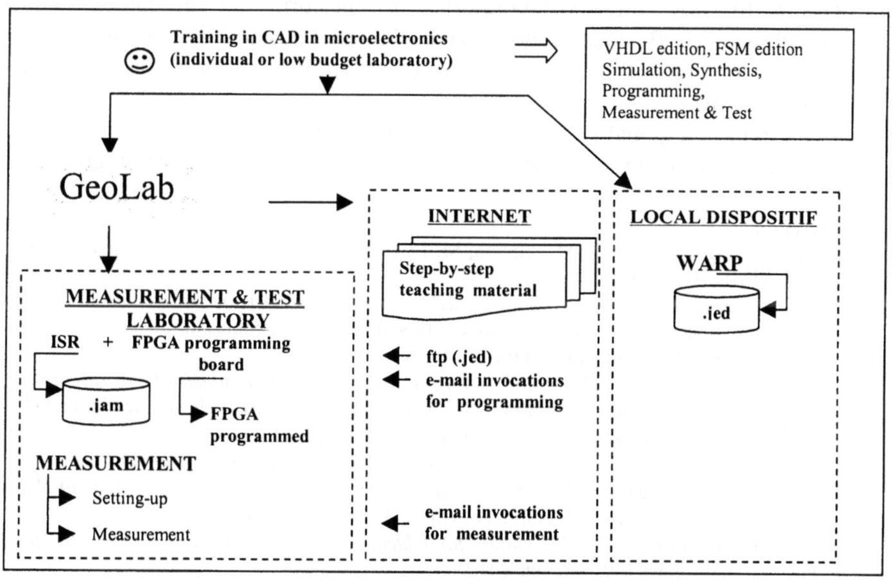


*Figure 3.* Training in GeoLab for CAD in microelectronics.

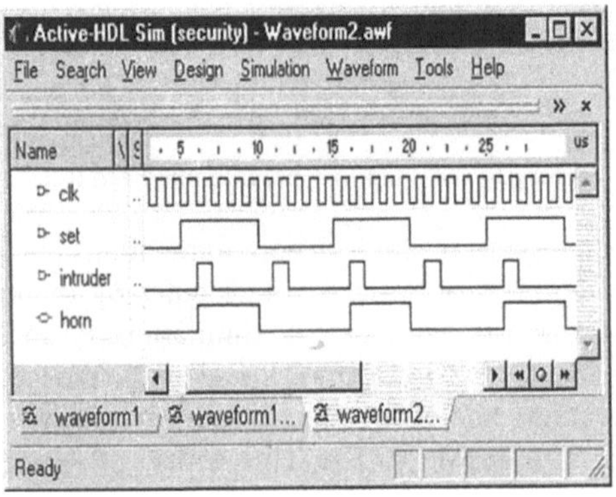


*Figure 4*. WARP simulation of a security system.

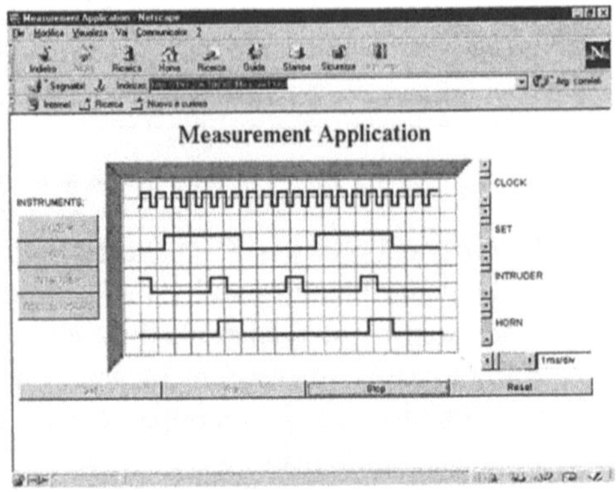


*Figure 5*. Distance measurement of a security system programmed on a FPGA.

GeoLab gives also the possibility for educational purposes to perform distributed design characterization and statistical modeling of MOS structures [6], distributed signal processing and system designs.

## 4. Conclusion

GeoLab has several advantages in teaching microelectronic design and testing. It permits to low budget laboratories to perform an acceptable education in microelectronics. Besides students have additional possibilities, as for example:

- to organize their own web-based environment for individual training with negligible expenses;
- to have a global view of the problems to be solved;
- to get accustomed to deal with advanced information technologies.

**References**


1. P.Foulk, M.Desmullier, L. Mackinnon, M. Ferreira, *The ASTEP educational multimedia framework*, 2nd EWME, Netherlands, 1998
2. Dave Grave, *On-line C-V and semiconductor parameter measurements*, Carnegie Mellon, [http://www.people.virginia.edu/~jcb6t/multimedia_microelectronics.htm]; [http://128002210165/captrans1.html]
3. T.Zimmer, F. Verdier, Y. Danto, *IC testing course performed in industrial and research environment*, Microelectronic education, Proceeding of the European workshop, Klewer Academic Publisher, 1996, pp.229-232; [http://www.ixb.u-bordeaux.fr/retwine.html]
4. C. Fernandès, *CAO des circuits électroniques*, CNAM/MEDIAS, Collection des cours 1998/1999
5. P.Arpaia, A.Baccigalupi, F.Cennamo and P.Daponte, *A remote measurement laboratory for educational experiments*, Measurement vol.21, No.4, August 1997
6. G. Marinova, *Computer-aided system for education in statistical design in electronics*, ISTET'97, Palermo, Italy, June 1997, pp.489-493
7. WARP 5.1, *Programmable logic design software*, "User's guide", CYPRESS Semiconductor, 1998
8. ISR 2.2, *In-system reprogrammable software for CPLDs*, "User's guide", CYPRESS Semiconductor, 1999

# MULTIMEDIA TOOL FOR SELF-LEARNING VHDL

S. PEREZ-LOPEZ, H. RAÑA AND J.L. PARADA
*Department of Electronic Technology. University of Vigo (SPAIN)*
*Phone: +34-986-812162   Fax: +34-986-469547   mailto:sperez@uvigo.es*
*Company Address: ETSI Industriales. Campus de Lagoas (Marcosende)*
*36280 Vigo.SPAIN*

Keywords: Multimedia, Tutorial, VHDL, Self-Learning

Abstract: We have designed this tool in order to make easier the study, learning and application of VHDL to our Microelectronic Design university students, both undergraduate and postgraduate.

First of all, we expose the syntax and all the elements language with plenty of examples and exercises to interactively try to resolve and evaluate their self-progress. This is addressed both to VHDL beginners and to those ones with basic knowledge.

Then we explain the kernel of a VHDL simulator, so they can understand the main differences between hardware description languages and general purpose high level programming languages (C++, TurboPascal, etc.) and how digital systems are to be modelled in order to take into account its inherent concurrency.

Finally we exposed a wide group of conventional MSI / LSI digital blocks, adequately selected and modelled in several ways, to allow them easily learning all of 1076.1 IEEE Standard possibilities to get digital systems be modelled and simulated satisfactorily.

In figure 1 we show the scheme for page links, which the proposed self-learning methodology is based on.

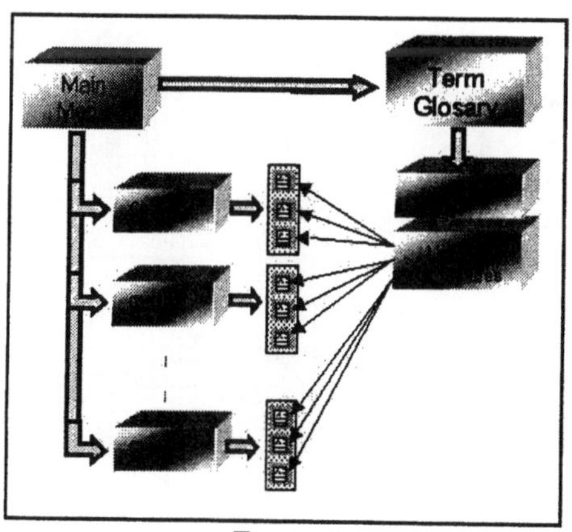


*Figure 1*

In figure 2 we can see the aspect of main menu showing chapter numbers. In this case CHAPTER 1 is selected and then its content appears in the big frame.
Down to the left side we can select the term and concept GLOSARY (see figure 3). The right oriented arrow allows entering the chapter pages and the option EXIT obviously allows leaving the tutorial.



*B. Courtois et al. (eds.), Microelectronics Education,* 135–138.

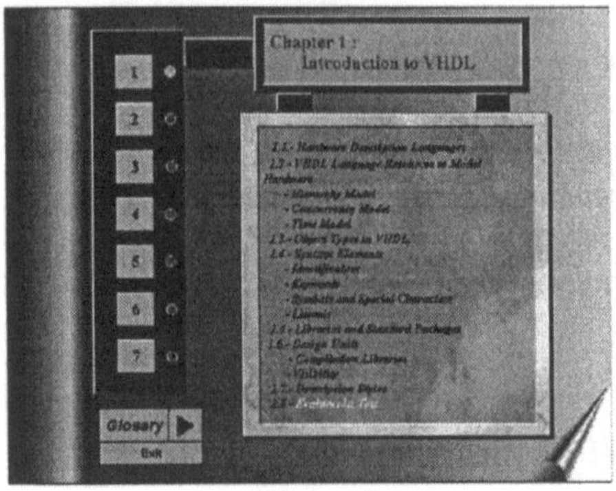


*Figure 2*

The glossary contains all the language reserved words and main terms and concepts for a quick consulting. You can always return to the page in which the glossary was called by clicking on <Link to page>.

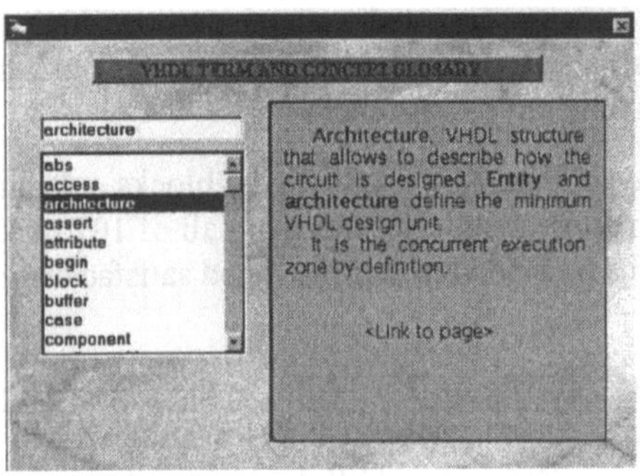


*Figure 3*

As an example of how concepts are explained, in next pages we show a few different cases. In figure 4 we can see a page that belongs to chapter 3. It explains one data type. When we click on <Example> we get more information about it, which is included in a new window, as it can be seen in figure 5. Clicking again on this window we get it disappear.

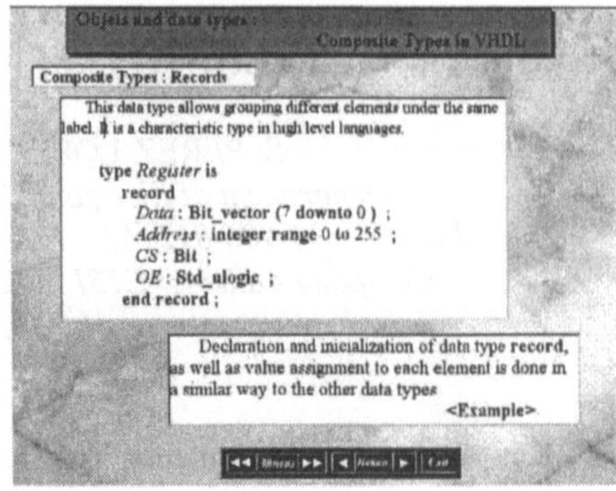


*Figure 4*

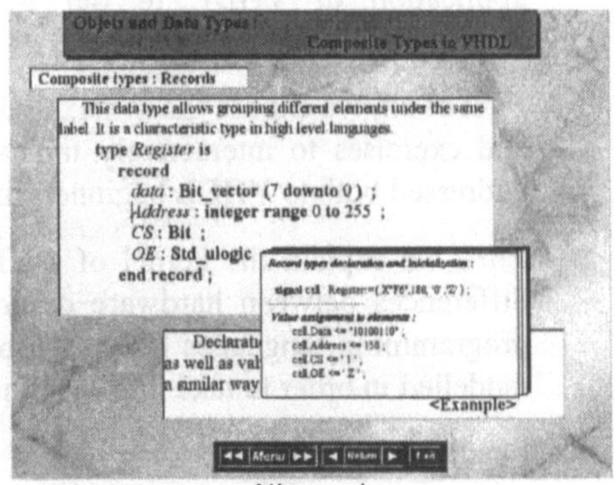


*Figure 5*

Figures 6 and 7 show another example, in which some Signal Attributes are explained. When <Timing Diagram> is clicked, two windows appear to show the diagram and the code that models it.

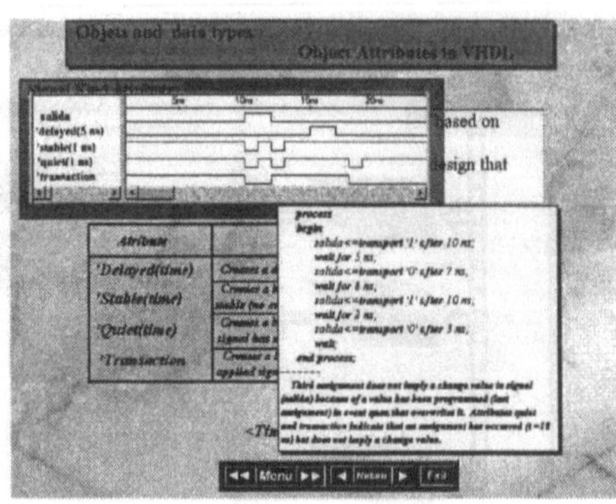


*Figure 6*

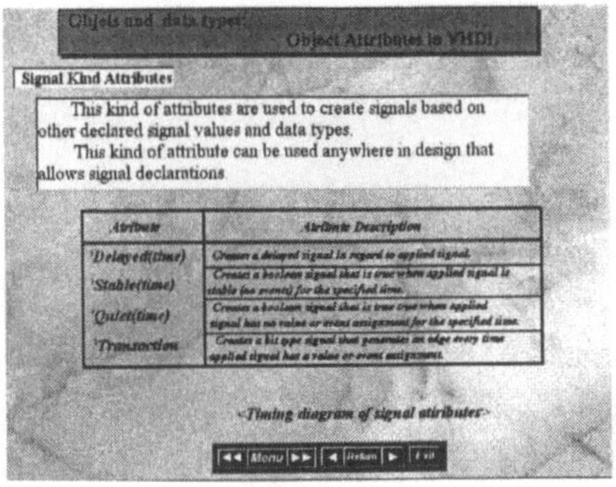


*Figure 7*

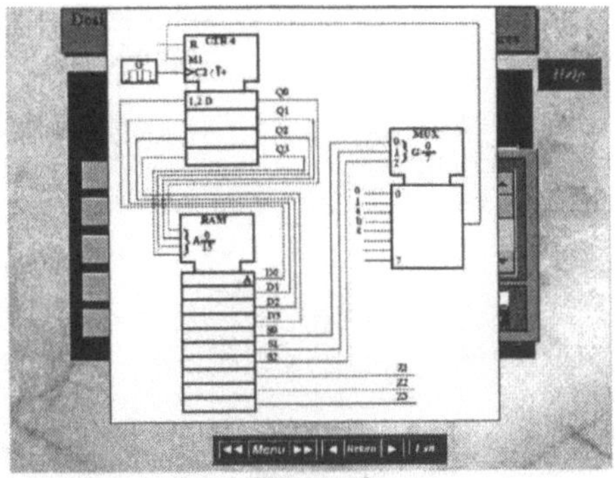


*Figure 9*

Figure 8 represent a page that belongs to last chapter (the seventh one). It shows a complete design example, in which one can select to see the schematic circuit (figure 9), its corresponding VHDL source code (figure 10), the test-bench code to verify the design, a link to a PC simulator (figure 11) and the list of libraries and packages used by the design (figure 12), which can be consulted by clicking there.

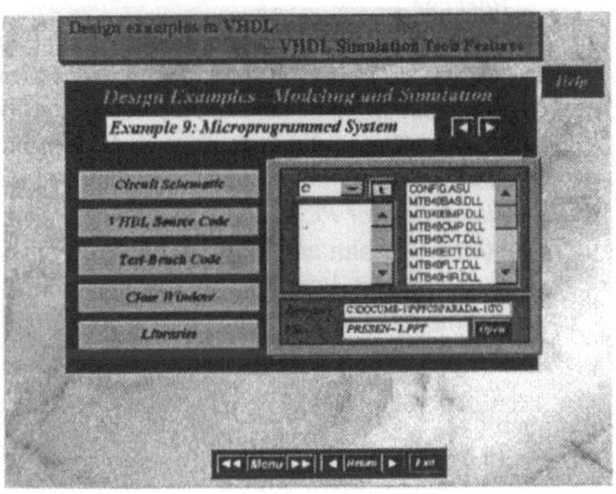


*Figure 8*

Finally, figure 13 shows the aspect of one of test questions included at the end of each chapter in order to evaluate the own progress as the student follows the tutorial. He/she has to mark all answers that apply. Then clicking on Verify button, one of these sentences will appear (in order from good to very badly and according to the number of wrong attempts):

OK! Correct. You can go on

Try it again

Please read once again the options in detail

Please study again chapter X

*Figure 10*

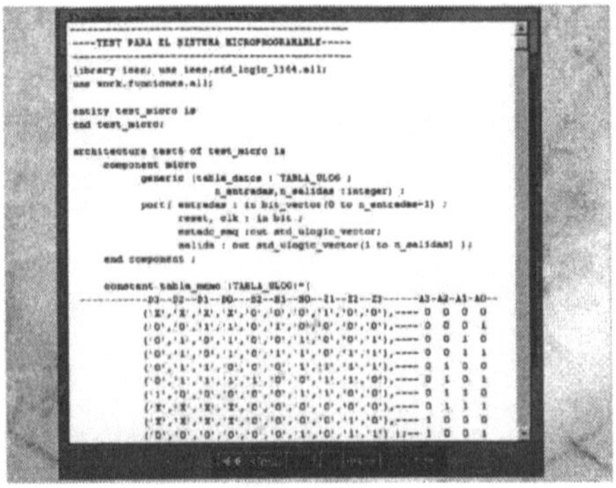

*Figure 11*

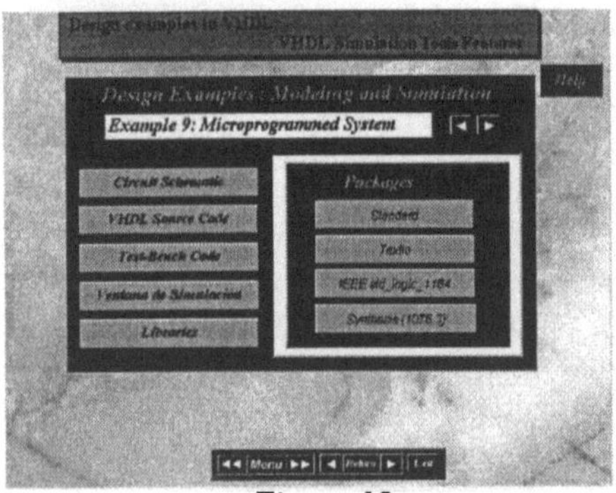

*Figure 12*

Presently, we are developing an extension to this tutorial in order to explain synthesis guidelines, model much more complex systems and ASIC design with IEEE 1076.4-1995 VITAL Standard.

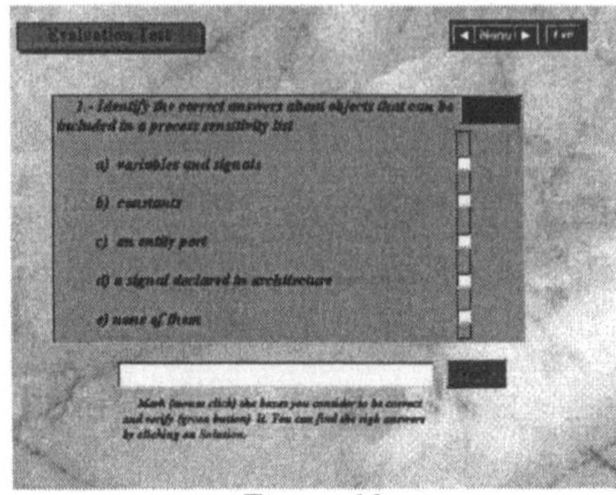

*Figure 13*

## References


- Baker, P. (1994) *Design Guidelines for Electronic Book Production*, Interface Design in Education, Berlin
- Bhasker, J. (1992) *A VHDL Primer*, Prentice Hall, Englewood Cliffs, NJ
- Bhasker, J. (1996) *A VHDL Synthesis Primer*, Star Galaxy Publishing, Allentown, PA
- De Bra P. (1996) *Hypermedia Structures and Systems*
- IEEE Standards Department (1996) *IEEE P1076.3-1996: Standard VHDL Synthesis Package*
- Institute of Electrical and Electronics Engineers (1993) *The IEEE Standard VHDL Language Reference Manual, ANSI/IEEE Std 1076-1993*
- Model Tech. Inc. (1995) *V-System/Windows (1995) User's Manual (Version 4), VHDL Simulation for PC's*
- Stefan Sjoholm and Lennart Lindh (1997) *VHDL for Designers*, Prentice Hall Europe
- Asymetrix Corporation (1994) *ToolBook User Manual*

# Using Multimedia in a Digital Design Lab

M. Serra, E. Wang and J.C. Muzio
*Dept. of Comp. Science, Univ. of Victoria, Victoria, B.C. V8W 3P6 Canada, mserra@csr.uvic.ca, jmuzio@csr.uvic.ca*

**Abstract:** We introduce a multimedia-based Virtual lab, developed with Macromedia Director, to enable students in Digital Logic Design courses to practice, exercise, learn in a simulated environment, easily integrated with more traditional physical lab boards.

## 1. Introduction

Digital design courses have, for many years, formed part of the fundamental core of both ECE and C Sc programs. While they usually include some low level construction of small scale hardware systems, it is important that they also reflect current industrial design styles. Certainly, in upper level courses in microelectronics system design, one of the goals is for students to acquire expertise in synthesis, capture, simulation, prototyping, VHDL, placement and routing, technological choice and programmable hardware. Normally, such courses require an introductory course in digital design which typically ranges from Boolean algebra, gates, transistors and flip flops, combinational and sequential systems through to programmable devices and controllers. There are many standard texts covering such material and we use Katz [2]. In some C Sc programs, it is the only course including any digital design. It is this course that is the primary focus of our Virtual Lab.

Further, it is increasingly important for students to make connections between intra and interdisciplinary areas, so that they are better equipped to negotiate the continually shifting boundaries of hardware and software design, as well as being exposed to such increasingly critical areas as embedded systems and hardware/software codesign. These advances in industrial practice illuminate the need for effective integration of software tools, but without compromising the real needs of a hardware design course.



*B. Courtois et al. (eds.), Microelectronics Education,* 139–142.

Traditionally, the introductory course includes a hands-on hardware lab, where students design and physically wire up small circuits. Some instructors feel this component is no longer critical, and the courses have migrated to being based solely on a simulation environment, (e.g. Logic Works), or on VHDL. Lack of resources has also, in some cases, dictated the abandonment (or sharp curtailment) of the physical laboratory. However, it is our view that the exposure to boards and wires still provides a very valuable pedagogic tool and an excellent learning experience for the students. Since testing and design for testability are also included within our course, it is important that students get real experience with actual blown chips and faulty wiring. However, it is clear there are real opportunities from the new paradigms of computer-assisted tutorials and the inclusion of good multimedia systems into the curriculum.

Our solution with the multimedia *Virtual Lab* (VLab) produced using Director [1], has not been to replace the physical hardware component, but to enhance its effectiveness by providing a simulation experience that reproduces many of the primary features of the board. The students see a lab environment including the VLab, physical designs on a board, Logic Works and VHDL. We think that, for any institution that does not include a hands-on lab, the VLab comes very close to giving the students this experience.

## 2. Primary Features of the Virtual Lab

The design strategy for VLab is to incorporate best practices from multimedia design and human-computer interface to ensure the final product is easy for the students to use, as well as effectively meeting its educational goals. We assume that all introductions, design preparation, and basic design simulation are done using the VLab platform on a PC. Consequently, we designed VLab for maximum flexibility , so it can be used across a wide spectrum from a totally 'virtual' mode, through to being an off-line practice tool prior to building the circuit on the physical board. The primary features include:

- VLab is an animated environment, incorporating textual tutorial links, demonstration movies, tutorial assistance and interactive practice modules.
- The multimedia platform of Director presents the whole system as a movie which can be scrolled in time, yet can be saved in a file and restarted at any point in an exercise.
- A simulator is incorporated into VLab, so that wired designs can be fully verified, and their actual behaviour observed within VLab.
- The design and implementation exercises take place on a mock board which is presented within a high quality and very realistic graphical environment. A typical screen shot is shown in figure 1 (it looks much better on a full screen, and in colour).
- VLab is not a substitute for a complete simulation environment, which is still needed for a detailed timing analysis and similar purposes.

## 3. Content, Implementation and Experience

Our course includes nine labs, with two based on VHDL. Of the rest, three are solely

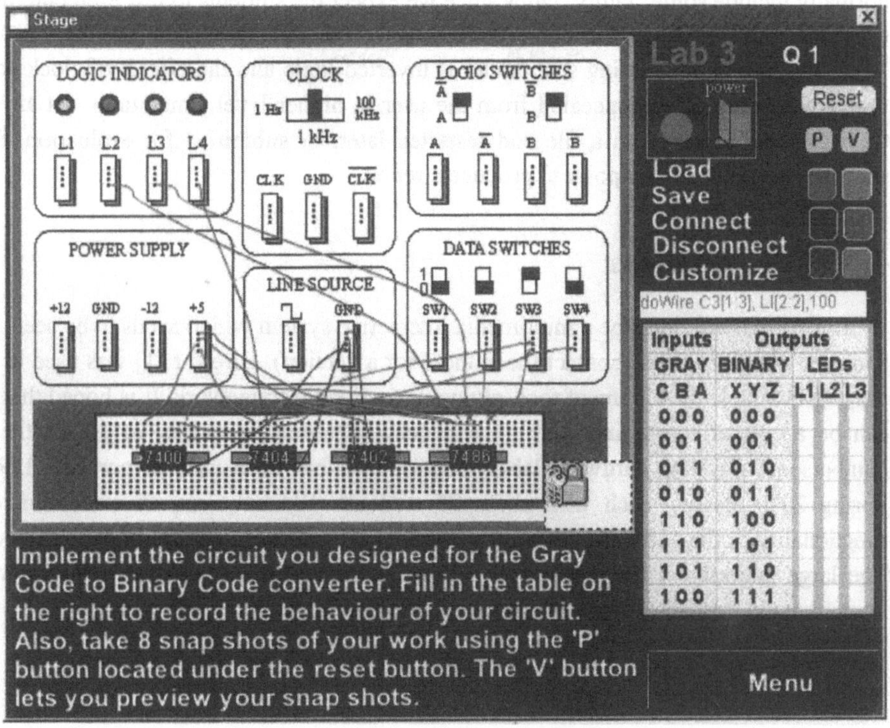


*Figure 1.* Wired lab exercise

in VLab, and the remaining five make use of VLab for their preparation and design, with the students then constructing the verified design on the physical boards.

VLab can really only be explained by showing it. Overall, it is in the manner of a cyber classroom, with carefully defined programmed steps and options for the students to follow within three main categories:

a) The appropriate background , with short demonstrations and tutorials leading up to the required pre-lab work (see Figure 2, including Karnaugh map tutorial).
b) The implementation of a design, its exploration and testing (similar to Figure 1).
c) The preparation of the final design, and its verification for submission.

To illustrate, we include a summary of work flow for Lab 3.

The design is described textually, with tutorials. The pre lab exercises take the student to the production of a gate level design, including a Karnaugh map tool. In VLab, the student has a choice of chips with schematics provided in a floating window (see Figure 2). Using the mouse, the chips are placed on the breadboard, and wiring added. At any time, the student can enter a 'test' mode to confirm the logic of each submodule. Connections can be edited, deleted, or reset at any point. Wires of different colours can be chosen, and the layout is automatically adjusted for clarity. There is a zoom feature to assist in the wiring and also a command line semantics, so that wiring can be done by a script language (it helps with people with little dexterity or visual handicaps). We normally include some 'faulty' chips since we expect the students to test each chip prior to use. This helps instill good design practices, as well as an understanding of the schematics. VLab is good for illustrating such areas as inverted logic and the effects of clock edges – things which are often concealed from the user in higher level simulations. At any point the work can be saved in a file and restarted later, or submitted for evaluation. Screen shots are available at any point to produce a report.

## 4. Conclusions

It is difficult to describe a multimedia interactive system which needs to be seen, tested and used directly (please contact the authors for a version). Director [1] was used as being the best available development tool, giving us a portable framework. It is hoped that VLab can be a cost effective and learning enhanced substitute for some, or all, of the more cumbersome physical hardware labs. At present, it has been used in our Fall 1999 and Spring 2000 classes, each with about 75 students. We have run an evaluation of the functionality of the software, its interface and its effectiveness (as perceived by the users). The large majority of the respondents (89%) stated that the Vlab is an effective way to learn, that the Pre-lab exercises were useful and that the practice was useful.

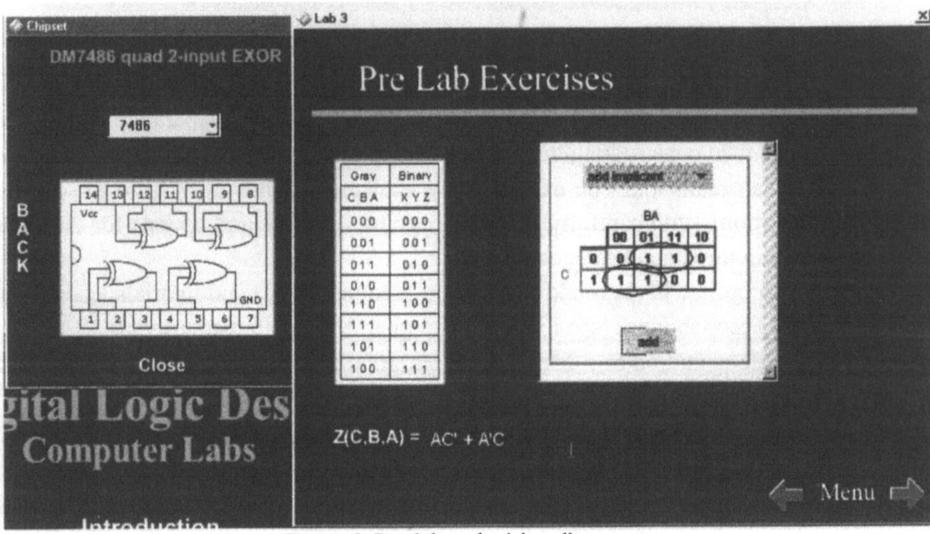


*Figure 2.* Pre-lab and wiring diagram

## 5. References

1. Macromedia, "*Director 7*", San Jose, CA, 1999.
2. R. H. Katz, "*Contemporary Logic Design*", Benjamin/Cummings Publishing, 1994.

## Memphis, a hypermedia system for learning microelectronics

Angel Salaverría* **, María J.Moure**, María D.Valdés**, Enrique Mandado**, Julio González***.
*.E:U.I.T.Industrial de Donostia.Universidad del País Vasco. (jtpsagaa@sp.ehu.es)
** Instituto de Electrónica Aplicada. Universidad de Vigo.iea@uvigo.es
*** State University of New York at New Paltz.(gonzalj@eelab.newpaltz.edu)

**Abstract:** Memphis is a hypermedia system [1] using a very easy interface, multimedia possibilities [2], hypertext and PSPICE simulation facilities [3].

### Introduction

Existent computer aided learning (CAL) systems only use hypermedia and they are not so useful for teaching electronic circuit design. This article describes an original CAL system that combines hypermedia and simulation facilities to improve microelectronics fundamental concepts and design learning. To do that it includes selected exercises simulating the student answer.

The present version of the system is dedicated to analogue microelectronic circuit design learning but a digital system is under construction.

### Memphis

Memphis is a hypermedia system for microelectronics learning using multimedia possibilities, hypertext and PSPICE simulation facilities. The system includes lessons, exercises, spoken comments and the possibility to simulate the circuits explained on theoretical and exercises parts.

Lessons and exercises are under html format (Figure 1). The main advantage of html files is the possibility to allow hyperlinks. By means of hyperlinks it is possible to go back to previous lessons in order to review forgotten concepts, or to display some additional information to reinforce student's knowledge. At the same time, html files are especially useful to display animated graphics and, in that way, to compose difficult circuits step by step, very slowly, including each step explanation. Memphis also includes a small glossary with the most common used words on electronics that the student can access by means of the corresponding hyperlinks.



*B. Courtois et al. (eds.), Microelectronics Education,* 143–146.

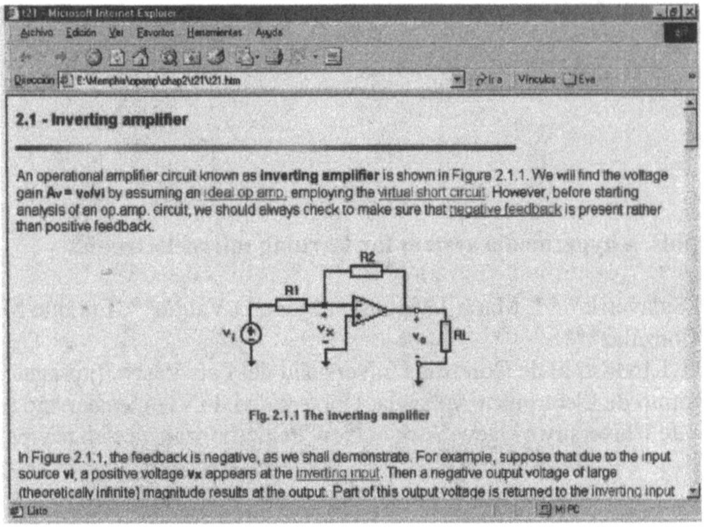


Figure 1. Inverting amplifier.

Including spoken comments (Figure 2), the students can listen the abstract of the theoretical part. This abstract shows the most important parts of the theory and it may be used at the beginning or at the end of a new part. When using it at the beginning, Memphis points out the attention of the student on important concepts. When used at the end, the student tests her/his knowledge.

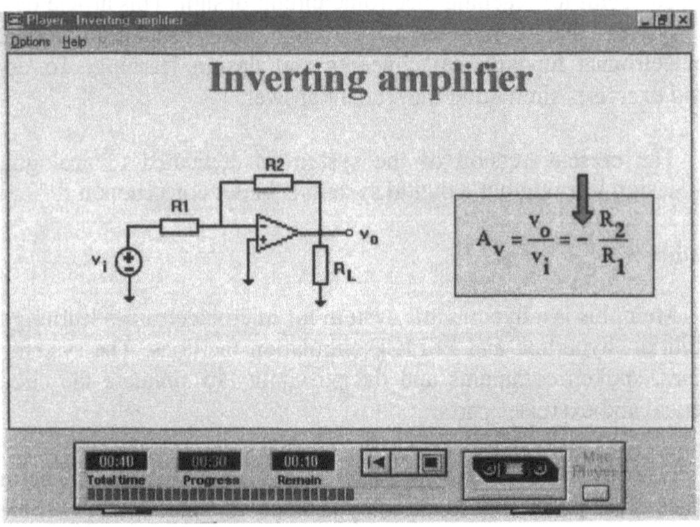


Figure 2. Multimedia player.

Simulation is a very useful activity before practical experimentation with real circuits. Working with real circuits the user needs an analogue to digital converter card (ADC), virtual instrumentation software and the corresponding circuit building experimental system [4]. Memphis uses MSIM evaluation version, a free version of PSPICE. Although the limitations of this free version, it is possible to simulate the most important circuits. The students learn the use of the most popular electronics simulator using pre-designed circuits and, step-by-step, she/he learns to modify them. At the end, the student has to design the whole circuit. Memphis has been designed for a single user, the student, working at her/his Institution or, mainly, at home where she/he has not the possibility to use ADCs and special hardware.

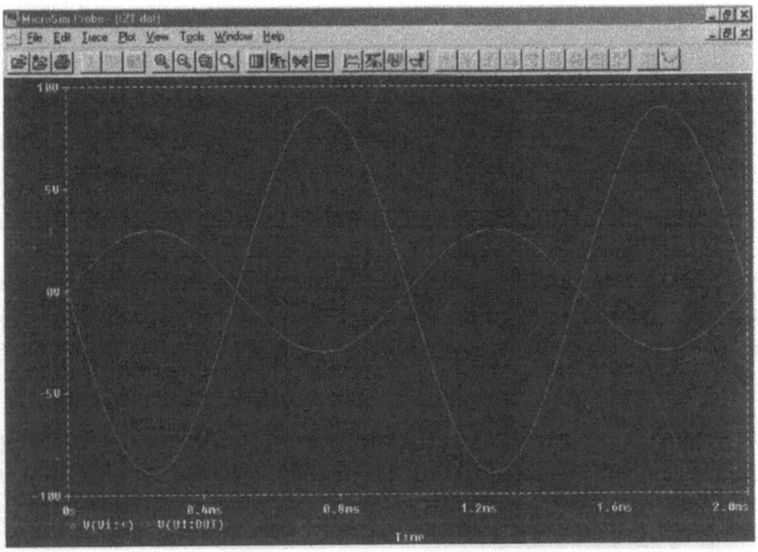


**Figure 4. Inverting amplifier simulation with PSPICE evaluation version.**

A disadvantage of simulation systems is to allow the student to use non-existing devices, such as a capacitor of 10 Farads or a 10000 GigaOhm resistor. With Memphis, when the user tries to do so, a splash window appears showing the mistake.

When the student solves an exercise and gives a wrong answer, the same window is displayed but Memphis allows the user to continue. The system generates the schematic file (Figure 5) with the wrong answer and simulates the wrong circuit. The idea is that the student could check her/his wrong answer and then compare it with the good one. The student can learn from her/his own mistakes.

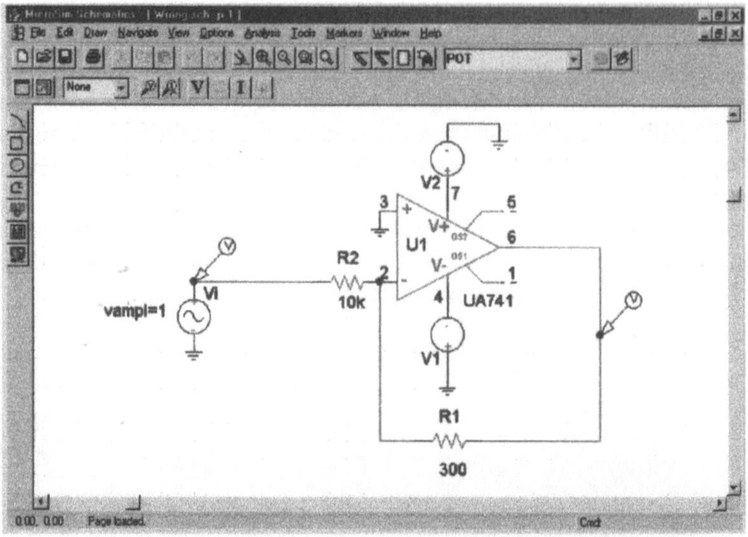


Figure 5. Schematic file generated with the wrong answer, in this case R1=300.

Using simulation the possibility to destroy an electronic device, a very common situation at the beginning of the studies, is avoided. Simulation gives the opportunity to use "software" instruments like oscilloscope and voltmeters at home with a very low cost.

Memphis needs a PC with 32 Mbytes of RAM memory, 32 Mbytes free on the hard disk and SVGA graphics card. The requirements are not high for a domestic user.

**Bibliography**

[1] Valdés & others. Using Hypermedia for Programmable Logic Devices Education. Proceedings of IEEE International Conference on Microelectronics System Education. USA 1997.

[2] Zysman E. Multimedia Virtual Lab in Electronics. Proceedings of IEEE International Conference on Microelectronics System Education. USA 1997.

[3] Gonzalez J.J. Toward an optimized computer assisted electronics laboratory. Proceedings of IEEE International Conference on Microelectronics System Education. USA 1997.

[4] Angel Salaverría & others. M.A.C. sistema hipermedia para el aprendizaje de la electrónica. Congreso CONIED'99. SPAIN 1999

# HYPERMEDIA APPLICATIONS AS INTERACTIVE DATABASE RESOURCES

M.D. VALDÉS, J.A. TARRÍO, M.J. MOURE, A. SALAVERRÍA AND E. MANDADO
*Institute of Applied Electronics. Electronic Technology Department. University of Vigo. Apartado Oficial de Correos. 36200 Vigo. Spain.*

## 1. Introduction

The struggle to obtain a stable market position promoted that each semiconductor devices manufacturing company develops their own products. As a consequence, the market is saturated with devices oriented to the same general purpose but with different structural or functional characteristics, making them more suitable for one particular application. This is the case of FPGAs where manufacturers offer thousands of devices, sometimes very similar, supporting digital systems rapid prototyping. From the design engineer's point of view the problem arises when deciding which device can be used to implement an application taking into account its particular characteristics.

This problem can be extended to the field of education. There are so many market solutions emerging, bringing about new nomenclatures and characteristics, that it is necessary to answer two questions: what to teach?, how to teach?.

## 2. What to Teach?

A very extended criteria is to focus education on the analysis of a limited set of devices from different manufacturers. The advantage of this solution is that students can be introduced to the analysis of a specific technology without great effort from the teacher. The disadvantage is that it gives only a very particular insight of the technology.

To make up for the previous mentioned disadvantage our research group developed a methodology oriented to favor the progressive analysis of the whole of the characteristics of complex technologies (the methodology has been published in different articles [1][2][3]).



*B. Courtois et al. (eds.), Microelectronics Education,* 147–150.

## 3. How to Teach?

The proposed methodology is based on a descriptive model where the characteristics (general and particular ones) of the devices belonging to a technology are classified and organized in a hierarchical way. The greatest advantage of a descriptive model is that it provides all the characteristics of a technology as well as the possible interactions between them. In this way any system or device can be described from it. Nevertheless, as the characteristics associated to each system differ from one to another, it is necessary to go through the model in an appropriate way.

The way in which the descriptive model is presented to the students is another important part of the methodology. In our opinion, hypermedia is the best support for transmitting this kind of information because students can navigate from one point where a concept is described to another linking related information. In this way the intuitive learning is favored. On the other hand, the hierarchical structure of the descriptive model makes the design of hypermedia applications easy.

The development of a descriptive model and the hypermedia application demands great effort from the teacher, however its benefits have been verified from the educational point of view.

## 4. Improving the Methodology.

It is interesting to use the descriptive model for the analysis of the actual characteristics of different real devices to achieve a practical knowledge of the studied technology.

A possible solution is to implement a commercial device database provided with links to those points of the hypermedia application where the concepts of the descriptive model are explained [1][3]. The database relates commercial devices with its characteristics and they in turn are related with concepts describing them. As result a feedback information system allowing the analysis of technologies as well as particular devices is obtained. If the multimedia advantage for transmitting information through powerful audiovisual resources is added, the outcome is an effective and sturdy educational system.

## 5. A Hypermedia Application

Combining the proposed methodology for the analysis of complex technologies with a hypermedia database we developed a CAL tool oriented to FPGA education with the following objectives:

- To support FPGA education of electrical engineers.
- To make the choice of suitable FPGAs for a particular application easy.

As a result the multimedia application “FPGA” was created (up to now in Spanish) [4]. Such application has two parts, the first one includes concepts related with the FPGA architectures and its resources. This part was developed following the methodology described in [1]. The second constitutes the commercial devices database itself. Links between both parts are established by means of hyperlinks.

To make the application the author system ToolBook 5.0 combined with sound and image edition software tools was used. It must be pointed out that although the database has been developed using a non-specific purpose tool, it is provided with typical functions for updating and introducing new data, in this case new devices and their characteristics.

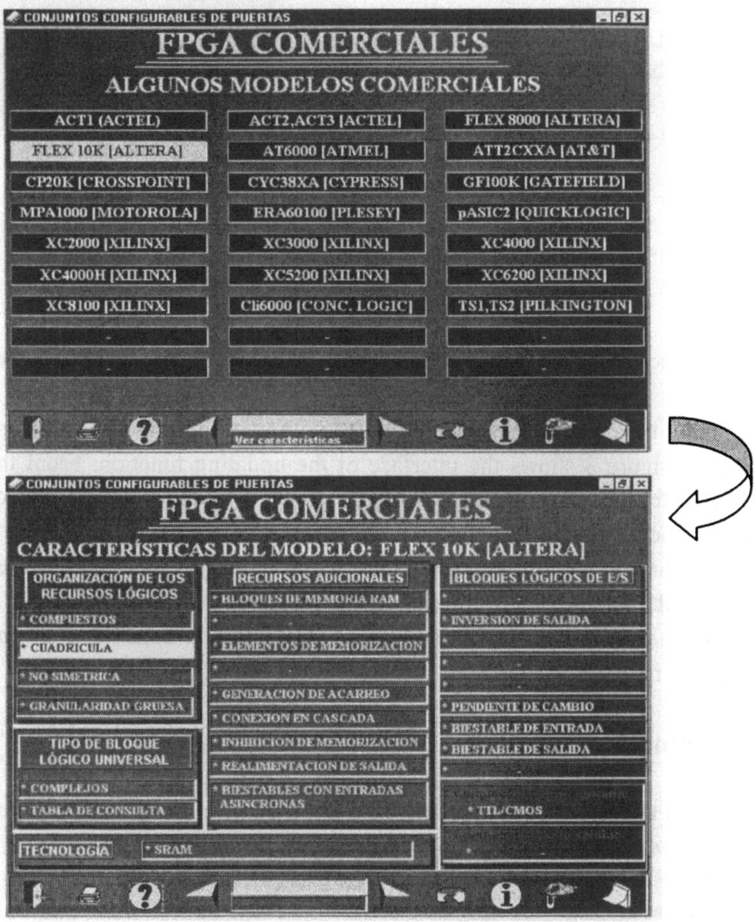


Figure 1. Accessing the database information.

Figure 1 shows two screens of the “FPGA” application belonging to the database. The first one presents the commercial devices included in the application, among them the FLEX10K of Altera. When clicking on its name the second screen of the figure, where characteristics of the FLEX10K are stored, appears. All pages containing device characteristics have the same structure. The access from any data field to one or more pages, where concepts associated with it are explained, is possible. For example, clicking on the outlined field "CUADRICULA" ("matrix" in English) we access to screen of figure 2. In this screen the "symmetric matrix architecture" is explained. The link between both parts of the application is established and the information is complemented making the understanding of related concepts easier.

Figure 2. A screens of the application "FPGA".

Figure 3. Updating the database.

Finally, the database has been provided with two important functions: searching and updating. Figure 3 shows the interface of the updating function. From this screen the user can introduce new commercial FPGAs to the database and their characteristics. The search function is oriented to those users interested in choosing the FPGAs corresponding to a set of characteristics. To execute this function the user must select the desired characteristics through different dialog boxes. As result the program generates a list of the available FPGAs. This function constitutes an important help when designing FPGAs based systems.

## 6. References


[1] Valdés M.D., Moure M.J. and Mandado E., "Using Hypermedia for Complex Technologies Education", *IEEE Transaction on Education* (USA), Vol. 42, No. 4, November 1999.

[2] Valdés M.D., Moure M.J., Rodríguez L., Álvarez J. and Mandado E., "Using hypermedia for Programmable Logic Devices Education", *Proceedings of IEEE International Conference on Microelectronic Systems Education* (USA), 1997.

[3] Valdés M.D., "Métodos de enseñanza y de diseño de sistemas basados en FPGAs", *Doctoral Thesis*, University of Vigo (Spain), 1997.

[4] Tarrío J.A., "Aplicación hipermedia para la enseñanza de FPGAs", *Degree Project*, University of Vigo (Spain), 1998.

# Chips on the Net
*An FPGA prototyping platform*

M. J. Smith and H. Fallside
*University of Hawaii and Xilinx*

Key words: FPGA, prototyping, platform.

Abstract: We describe a new FPGA prototyping platform for educational use.

## 1. INTRODUCTION

The new US ABET 2000 accreditation requirements demand the "ability to design a system, component or process" [ABET1]. Systems are becoming more complex and engineers and education are not keeping pace. There is no easy solution to this problem. Capstone design courses are often too difficult for the average student and too expensive for the average school.

## 2. PROJECT DESCRIPTION

One way to help students learn system design is to provide a framework or platform. In the early 1990s, working with the NSF and Stanford (Prof. Abbas el Gamal), we distributed the first FPGA prototyping systems, the Protozone prototyping boards [Protozone]. We learned a lot from Protozone. It was a hard for students to learn new software tools, program the PC interface, and design something in one class. The PGA socket and surrounding circuits (which were not wired in order to permit use of different FPGAs) compromised by being a bit too flexible. The two-card design attached to the PC restricted portability and required opening the PC. Putting the prototyping areas on the boards made them difficult to re-use.



*B. Courtois et al. (eds.), Microelectronics Education,* 151–154.

The board was a little too big, complex, and expensive. We distributed boards to a dozen schools to gather feedback. We learned that it is hard to get students and even harder to get professors to write reports. We did prompt Xilinx and Altera to distribute boards as part of their education programs and as a result this did have an impact on microsystems education around the world.

The development of FPGA technology has generated a large number of commercial and university research projects involving FPGA boards. Currently Altera uses the UP-1 board [Altera1] and XESS designs and distributes Xilinx-based FPGA prototyping boards [XESS1]. Both companies have donated hardware (and software) to schools worldwide. XESS currently distributes 5000-7000 XS40 boards per year. XESS, Xilinx, and Prentice-Hall have placed more than 10,000 boards into over 1000 schools worldwide since 1992. Both boards can handle designs of 10k-gates, but are small for complex system design.

There are comprehensive lists of commercial FPGA platforms on the web [Boards]. We have looked at these carefully to determine our own direction. The Annapolis cards are representative of the cards that plug into PCs, eminently suitable for research, but less so for undergraduate projects [Annapolis]. The Associated X240 is an FPGA surrounded by connectors that costs $500 with a Spartan FPGA (30k gates). You can add an Ethernet module for $500, but there is no direct Internet connection to the FPGA. [Associated]. The Biel Scool of Engineering ProTest, which uses a programmable base station for prototyping, may not flexible enough for class project needs [Biel]. Virtual Computer produces a board using Xilinx Virtex FPGAs that costs about $1000, which is expensive for undergraduate use [Virtual]. We are working on the next-generation Xilinx educational board. Figure 1 shows our development platform, the Demonstation.

Demonstation contains a socketed Virtex FPGA. We used the HQ240 package, which contains a wide range of device sizes. This is important as the student version of the software may only have the capability to program lower capacity devices, while those with full versions of the software, may wish to use larger members of the Virtex family. We use a small nonvolatile CPLD device to initialise the volatile Virtex device. A ROM holds this bootstrap code (we can use wither a ROM emulator or Xilinx flash ROM). We also included an LCD panel and parallel port for debugging purposes.

The most novel feature of our project the ability to (re)program the Demonstation over the Internet. An Ethernet PHY chip connects directly to the Virtex FPGA. The Virtex device contains a complete TCP/IP stack in order to receive an incoming bitstream file. The bitstream is assembled in SRAM before being used to configure, reconfigure, or partially reconfigure the Virtex device. This allows us to separate students, teachers, and hardware. Our aim is to allow a student to design at home, and download, using a Java API [XPRESSO], to a remote Demonstation. She then remotely debugs her design with a virtual logic analyser that can "see inside" the FPGA. If she runs into problems, she can email the instructor and they can debug the design together.

The XS40 boards cost $100, but Demonstation (and boards based on it) presently cost around $1000. While this cost will fall with time, we can reduce the effective cost by partnering with a portal, DACafe, dedicated to systems design. As an experiment DACafe placed an ASICs textbook online [Smith99]. 'ASICs... the book' generates about 10-15% (50% from .edu domains) of DACafe's 4 million hits per year. We wanted to answer the question: "Does a free online book increase book sales and Internet advertising revenue?" The answer to both questions is "Yes." Now it makes sense to bring together a publisher (Prentice-Hall), Xilinx, and DACafe. DACafe will host educational material, Xilinx will host design software and debugging tools [WebFITTER], and both (together with others) can host hardware platforms based on Demonstation. Schools can access all of these for free.

*Figure 1.* The Demonstation, development platform for the third-generation Xilinx educational board (the size of an A4 sheet of paper). Ethernet (10/100Mbps) enters at the top left. Middle right is a 240-pin socket that accepts Virtex FPGA devices with up to 1Mgate capacity. Bottom left is a prototyping area. Above and below the FPGA are two banks of SRAM. The configuration ROM is between the prototyping area and SRAM. At the bottom is an LCD panel for debug purposes. To the right is 5V, 3.3V, and 2.5V power conditioning. Underneath the board is a CPLD that contains a bootstrap circuit to load the FPGA from either ROM or RAM at power-on or during normal operation

We started this project, Chips on the Net or XCoNET, in July 1999 [XCoNET]. Hardware was tested and debugged in a few months. We are now writing code for the TCP/IP stack and other functions, all of which are completed in hardware. Demonstation relies purely on hardware, but the same functions could also be implemented using a soft microprocessor core in the FPGA. One of our project aims is to compare the two approaches. We now have Demonstation transmitting and receiving Ethernet frames. Our next goal is to demonstrate remote configuration. We have built a handful of Demonstations and XESS will distribute a production version that adds the following components: a video decoder to digitize video signals, a RAMDAC for generating video, and a stereo audio codec [XSV].

# 3. REFERENCES

[ABET1] ABET, University accreditation organization http://www.abet.org

[Altera1] The Altera University Program and UP-1 board http://www.altera.com/html/univ/features.html#package.

[Annapolis] Wildfire boards (Wildforce and Wildone) http://www.annapmicro.com/.

[Associated] See, for example, http://www.associatedpro.com/.

[Biel] Biel Scool of Engineering, http://www.isbiel.ch/I3S/microlab/products/protest.

[Boards] (FPGA boards) http://www.io.com/~guccione/HW_list.html/ http://www.optimagic.com/boards.html and http://www.xilinx.com/products/protoboards/protoboards.htm

[Protozone] MIP-8915229 Prototyping System for VLSI Education. http://www.nsf.gov/cgi-bin/showaward?award=8915229

[Smith97] Smith, M. J. S., 1997. Application-Specific Integrated Circuits. Addison-Wesley VLSI Design Series, pp. 1040, ISBN: 0-201-50022-1, TK7874.6.S63. Reading. MA: Addison-Wesley.

[Smith98] Smith, M. J. S., 1998. "ASICs... the website" at http://spectra.eng.hawaii.edu/~msmith/ASICs/HTML/ASICs.htm.

[Smith99] "ASICs... the book" at DACafe http://www.dacafe.com/ASICs.htm.

[Virtual] See for example, http://www.vcc.com/index.html

[WebFITTER] http://www.xilinx.com/sxpresso/webfitter.htm

[XCoNET] http://www-ee.eng.hawaii.edu/~msmith/XCoNET/XCoNET.htm

[XESS1] The XESS FPGA prototyping website at http://www.xess.com/FPGA/homepage.html.

[XPRESSO] Xilinx Silicon Xpresso initiative including IRL, Jbits, and Chipscope. See, for example, http://support.xilinx.com/products/software/sx/sxpresso.html#IRL

[XSV] http://www.xess.com/prod014.html

# WEB BASED TUTORIALS FOR TEACHING ELECTRONICS

D.G. BAILEY
*Institute of Information Sciences and Technology,*
*Massey University, Palmerston North, New Zealand.*

***Abstract.*** With a wide range of student abilities in a class, it is difficult to effectively teach and stimulate all students. A series of web based tutorials was designed to help weaker students and stretch the stronger students. The approach taken is illustrated with a tutorial showing the relationships between a truth table, Karnaugh map, logic circuit and Boolean algebra representations of a logic function. Use of the tutorial has resulted in significant benefits, particularly with weaker students.

## 1. Introduction

A second year Electronics and Design paper is one of the core majoring requirements at Massey University in a number of degree programs: BSc (Electronics), BInfSc (Computer Science and Electronics), BTech (Computer Systems Engineering), BTech (Information Engineering), BE (Information and Telecommunications Engineering). Other students with an interest in electronics may also take this paper. With the medium to large class size, there is a wide range of student ability and previous experience in electronics. Coupled with this is the fact that different students respond differently to different methods of teaching, depending in part on their prior understanding of the subject. At one extreme, there are students who have had little prior electronics experience, and at the other extreme there is a significant group of students who have electronics as a hobby. This wide range of ability and experience provides challenges for conventional teaching methods. For example, to reach most of the students in the class, lectures must be aimed at students in the middle of the class in terms of ability. Unfortunately, this approach fails to stimulate and stretch the better students in the class, or help the weaker students to keep up with the material.

Both of these needs may be met through tutorials tailored to different levels of ability. Running such tutorials to cover the range within a class is time-intensive. Another approach is to make available web based tutorials that allow students to work at their own pace. Using the course notes as a starting point, a series of interactive tutorials is being developed. The tutorials are made available over the web and also within the Electronics teaching lab (for students who do not have web access from home).



*B. Courtois et al. (eds.), Microelectronics Education,* 155–158.

## 2. Tutorial Design

The tutorials consist of a series of HTML web pages with embedded Java applets. The applets are mini programs that are executed by the Web browser within the context of the web page containing the applet. This combination of web pages and applets is particularly powerful for providing interactive demonstrations because any textual content (for example a description of the problem, or instructions) may be easily provided within the web page. The applet is able to be a complete working program that dynamically illustrates the concept, or provides a working environment for the student to experiment and work through their solution. The applet is therefore effectively like a figure in an ordinary textbook with the difference that it is dynamic, and responds to the student.

Applets achieve this interaction with the student through both mouse clicks and keyboard entry. These allow the student to adjust parameters, make selections, and affect the way the program is run or information is displayed. Such interaction allows each applet to provide a mini demonstration or experiment (exercise) to help the student understand a particular concept or technique. The content of the tutorials closely parallels that of the lecture course, allowing weaker students to pick up on details or concepts that they may have missed during the lectures. The exercises also range in difficulty, with some more advanced exercises available that will stimulate the more able students.

### 1.1 EXAMPLE: KARNAUGH MAPS

As the result of an initial proposal, a pilot series of tutorials was developed to illustrate the operation of Karnaugh maps for digital logic minimisation. From previous years experience in teaching this section of the paper, one concept that students initially struggle with is the ordering of the cells within the Karnaugh map compared with that within a truth table. A common error was to use a standard binary ordering of the rows and columns in the Karnaugh map. Since the purpose of using a Karnaugh map is for circuit minimisation, it is also useful to illustrate the complexity of the resulting circuit. This was accomplished by also showing the circuit function as a two level logic diagram, and representing the output as a Boolean algebraic expression.

The Karnaugh map applet therefore shows four different but equivalent representations of a logic function (see figure 1): a truth table, a Karnaugh map, a two level logic circuit diagram, and a Boolean algebraic expression.

Selecting one element in any representation (a line in the truth table, a cell in the Karnaugh map, a gate in the logic circuit, or a term in the Boolean algebra expression) automatically highlights in yellow the corresponding elements in the other equivalent representations. This allows students to explore the relationships between the different representations, guided initially by a series of leading questions in the accompanying instructions.

When the student selects multiple cells within the Karnaugh map, the corresponding multiple elements are also highlighted in the other three views. If the cells include both 0 and 1 output, the grouping is illegal, this is indicated to the students

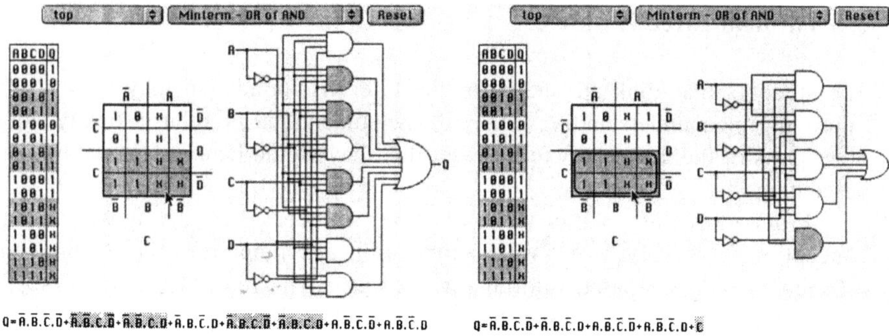


*Figure 1.* Applet view before and after releasing the mouse. A group of cells has been selected in the Karnaugh map and these are combined as part of a minimisation. The logic circuit representation and algebraic representation automatically update to represent the simplification that has taken place.

by highlighting the selected cells in bright red. Thus even if the student is unsure what they are trying to do, they are given very clear visual feedback as to whether or not an action may be performed. When the student releases the mouse after making a legal group selection, the group is shown on the Karnaugh map with a ring around the terms. At this point, the logic circuit and Boolean algebra representation automatically change to show dynamically the effect of the simplification (as shown in figure 1). In this way the students see very quickly and graphically the effects of the minimisation they have made.

The particular two-level representation may be selected from a menu as either minterm (OR of AND, or NAND of NAND) or maxterm (AND of OR, or NOR of NOR). The selection changes the circuit representation, the Boolean algebra expression, and the way that grouping is performed within the Karnaugh map. This allows the students to explore the similarities and differences between the different representations, again guided by leading questions in the accompanying web page.

With this particular tutorial, the following questions are used to guide the students while they experiment with minimisation using the Karnaugh map. These questions were chosen to reinforce what the students have been taught previously in lectures.

1. What is the relationship between the ordering in the truth table and the ordering in the Karnaugh map?
2. What is it about the Karnaugh Map that makes simplification easier?
3. Why does the maxterm grouping on the Karnaugh map group the 0s?
4. The maxterm grouping works by excluding 0s, rather than by including 1s. True or false?
5. Are the simplified minterm and maxterm expressions equivalent? Explain why or why not.
6. In the simplified minterm representation, the $\overline{A}$.B.C.D term is included in two groups. Does this matter? Why or why not?
7. For the minterm representation of a circuit, what is the difference between the OR of AND and the NAND of NAND circuits?
8. For the maxterm representation of a circuit, what is the difference between the AND of OR and the NOR of NOR circuits?

## 3. Tutorial Effectiveness

The students that took advantage of the tutorials found that they were useful for improving their understanding. Many of the students have enthusiastically suggested other parts of the course where they felt that they would benefit from additional web based tutorials.

From a teaching point of view, whenever a student is struggling with a particular concept, they are given an initial explanation and help from the lecturer and are directed to the web-based tutorial for exercises. The tutorials provided so far are effective at reaching the weaker students in the class, but as yet do not directly stretch the better students. However, there has been an indirect benefit for the more able students in that it has enabled topics to be covered to a greater depth within the lectures because of the availability of remedial help for the weaker students.

## 4. Implementation Issues

It is important that the interaction between the student is as seamless and natural as possible. Otherwise, learning how to use the applet can get in the way of learning the concept or technique that the applet is trying to demonstrate. To facilitate a smooth interaction in the Karnaugh map example, each of the four representations of the logic function was implemented as a separate object, with its own internal data structure. When items were selected or grouped in one view, the selection or group was passed to the other views which then updated and redrew themselves to represent the new state. While this meant maintaining multiple different copies of the same logic function, keeping the data for each view independent of the other views meant that data within a view was kept in a consistent state. Message passing between the views on user events was used to maintain consistency between the different representations.

The use of Java gave some degree of platform independence. One issue that became apparent was the Java support provided by the web browsers differed significantly from browser to browser, and platform to platform. This meant that the lowest common denominator, Java 1.02, was used even though much of the required message passing is easier to implement in the later versions.

## 5. Conclusions

Web based tutorials have proved to be a useful addition to the regularly scheduled contact with the students. As a result, the initial trial program is being expanded to other parts of the course.

One problem identified with the web based tutorials is the considerable time required to write the underlying applets to give a seamless and natural interaction with the problem. This problem will be the hindering factor to their widespread use without the development of higher level tools for applet creation.

# A WEB APPLICATION EXPERIENCE IN ELECTRONIC EDUCATION IN THE ENGINEERING SYLLABUS

M. MARTINEZ, F.J. AZCONDO, S. BRACHO
*Microlectronic Engineering Group.*
*Electronics Technology, Systems and Automation Engineering Department*
*Escuela Técnica Superior de Ingenieros Industriales y de Telecomunicación.(E.T.S.I.I.T.*
*University of Cantabria*
*Av. de los Castros s/n. 39005 Santander (SPAIN).*
*Tfno: 34 942 20 14 81. Fax: 34 942 20 18 73. e-mail:* bracho@teisa.unican.es

ABSTRACT
The University of Cantabria is promoting the use of Internet resources to improve the academic results in the syllabus of Industrial and Communications Engineering. The Microelectronics group participates in this program by introducing the teaching material of the basic electronics subjects using multimedia format. The aim is to analyze the impact of these techniques on the academic results of the students in the first years of Electronics subject within these Engineering, in comparison to the previous results of the last few years.

## INTRODUCTION

Education in Microelectronics using multimedia techniques in the University of Cantabria has a solid background that makes possible to undertake new teaching methodologies that improve the efficiency and reduce the cost of training activities. In this way a multimedia software that contains an explanation for self study of the design-flow has been developed, simulation and test of digital integrated circuits based on the CADENCE framework for the design with standard cells, Synopsys for description and synthesis and Altera for implementations in FPGAs [1].

During the 1999-2000 academic year, we have extended these methodology to the undergraduate courses. To do so, we have chosen the subjects of Electronic Systems and Industrial Electronics from Industrial Engineering and Basic Electronics from Communications Engineering to evaluate the introduction of these methods in the teaching of Electronics in the first years of Engineering.

## CONTENT OF THE SUBJECTS

The subject of Basic Electronics of the second year of Communications Engineering introduces to the initial concepts of analog electronics. Single-stage and multi-stage BJT and MOS amplifiers are studied, these contents are extended to BJT and MOS differential amplifiers, the frequency response of the amplifiers and linear and non-linear applications of operational amplifiers.



*B. Courtois et al. (eds.), Microelectronics Education,* 159–162.

Currently, the students have the teaching material used in the theoretical classes available as well as the problems and exercises solved in classroom. Furthermore, the subject has complementary practical classes given in the Basic Electronics Laboratory.

The subject of Electronic Systems in the third course of Industrial Engineering incorporates an analysis of the principal electronic devices, diodes, BJT and MOS as well as applications focused on solving industrial problems.

The laboratory work allows the student to experiment with the design of the most representative circuits of each chapter and to carry out measurements to check their right operation. The information about the function of electronic devices and the design of amplifiers, both DC and AC, is clearly increased thus improving the masters` classes. A self study course of these topics is introduced, incorporating the basic principles of electronic devices and exercises that help in the understanding of the most common electronic design techniques and the simulation of the circuits, thus re-enforcing the learned topics.

The method will help not only in the training on theoretical topics but also in reducing the time expended in reaching a satisfactory skill with the laboratory instrumentation.

The subject of Industrial Electronics is dedicated to the components used in Power Electronic converters. The subject has evolved substantially in the last few years thanks to the development of practical classes that allow the students to acquire real experience about the use of power components, the design of converters, their protection and control.

The desired improvement is achieved by facilitating the access to updated information related to power devices and converters. Within this initiative the basic bibliography, photocopies of transparencies, solutions to exercises and the guide to practices are given to the students via Internet.

## ELECTRONIC BOOK ON BASIC ELECTRONICS, ELECTRONIC SYSTEM AND INDUSTRIAL ELECTRONICS

Between the objectives for the first courses of electronics in engineering is to improve the student's access to information by the production of an "electronic book". To make the information accessible a multimedia format is given to the available material. The electronic book is intended for re-enforcing the knowledge acquired in classroom, in such a way that the student can access, from home or from the computer room, the transparencies used in class, exercises and practices and any other material which helps in the preparation of the subject. The material supplied to the student is considered to be complementary to the explanations given in class. The initiative is intended to improve the quality of the teaching and to reduce the time that the student expends in the subject. Part of the teaching time of each teacher that participates in this activity is substituted by the effort of preparing all necessary materials, transparencies explaining theory and exercises with solutions for each subject.

Figure 1 shows the site dedicated to Electronic Systems. The icons that show the program and references, laboratory, simulation software and educational material are links to new web pages. The first icon leads to the subject program and the basic selected books. The manuals appear in the simulation software page, with tutorials and direct links to the free student version of IS-Spice and PSpice simulators.

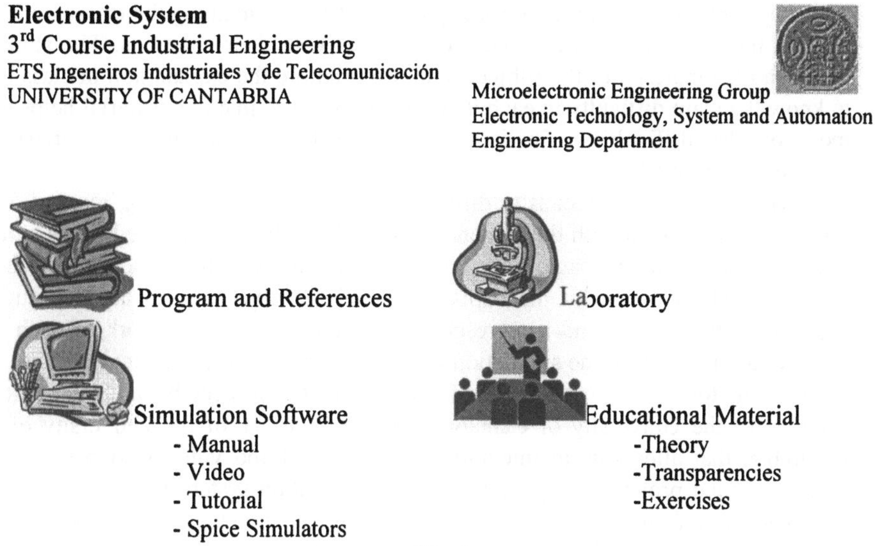


Fig 1

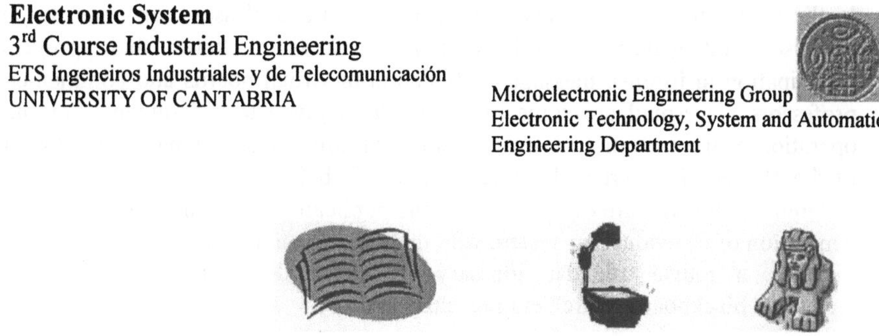


| | Theory | Transparences | Exercices |
|---|---|---|---|
| Electronic Circuits | ✘ | ✓ | ✓ |
| Diodes Circuits | ✘ | ✓ | ✓ |
| Bipolar Junction Transistor | ✘ | ✓ | ✓ |
| MOS Transistor | ✘ | ✓ | ✘ |
| BJT Amplifier | ✘ | ✓ | ✓ |
| MOS Amplifier | ✘ | ✓ | ✓ |
| Frecuency Response | ✘ | ✓ | ✘ |
| Operational Amplifier | ✘ | ✓ | ✓ |

Fig 2

The figure 2 is the web page that give access to the educational material composed by notes of theory, transparences and solved exercises of each item of the subject.

This page links to the Adobe_Acrobat pdf format files containing the transparencies that support the theoretical classes and the exercises with their solutions. The student can download the material of the subjects and use it before the explanation in class in order to know the main difficulty in each item. After the class time this material helps to solve most of the difficulties, finalizing the understanding using the exercises as a complementary work.

The hyperlink laboratory leads to different samples of practical work, most of them are prepared to be solved with the student version of the above mentioned Spice simulation tools. In this laboratory page, now under construction, we are including the laboratory work manuals that content the explanation and the objectives to achieve during the progress of the subject and a representative practical simulation work of each chapter which is a reference for the simulation work to be done in each practical exercise.

A common tool to be used in the future within the project by all the participating teachers of the University of Cantabria will be Web CT of the University of British Columbia, that allows us to integrate graphic, sound and video resources through an object oriented programming technique. The navigation system is based on a two-way access to the information as is shown in fig. 1 either directly or via a menu. Different automatic tutorial has also been included, in which there is no interaction between the application and the user and the learning is achieved by mere observation of the screens.

In the next future the students will be able to attend classes, carry out the proposed exercises and generate a self-evaluation process to know their progress. The participation in forums, messages publication in virtual boards and communications to professors via e-mail are some of the facilities given to the student. For the proper operation of the system an instructor is responsible for observing on-line the evolution of the classes via internet. By means of the Web CT software a set of web pages of different complexity are designed: from the publication of the notes of the subject to the generation of an evaluation system with multiple questions and answers. The design tool includes: a course calendar, glossary of key words, exercises for self-evaluation, electronic blackboard, search engine, chats, etc.

## CONCLUSIONS

The publication of educational materials of undergraduate courses introduces new teaching and evaluation methods that up to now was mainly oriented to open universities. The academic activities are enhanced as the material is acquired from any computer with direct communication with the author. The system is transparent so that comments on the contents may be received from colleagues worldwide. New capabilities are introduced for the practical classes because the traditional home exercises are become into simulations based on free evaluation software. It is expected that the academic results improve because the documentation is clearer and the use of internet technologies means an extra motivation for the students

## REFERENCES

[1] R. Casanueva, F.J.Azcondo, M. Martínez, S. Bracho,. Multimedia Training of Integrated Circuits Design with low cost hardware. Microlectronics Education. Kluwer Academic Publishers, 1998.

# BASIC MICROELECTRONICS SELF INSTRUCTOR

Eros Pasero
Electronics Dept. – Politecnico of Torino
c.so Duca d. Abruzzi 24 – 10129 Torino – Italy
ph: +39 0115644099 Fax: +39 0115644099
Email: pasero@polito.it

## 1. Abstract

The re-organization of the Italian university system caused the need for new forms of teaching. The worker-student, for example, needs to conciliate his available time with the study. New universities, located far from the original center, don't have a local teaching staff. The "continuous learning initiative" [1] uses videotapes to teach basic and advanced electronics to adults. All these approaches suffer from the same problem: these students don't have a university structure available for their requirements. They don't have the possibility to remove a doubt by means of an assistant, they don't have the books you can find in a library, they don't have data sheets for electronic components, and so on. In a few words: they don't have an assistant available during their learning period.
New multimedia technologies can open up new horizons for education. Everyone in his own personal computer can easily use a CD-ROM. Hyper textual techniques can help to organise new approaches to a book reading. Recorded voices can stimulate the student attention. Animations can better explain some electronic aspects. CAD tools can be used as virtual laboratories.
This paper presents an approach we are experimenting with the first level of degree (diploma) of Electrical Engineering in the Politecnico of Alessandria, Italy.

## 2. Introduction.

The recent introduction of a first level of degree in engineering in Italy, without an increase of the teaching resources, has aggravated the already heavy situation of the poor contacts between students and professors in the university. The difficulty for students to have explanations from their super busy teachers, often involved in three or four different classes, is becoming a serious problem. The percentage of students able to join the final degree is today very low. We decided to experiment a new method of learning for the students of the electronics classes of a decentralized site of degree. In this site no teacher is available for further explanations after lessons and therefore the problem of poor contacts is particularly present. The same approach is experimented in a distance learning degree of electronics, where students live far from the teaching site and only videotape lessons are available.



*B. Courtois et al. (eds.), Microelectronics Education,* 163–166.

The aim of this project is to give a tool that helps students to self organize their study cycle. This tool is a CD Rom where several topics are available. It enables students to read a book of basic electronics, have hyperlink key words to the most significant points, find self evaluation exercises with different help levels, able to indicate the weak points in their preparation level. Moreover the CD Rom has a virtual laboratory where each student can use the Pspice simulator as a pure tool, without spending hours to learn how an electronic simulator works. This engine gives the students the possibility to verify the proposed circuit with a pre-configured version of Pspice. All the verification analysis is performed in automatic way but students can interact changing some parameters or some parts of the circuit, testing their level of preparation. Intermediate evaluations will give indications about the part of the program which need to be better examined.

An Internet section [2] will allow the students which have an Internet connection to use the official Politecnico electronic web site, to download new material, to mail and videoconference with professors.

## 3. The course organization

The two typologies where we decided to intervene are different but show similar problems. The first one is the so-called "day-degree" in electrical engineering. This is a sort of "shrinked" (3 years) degree of the original 5 years long "ingegneria elettrica" degree. All the teaching staff come from other universities (no local staff is available) and is present only during lessons. Electronic laboratories are poorly equipped and no library is available. Students choose this form of study because they think is easier and shorter than the original 5 years long: their motivations are not so strong. Teachers find students with a short qualification and are not motivated. Courses contents are rather a shrinking of the original modules than a complete re-organization of the matter. Books are the same ones used in the major degree level: too difficult.

The second case of our experiment is the tele-teaching degree in information technology. Again three years but students live far from the university (also hundreds of miles) and use videotapes to learn electronics.

The typologies are different but they show the same difficulties. Students and teachers were requested to fill in a questionnaire where they indicated main difficulties during their university life. These data were matched with the exams results and other meaningful information. Therefore we had some directions where to operate. Major problems can be here summarized.

*Difficulty to organize the study activities.* In the first case students coming from secondary school are not used to self organize their student life. The lack of clear indications joined together with their short motivations is reasons for serious difficulties. Moreover they feel the lack of contact points. They need to have a point of reference, able to give them indications, suggestions, answers and so on.

*Poor books and course programs.* These courses are often only a minor part, or a summary or a sort of "digest" of major courses. Students are also younger and less prepared than their "five years long cousins". It follows that they also have difficulties to auto-evaluate their own level of preparation. This is a problem shared with the higher level of degree. The lack of tools and tutors, which can monitor student activities, is often reason of faults in final tests.

Owing to the impossibility to have more teachers, tutors and resources we decided to verify the efficiency of a hyper textual multimedia tool.

## 4. The project topics.

Following the considerations emerging from the tests described in the previous section we decided to implement several strategies. The difficulty to pre-evaluate the best approach from the student point of view suggests in fact to experiment different aspects. Here we show the main lines we developed.
*Interaction.* A passive way of utilization is one of the main reasons of failure of a multimedia product. The student should interact with the CD book and not to undergo a pure passive reading: electronics sometime can be very tedious! The aim of this interactive book is to stimulate the student attention with personalized reading courses, self-explaining exercises, and virtual electronics labs.
*Multimedia.* Multimedia means sounds, colors, animations and other stuff, which can give new life to a gray electronic book. Often a short movie can better show a physical phenomenon usually described by one or two pages of written words. Also a formula can "live" when explained by an impressive animation.
*Reading keywords.* A traditional book can only be read in a sequential way. At most you can select some more interesting sections. In this CD book the student can select his goal. A pure reading will allow the reader to consult the full CD. If he's preparing his final examination only resuming tables will be shown. If he's trying to verify his level of preparation a selected number of exercises without solutions will be proposed. A final evaluator will give him a score of his confidence with the subject.
*Practice.* We propose two ways of testing. The first one is a typical numerical approach were the student must find a numerical solution. Here the student finds several exercises, divided by topics and level of difficulty. For every exercise he has access to different levels of helps. The first level is just an indication, in the case he is not sure about the way to begin to solve the exercise. The second level contains a first level of solution. Here the student finds a major help in solving the exercise but must look for the final solution in an autonomous way. The third level is the complete solution of the exercise. During the solving phase the student has available several other tools such as a guided textbook with direct index to the parts significant for the exercise, a data book for specific electronic components, a scientific calculator and a block notes. The second way to the test section requires a major skill. Some exercises propose an interactive solution. They suggest a way of solution, with different alternatives. The student chooses a way and another series of choices is again presented. Here the student has to make another decision. This approach is intended to verify the guidelines of the student preparation. Different choices can indicate different strategies in the problem solving skill of the student.
*Virtual Lab.* CAD tools are today mandatory to analyze an electronic circuit. Several pc programs are available but Spice (or Pspice) [3] is surely the most used in industry. Owing to the poor number of hours available for the electronics basic classes, students must choose whether to learn to use Pspice or to learn how electronic basic circuits work. The approach of the self-instructor is completely different. Here the virtual laboratory presents a possible approach in the simulation and analysis of a circuit. The student is guided up to the simulation phase from a pre-programmed version of Pspice, with several self explaining guides and already set parameters. The student can use

Pspice as a powerful analysis tool, without to loose time to learn its commands. After the analysis he can change the characteristic of some parameters of the circuit, such as voltages, currents, resistors values and so on. Pspice can show the changes in the circuit behavior, without any effort by the student. Using this approach he can benefit from the Pspice simulator without to be scared by its complexity. On the contrary the help windows to learn more about the simulator and to know how to use it stimulate him.
*Internet.* Here students can find new circuits, new data sheets and other stuff. They also have the possibility to contact the teacher by Email or by Videoconference. This added benefits stimulate the students to have more contacts with the professor, overcoming the difficulty of a face-to-face contact, sometimes critical for young students.
The self-instructor is today available as a preliminary release, with several pages of electronics theory and about 30 exercises. In this first release only the Italian version was implemented to overcome the language difficulties of our students.

## 5. Conclusion

The self-instructor is today an experimental platform to verify the real feasibility of these techniques. Students of basic electronics classes are experimenting these tools in the electronics labs of the Politecnico and will give useful indications about the utility of this approach. First impressions are that they are not really interested in the tool but they are only looking for an easier way to pass exams. Only next year fallbacks will be useful to have an evaluation of our approach.

## 6. References

[1] M.A.Garito (1994) "Multimedia and Distance Learning for Science and Technology", Garamond Pub.
[2] http://neurosrv.polito.it/neuronica/didattica/corsoelettronica/telehome.htm
[3] MicroSim Pspice A/D Student Ed., MicroSim Corporation, Irvine, CA

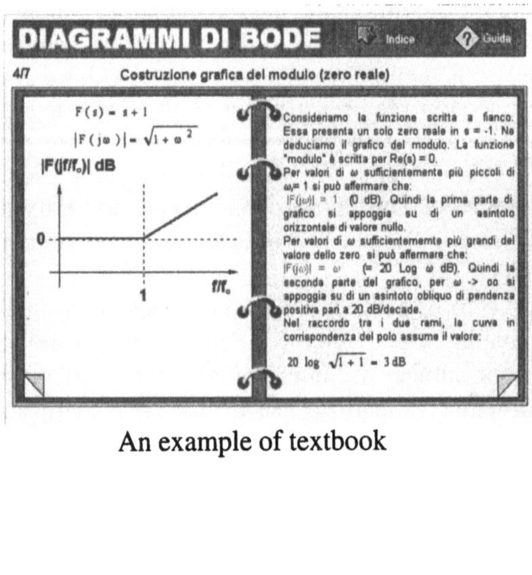

An example of textbook

# MREL : METAREGISTER OF EXPERT LINKS

*The development of a prototype for an electronic course on the design of analog integrated circuits and systems*

PROF. DR. W. SANSEN

*KULeuven ESAT-MICAS*

*Kardinaal Mercierlaan 94*

*B-3001 Leuven-Heverlee*

*BELGIUM*

PROF. DR. G. DE CORTE & DR. PH. YDE

*University Antwerp – Department Didactics*

*Universiteitsplein 1*

*B-2610 Wilrijk*

*BELGIUM*

## Introduction

New learning technologies provide for interactive and multimedia tools that facilitate teaching and help students acquire self-study skills. Although these technologies can be powerful educational tools, their introduction in academic teaching and learning is slow and fraught with problems. Lectures or seminar presentations remain the preferred teaching methods. Lecturer-independent materials – scarce as they are – rely rather heavily on printed textbooks. New learning technologies as such are insignificant without effective and efficient learning materials. In our contribution we want to elaborate on some personal experiences while developing courseware for adults in regular higher education. Our main research question is to determine to what extent and



*B. Courtois et al. (eds.), Microelectronics Education,* 167–170.

in what respect an electronic version of a text-book can realize a didactic added value for learners as well as for lecturers. What works, what doesn't and what need to be changed ?

With Kenneth Laker as co-author Willy Sansen (1997) published a standard text book on 'Design of Analog Integrated Circuits and Systems' with a supplement 'Solutions Manual to Accompany'. The textbook aims at learners who use it in combination with lecture-notes, but also at those who don't follow lectures. Sansen does not use the textbook primarily during his lectures at the university. The main function of the book is to support students when they assimilate what has been treated during the lectures (for reference purposes) but in particular to help students when they work on projects with regard to the design of analog integrated circuits and systems. At final examinations – which focus on a particular project, not on the lectures – students are allowed to consult the textbook.

**The development of a computer-based course**

For a pilot-project we developed a CD-ROM on a particular topic of analog integrated circuits and systems' : CMOST OTA (De Corte and Yde, 1999) with Willy Sansen as content specialist. First we organised the topic in short manageable topics that provide for self-contained and coherent chunks of learning. Since courseware must be grounded in an effective model of teaching and learning, we then developed the electronic course according to SITO – an acronym for the learning routes Study, Inform, Test, Overview.

**Study** : basic contents, 'need to know'. Learners are encouraged to go through all the contents of this route.

**Inform** : enrichment program with background information, 'nice to know'. Learners are encouraged to explore the topics they are interested in.

**Test** : self-evaluation, yes/no questions with immediate feedback. This route enables learners to detect eventual areas of weakness.

**Overview** : tools that enable students to orientate themselves within the course, search engine, overiews of all the topics treated in Study – Inform – Test.

For learners who use the CD-ROM in combination with lecture-notes, but also for those who don't follow lectures, the CD-ROM provides supplementary learning materials.

**mREL as a self-study tool**

Many existing electronic courses stimulate learners to follow a predetermined, linear path and hardly involve them in active learning activities. To make our computer-based course an even more powerful educational tool, we developed a self-study tool: mREL (metaRegister of Expert Links). The major objective of mREL is to stimulate students to relate different course contents, thereby acquiring a cognitive network of the course (Prawat, 1999). From a chapter of 'Design of Analog Integrated Circuits and Systems', Willy Sansen as an expert selected 23 main topics. mREL therefore contains 23 windows, each presenting a main concept in a yellow rectangular, surrounded by other terms which Sansen considers to be closely related to the main term.

An example. The main term *current mirror* is surrounded by four related terms : *active load, differrential amplifier with active load, width W, length L.* The relationship 'main term–related term' is visualized by an arrow. Before students click on any arrow, they are prompted to define in a few sentences for themselves the relationship they perceive between main term and related term (e.g. between *current mirror* and *active load*).

To compare their perceptions as learners with the expert-perceptions (i.c. Sansen's perceptions), students then click on the arrow. A popup appears, describing the relationship as perceived by the expert often followed by relevant formulae :

In our example the popup contains the following text: « *Current mirror – active load* : *A MOST differential amplifier achieves a high gain because the differential pair is loaded*

*by a current mirror which acts as an active load. The output resistance is now high and so is the gain.* » Next to each related term, there is a hot spot. Once learners click on it, they jump immediately to a window where the related term becomes the main term.
An example. When students click on the hot spot next to *differential amplifier with active load*, they jump to the window where *differential amplifier with active load* is the main term (in a yellow rectangular), surrounded by three other related terms : *operational amplifier, differential pair, current mirror.* So concepts are constantly presented in different contexts.

## Conclusion

We consider mREL as a very powerful and flexible learning-tool for computer-based learning materials. Our poster session will demonstrate the effectiveness and efficiency of mREL as a self-study tool to develop cognitive networks, in this case with regard to a topic of the design of analog integrated circuits. Our presentation will also include a demonstration of how mREL works in practice. Finally we want to illustrate how the development of CMOST OTA, and in particular of mREL, helped us to delineate issues that need to be considered in greater detail with regard to the ideal mix of lecturer guided- and computer based-teaching.

## References

De Corte, G. and Yde, Ph. (1999) *Adult learning and courseware. In search of an effective interface*, Scientia Paedagogica Experimentalis, 1, 57 – 71.

Laker, K. and Sansen, W. (1994) *Design of analog integrated circuits and systems*, MacGrawHill

Prawat, R. S. (1999) Peirce and the learning pardox, *American Educational Research Journal*, 1, 47 – 76.

# NEW CONCEPTS IN

# TEACHING AND

# INDUSTRIAL PROJECTS

# EuroPractice Training – a Pan-European Infrastructure for Microelectronics Training

Prof. Nihal Sinnadurai. UK
Prof. Ole Olesen, DTU, Lyngby, Denmark
Prof. Pierluigi Civera, COREP, Torino, Italy
Dr Malcolm Wilkinson, TFI, Cambridge, UK
Ivan Ring Nielsen, TechnoConsult, Copenhagen, Denmark
*[Principal Author: Prof. Nihal Sinnadurai: Fax: +44 1473 253649, Phone: +44 7803 182769
Address: 44 Anglesea Road, Ipswich, IP1 3PJ UK]*

**Extended Abstract**

EUROPRACTICE is an European initiative on technology transfer to benefit European industry, which started in 1995. The focus was on the three technologies: Integrated Circuits (ICs), MultiChip Modules (MCMs) and Microsystems together with two supporting activities which were CAD software, and Continuing Professional Development (CPD) Training. EUROPRACTICE Training was established under the title "Training and Best Practice Service (TBPS)". The TBPS goals are to identify the training needs of European industry and academia in ICs, MCMs and microsystems, and to identify related good calibre training programmes provided by reputable Course Providers (CPs) to match the identified training needs. TBPS has set up an extensive Course information base together with the marketing, handling and helpline infrastructure to reach those seeking training. The user friendly web site (http://cie.it.dtu.dk/tbps) has course data and the electronic Training Registration On-Line (TROL) system. Over four years, TBPS has penetrated through to many Course Providers in Europe and enabled training directly for over 900 attendees from industry, plus over 230 attendees at customised courses delivered at company premises. Where gaps in training have been identified, the TBPS has organised training at two levels: Awareness Training to raise awareness amongst novices to the field, followed by Advanced Training to help build up the skills of those who then seek in-depth knowledge. Further help is provided on a consultancy basis. This enables more rapid learning than following a conventional route (Figure 1) and has proved to be very popular.



*B. Courtois et al. (eds.), Microelectronics Education,* 173–175.

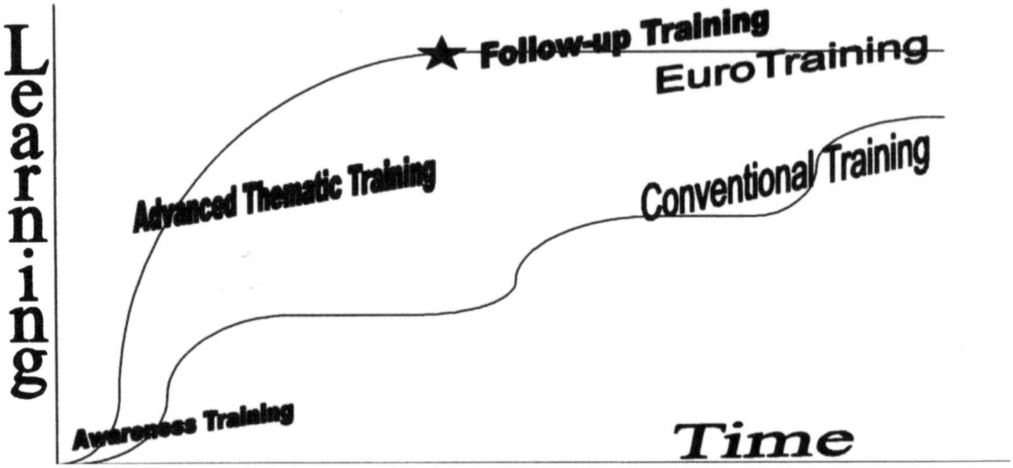


**Figure 1. The learning process with and without EuroTraining**

The training programme has progressively grown in strength in the take-up of training (Figure 2).

**Figure 2. Course Participants for Phase II**

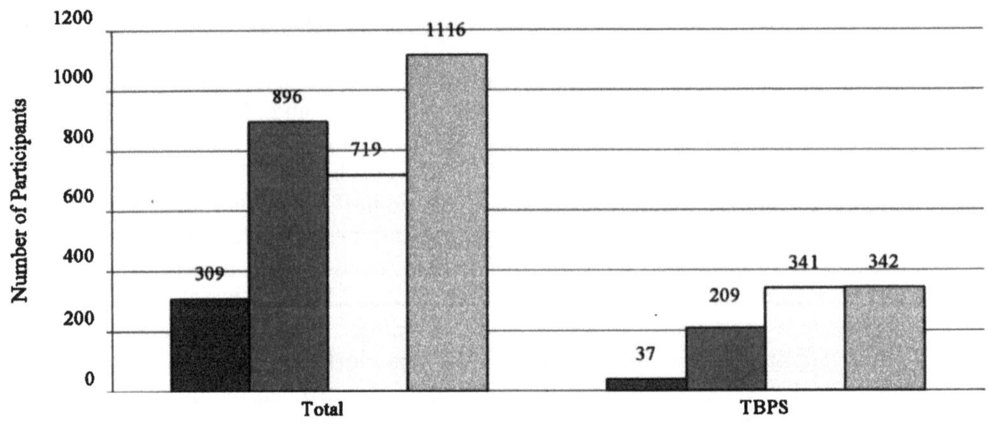

In addition, specific courses have been tailored (customised to meet the requests of a number of companies, and this interest has also grown progressively (Figure 3)

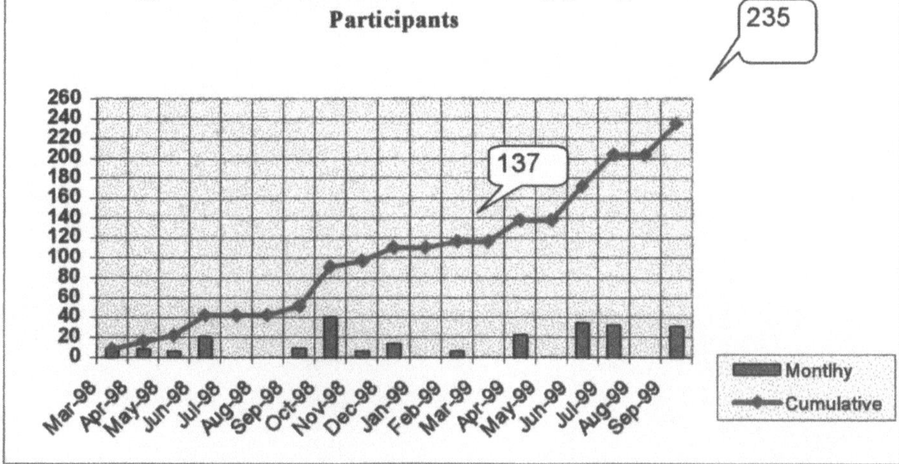


**Figure 3. Customised Training progress**

TBPS maintains a database to analyse Course patterns for attendance, cancellation and demand, and such analyses are regularly reviewed to help CPs target the potential attendees. A further step is the Quality Accreditation of Course Provider programmes to ensure they meet a carefully considered set of evaluation criteria [Ref: "Search for Quality and Quality Assessment in Continuing Education and Training", R Castagne and C Alfandari, EWME 2000]. By the end of 1999 almost half the CPs will have been evaluated and those meeting the criteria will have received the Quality Label (QL).

The benefit of the EUROPRACTICE Training service is now observed to be valuable both in terms of the quality of the service to industry as well as the quantitative benefits that are tangible. This has led to the new EC initiative "EuroTraining", which can be reached at the web domain "Eurotraining.net"

# AN EXPERIMENTAL APPROACH TO THE STUDY OF SEMICONDUCTOR LASER SOURCES

P. ARGUEL
*Université Paul SABATIER - Toulouse III*
*LAAS-CNRS*
*7 Avenue du Colonel Roche*
*31077 Toulouse Cedex 04*
*France*

## 1. Introduction

The development of semiconductor lasers is a good illustration of the major progress that has been recorded in terms of basic concepts and the mastery of manufacturing processes. Owing to their maturity and diversity, laser diodes are becoming highly popular in a number of applications ranging from the general public to leading edge industrial processes. Nevertheless, courses derived from this area of microelectronics are still too few and far between in postgraduate studies and remain, on the whole, essentially theoretical.
The approach retained in this paper has been followed over the last two years for students completing their postgraduate studies at the University of Toulouse III. The whole course is planned over six hours. It relies on an experimental approach in several steps allowing students to progressively get to grips with the basic concepts briefly dealt with during a "conventional" lecture. Each step gives rise to discussions with the teacher and the main conceptual or technological innovations specific to that step as well as the remaining challenges in the field are detailed.

## 2. Progress of the study

The experimental approach can be broken down into three steps that correspond to the main families of semiconductor laser source characteristics. The experiment involves the use of standard equipment including laser diodes ($\lambda$ = 660 nm or $\lambda$ = 780 nm), amplified photodiodes, a Peltier effect mount, a spectrometer and a goniometer.

### 2.1. FIRST STEP: V(I) AND P(I) CHARACTERISTICS

The aim of this first step is to allow students to get to know the basic characteristics of a laser diode and to establish the link with electronic diodes they already know. Therefore, we consider the evolution of the voltage V across the device and the



*B. Courtois et al. (eds.), Microelectronics Education,* 177–180.

evolution of the optical output power P as a function of the value of the biasing current I.

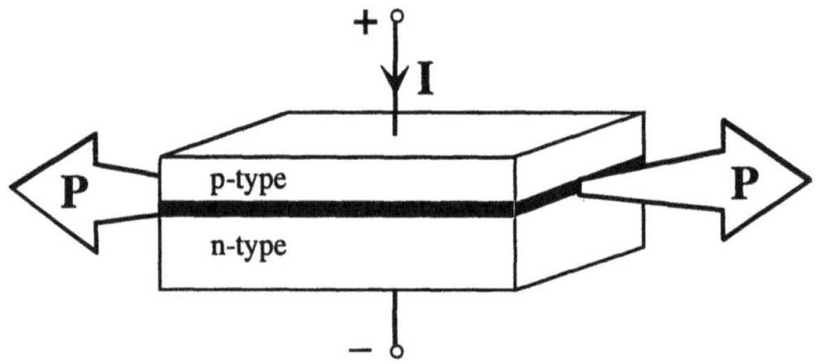


Figure 1 : Schematic illustration of a laser diode.
When forward bias is applied to the p-n junction (current I), optical power P is emitted.

The experimental plot of V(I) highlights the presence of the turn-on voltage $V_c$ and allows the device's series resistance $R_s$ to be determined. Through a comparative trial we can specify typical values of $V_c$ for different materials.
The experimental plot P(I) shows the evolution from an electroluminescent operating mode to a laser regime and the presence of threshold current $I_{th}$ whose definition is given (see Figure 2). The notions of differential and external quantum efficiency for one or two emitting facets are equally specified. Then a brief history of the main technological innovations and their related performance in terms of threshold current density is presented.

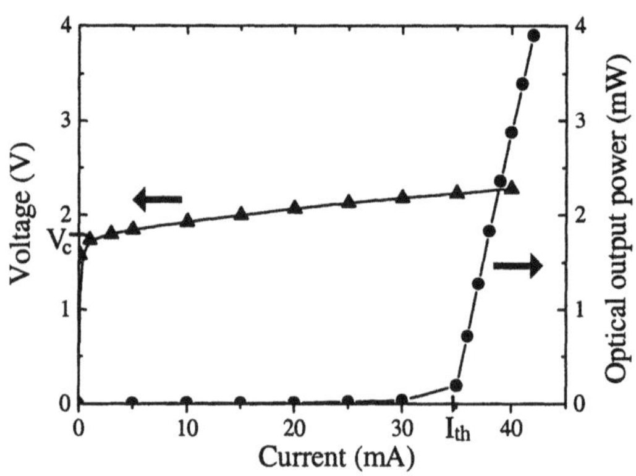


Figure 2 : V(I) and P(I) characteristics.

## 2.2. SECOND STEP: RADIATION PATTERN AND EMISSION SPECTRUM

Here, the spectral and spatial distributions of the optical power emitted from the laser diode are investigated. These characteristics are then used to determine the physical and geometrical parameters inside the device.

The determination of the emission spectrum for the two aforementioned operating modes allows the students to assess precisely the spectral narrowing generated by the laser effect (see Figure 3). Such concepts as bandgap engineering are tackled and major families of semiconductor alloys are described. Likewise, distributed mirror laser structures and their related technological constraints are then presented.
The experimental plot of the radiation pattern given in Figure 4 along the axes perpendicular and parallel to the plane of the active region ($P(\theta_{\perp})$ and $P(\theta_{//})$ respectively) supports the determination of the lateral and transverse dimensions of the optical guide inside the device. The structure based on a stack of semiconductor layers as well as the optical and electrical confinement criteria are then considered.

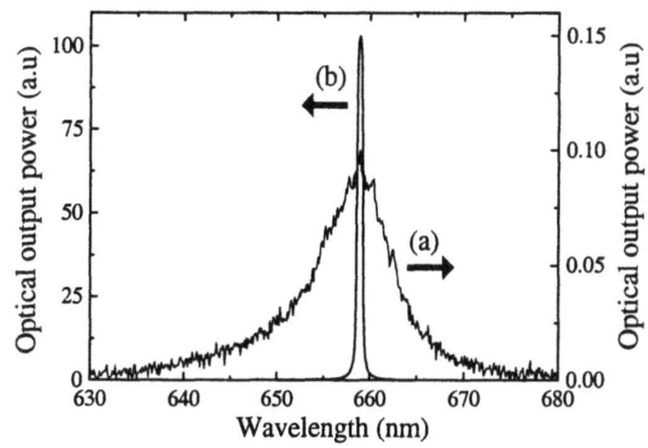


Figure 3 : Emission spectrum for the electroluminescent (a) and the laser (b) operating mode.

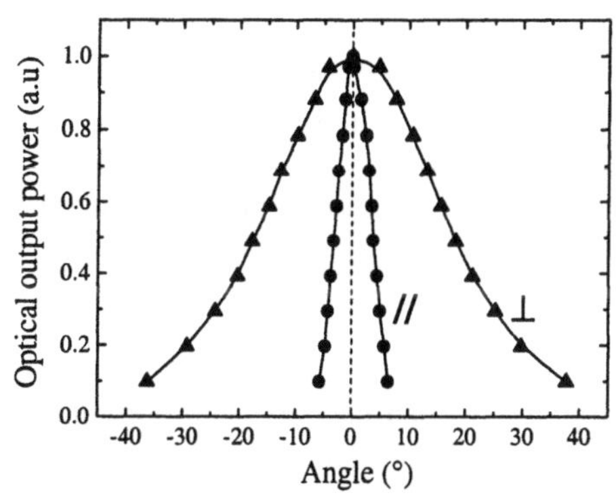


Figure 4 : Radiation pattern perpendicular (⊥) and parallel (//) to the plane of the active region.

## 2.3. THIRD STEP: OPERATING TEMPERATURE

This step is used to highlight and characterise the impact of the operating temperature on the laser diode performances.

In a first step, the evolution of the emission wavelength is recorded when temperature varies from 10°C to 50°C. We determine the device's spectral drift coefficient and after interpreting the phenomenon from a physical point of view, we discuss the applications and constraints associated with this property.
Then the evolution of the threshold current value is noted as a function of the operating temperature. This evolution is usually formulated as follows:

$$I_{th}(T_2) = I_{th}(T_1)\exp(\Delta T/T_o) \qquad (1)$$

where $T_2$ and $T_1$ are two operating temperatures, $\Delta T = T_2 - T_1$, and $T_o$ is the temperature coefficient expressed in Kelvin.
The corresponding plot (see Figure 5) naturally leads to the definition and determination of the coefficient $T_o$ which is an important parameter of the laser diode.
The temperature coefficient is then taken up in a comparative trial for different material and technological processes, emphasis being laid on the process specific to optical telecommunications and the important issues at hand as a result.

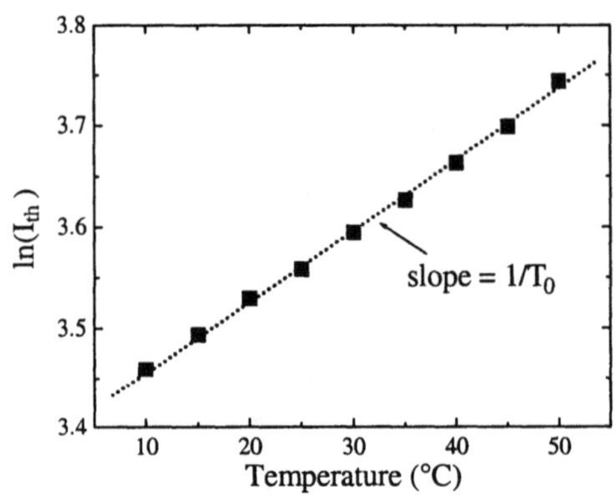


Figure 5 : Threshold current variation as a function of temperature.

## 3. Conclusion

The experimental approach which has just been presented is used to allow students to become familiar with the basic concepts that govern the study and design of semiconductor laser sources. Through the discussions thus generated, the approach supports a deeper acquisition of knowledge tailored to each one's concerns.

## 4. References

Saleh, B.E.A. and Teich, M.C. (1991) *Fundamentals of Photonics*, Wiley Interscience, New York.
Suematsu, Y. and Adams, A.R. (1994) *Handbook of Semiconductor Lasers and Photonics Integrated Circuits*, Chapman & Hall, London.

# LABORATORY OF MICROELECTRONICS: AN ADVANCED PRACTICAL COURSE ON ANALOG ICs DESIGN.

MAURIZIO VALLE

*Department of Biophysical and Electronic Engineering*
*University of Genoa*
*Via All'Opera Pia 11/A, I16145 Genoa, Italy*

**Abstract:**
In this paper, we give a detailed overview of the content and of the educational approach followed in the course of Laboratory of Microelectronics held at the University of Genova, Italy. The course follows an original approach in teaching analog integrated circuit design to students at the fourth year of their five years curriculum in Electronic Engineering; such approach is mainly based on teaching to students practical and reliable design methodologies and on intensive laboratory exercises.

## 1. Introduction

This paper introduces the content and the teaching approach that have been adopted and implemented in the Course of "Laboratory of Microelectronics". The course aims at teaching CMOS analog integrated circuit design to students at the fourth year of their five years curriculum in Microelectronic Engineering (Italian University Master Degree, Laurea Degree, in Electronic Engineering) at the University of Genoa (Italy). Students who attend the course have a background on device physics, integrated circuit technology, electronic circuit design and basic microelectronic circuits.

The main objectives of the course are:



*B. Courtois et al. (eds.), Microelectronics Education,* 181–184.

a) to give to students an operational knowledge and expertise of main issues, design approaches and solutions in the design of the analog subsystem of mixed-mode ICs;
b) to introduce basic and most effective design methodologies;
c) to make students to learn through intensive laboratory exercises and through the development of a design project performed at the end of the course.

## 2. Course organisation and content

The course is based both on classroom lectures and laboratory exercises. The practical part of the course is considered vital for the education of graduated engineers in microelectronics. The course is organised as follows:
a) a set of classroom lecture-based lessons (about 60% of the overall time);
b) a set of supervised laboratory exercises (about 30% of the time);
c) a set (about four) of invited talks (about 10% of the overall course time) given by experts of the academia and/or industry covering few well defined design case studies (e.g. very high speed AtoD converters, ΔΣ converters for sensors data acquisition).

Each set of classroom lectures dealing with a specific topic (see the next Section) is followed by a laboratory exercise during which students put what they have learned into practice. The laboratory exercises are subdivided into a first part, held with the supervision of a tutor, and a second part without tutor; in this case the tutor only verifies the results of the work done by the students.

The final examination is based on the design of an analog full-custom circuit starting from the specifics (given by the tutor, e.g. the circuit basic topology, the power consumption, the GBW, etc.) till the layout design. The design must be fully verified: the students must run the DRC, LVS, etc. i.e. all standard design verifications. The design topics are based on the arguments illustrated during the course; the circuit complexity is limited to at most twenty transistors.

The course deals with the fundamental aspects of analog sub-system design in modern submicron mixed-mode ICs. The course topics are the following:
a) Physical (e.g. layout) design of ICs. Layout design rules and guidelines for active and passive integrated devices (e.g. design rules for good matching of devices, for interference reduction, etc.). Physical design of mixed-mode integrated circuits. Physical design verification (e.g. DRC, ERC, LVS). Examples of industrial CMOS technologies are introduced. Layout design methodologies. [1] [2].
b) The MOS transistor: regions of operation and modelling [1].
c) Elementary transistor stages and operational amplifier design [3].
d) Basic design methodologies: Gate Voltage Overdrive (GVO) and $g_m/I_d$ based methodologies [3] [4].

e) Design and implementation of integrated active filters [3] [5] [6].
f) Integrated A/D and D/A converters [7] [8].
g) Translinear circuits and synthesis of non-linear functions [9].

## 3. Educational approach

The students learn to apply and to develop design methodologies: at the beginning they analyze the given circuit, write down and solve the circuit equations and then implement a design methodology (usually based on an iterative loop) by hand or by using Matlab. So the students are disouraged from using SPICE as design tool but they learn to use it only for the design verification. The rationale of this approach is to try to take into account into an effective way of all design variables; morevoer, following this approach, the circuit can be easily resized when changing the specifications or when moving to another technology. The MatLab implementation of the design methodology is of basic importance because it makes the design space exploration much more effective and fast.

The educational approach is also based on the following guidelines:

a) intensive practical laboratory activity;
b) to avoid to make students to "struggle" with complex CAD tools and to make them to concentrate on design issues. The practical part of the course is based on "easy-to-use" CAD tools such as L-EDIT (Tanner Tools) and HSPICE (AVANT!);
c) usage of MATLAB to implement the synthesis methodology;
d) design methodology based on all regions of operation of the MOS transistor (i.e. not only strong inversion but also weak and moderate inversion);
e) all the course material and laboratory exercises are going to be pubblicated on the WEB (http://melserver.dibe.unige.it/micro/) and will be directly accessible to the students;
f) to make more effective the teaching activity, throughout the course we refer to a real, industrial technology: the AMS CMOS CYE 0.8 μm minimum channel length, available through EUROPRACTICE.

## 4. Practical laboratory exercises

One of the main educational goal of the course is to make students to acquire an operational design expertise through a set of design laboratory exercises. This part of the course is meant not only to teach to students the usage of CAD tools and design environments, but to make them to face simple but real design cases; the students are then forced to put into practice the design methodologies learned during the classroom lessons.

The laboratory exercises given in the present academic year were the following:

a) layout design; the students must design the layout of some basic structures such as: matched resistors, matched differential pairs, common centroid structures, etc.
b) introduction to features, drawbacks and limitations of SPICE models: i.e. LEVEL 2 and LEVEL47 (BSIM3V2). The students must run some benchmark test simulations [1] for evaluating the MOSFET models for analog design.
c) GVO based synthesis of a cascode CMOS OTA and design verification through circuit simulation. During some classroom lectures, the students learn how to size the transistors and to design the bias network. During the laboratory exercises, the students must perform all the SPICE simulation verifications (.DC, .AC, .TRAN, .FOUR, .NOISE, etc.).
d) $g_m/I_d$ based synthesis of a CMOS MILLER OTA using MatLab. The $g_m/I_d$ synthesis methodology [4] is a very effective approach for analog circuit design in particular for current mode circuits. Moreover, it is a very general synthesys methodology and is usefull for all regions of operation of the MOS transistor. The effectiveness of this approach is much more augmented if it is implemented through MatLab. The MatLab implementation of the MILLER OTA synthesys methodology (see for instance [4] and [10]) is reported in [11].

## 5. Acknowledgments

The author would like to thank very much Dr. Francesco Diotalevi for its valuable help in implementing the practical laboratory exercises.

## 6. References

1. Tzividis, Y.P. (1995) *Mixed Analog-Digital VLSI Devices and Technology: an Introduction*, McGraw-Hill.
2. Vittoz, E. Notes of the Lectures on IC Layout Design, EUROCHIP Advanced Analog Digital Design Course, Nov.1992, ESAT, KUL, Leuven (B).
3. Laker, K. R. and Sansen, W. M.C. (1994) *Design of Analog Integrated Circuits and Systems*, McGraw-Hill Book Co.
4. Flandre, D. et al., (1997) Improved Synthesis of Gain-Boosted Regulated-Cascode CMOS Stages using Symbolic Analysis and $g_m/I_d$ Methodology, *IEEE Journal of Solid State Circuits*, **32**, 1006 - 1012.
5. Ismail, M. and Fiez, T. (1994) *Analog VLSI: Signal and Information Processing*, McGraw-Hill.
6. Tzividis, Y.P. (1994) Integrated Continuous-Time Filter Design - An Overview, *IEEE Journal of Solid State Circuits*, **29**, 166 - 176.
7. Van der Plassche, R., (1994) *Integrated Analog-to-Digital and Digital-to-Analog Converters*, Kluwer Academic Publishers.
8. Razavi, B. (1995) *Principles of Data Conversion System Design*, IEEE Press.
9. Tomazou, C., Lidgey, F.J. and Haigh, D.G. Analogue (1990) *IC Design: the current mode approach*, IEE Circuits and Systems Series 2.
10. Jespers, P., (1998) Notes of the Lecture on "CMOS Operational Amplifier Design", DIBE, University of Genova.
11. http://melserver.dibe.unige.it/courses/valle/material/eserc_Matlab.pdf

# A WORLD WIDE WEB TEACHING AND LEARNING ENVIRONMENT APPLIED TO EDUCATIONAL COURSES IN MICROELECTRONICS

N. Speciale, F. Franzè and G. Masetti
*DEIS, Università di Bologna*
*Viale Risorgimento 2, 40136 Bologna, Italy*
nspeciale@deis.unibo.it

**Abstract** This work describes a user-friendly interactive system developed for the World Wide Web to help students to deepen and verify basic concepts and notions in microelectronics. In particular, the presented tool enables to follow both the logic schedule of a course established by the teacher and an own reading sequence, overcoming classical approaches. This was done by exploiting the complex hypermedia nature of the devised text, that can be explored through suitable navigation panels and hypertext links to multimedia resources and several Internet services, mainly based on Java technology.

## 1. INTRODUCTION

Java is a programming language that represents a powerful tool for the development of platform-independent interactive software, which can be used on the World Wide Web (WWW) through a Java-enabled Web browser [1]. One of the main features of WWW is the fact that informations in hypermedia format can be easily created and disseminated, combining hypertext and multimedia. The introduction of Java extends the capability of Hypertext Markup Language (HTML) documents allowing developers to freely create interactive content using new data types and methods to operate on them. In particular, exploiting pieces of compiled code that are referenced within Web pages (Java applets), it is possible to display still or animated graphics, play audio files and receive user input from the mouse and keyboard.

One of the main goals of this approach is the possibility given to students to develop by themselves models, mathematical formulas and design concepts of electronics, overcoming classical approaches that are merely static and offer no significant advantages over textbooks. Instead of clicking on a passive screen full of images and text, the student becomes an active participant that can be



*B. Courtois et al. (eds.), Microelectronics Education,* 185–188.

prompted for a response by offering a real-time interactive menu and figures. By so doing, important concepts can be intuitively highlighted and students can develop proper ideas or correct physical pictures to grasp concepts.

Theoretically, there are no specific limitations on what the Java language can be used to do, apart from security restrictions on how Java code can access the computer on which it is running. In our project, these problems have been solved by using the technique of digital signature of the applets [4]. The proposed tool enables the students to follow either a pre-defined logic schedule of a course established by the teacher or an own specific reading sequence, warranting constant feedback and interaction with the tutor.

As shown in Fig. 1.1, this purpose is achieved by fully exploiting the complex hypermedia nature of the devised text (theory, interactive exercises and examples, simulation programs), that can be explored through suitable navigation panels, hypertext links to different multimedia resources, Internet services and information retrieval based on Java. Particular efforts were given to get a

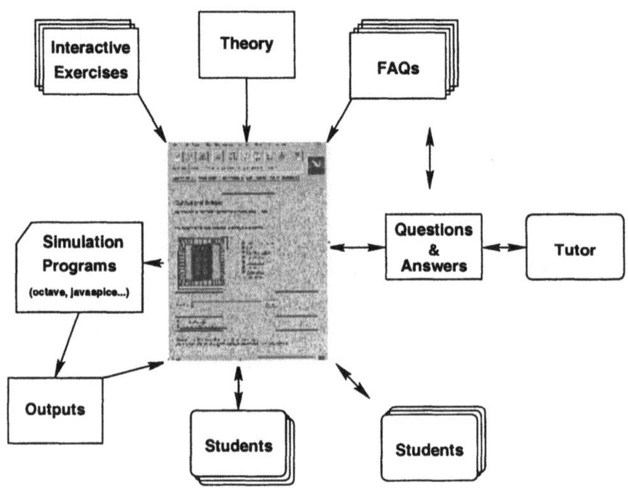


*Figure 1.1* Schematic organization and interactions between different resources and Internet services for the proposed environment.

system very user-friendly also for tutors, which, in turn, can in a very easy way draw interactive figures and update text, graphs, schematics, links, by using high-level automatic procedures to write reusable and easily exportable HTML and/or pdf texts.

The access of all the above features and materials is provided through standard WEB browser such as Netscape or Microsoft IE, making the designed software fully-portable (Linux, Solaris, Windows).

# 2. THE PROPOSED INTERACTIVE ENVIRONMENT

The defined environment allows an interactive use of different programs such as numeric and graphical tools and circuit simulators as SPICE [2], a well known general purpose analog simulator that represents the most commonly used software both in universities and electronic industries.

Our implemented version, named JavaSpice, is a program that exploits Java Native Interface (JNI) to integrate both the front-end in Java and the back-end in C and can used transparently with a browser like Netscape on different machines. Version for Linux and Solaris have already been tested, and also a MS Windows is under development. Its kernel, called TinySpice, is based on a modified version of Berkeley Spice3 code in which all standard analysis (DC, AC, TRAN, location of numeric poles and zeros for different kinds of transfer functions, small signal sensitivities, Fourier, noise and distorsion) are supported. Additionally, new features like device models, circuit parameters extractions and a simple monte-carlo analyses, have been added.

Fig. 1.2 shows an example of a graphical interface, represented by a simple editing frame used to handle the load netlist and simulation results. The JavaPlot window is a java class that implements useful functions to display and analyze the curves, such as scaling and scrolling in all directions, but it can also be used as a stand-alone program to display simulated and/or measured data files written in a common ASCII two-column file.

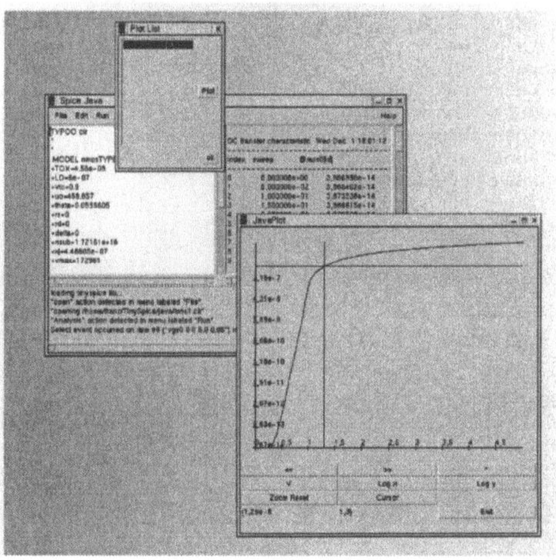

*Figure 1.2* JavaPlot window, the graphic interface of JavaSpice program.

Overworking the possibilities offered by the Common Gateway Interface (CGI), Java language and a WWW server, the students can carry out helpful (either pre- or runtime defined) circuit simulations to learn the correct use of the program and understand main aspects related to typical analog and digital designs, varying both designable and device parameters to explore the main features of the integrated circuit design environment. Moreover, to allow a real-time understanding of the system evolution, simulation results can be displayed in progress, by graphically plotting model parameters or circuit variables during the simulation. As an example, Fig. 1.3 shows the DC transfer characteristic of a CMOS inverter. In this case, the influence on the I-V shape induced by variations in threshold voltage, operating temperature and device dimensions can be interactively investigated selecting different values of the transistors conduction factor $\beta_n, \beta_p$.

Finally, the possibility given to students to contact the tutor via email to achieve explanations to doubts and to answer to questions, allows both a fast feedback to students and the collection of Frequently Asked Questions (FAQ) in an organized database that represents an important self-checking opportunity for students, mainly for those having minor changes to meet their own tutor and colleagues (workers-students, far-away-students).

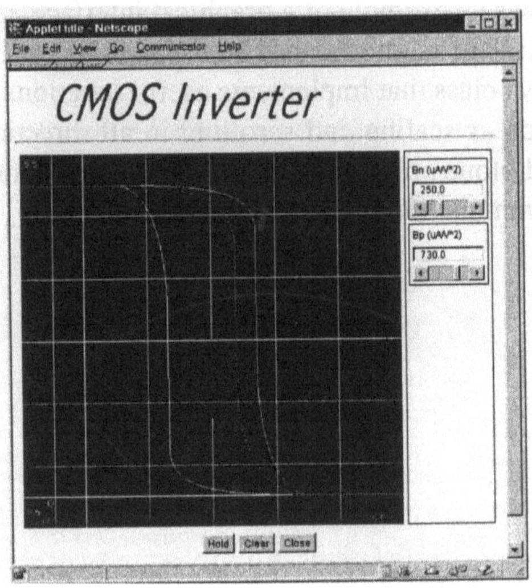


*Figure 1.3* Interactive DC transfer characteristic of a CMOS inverter.

## References

[1] JavaSoft *"Downloading the Java Developers Kit"* `http://www.JavaSoft.com`

[2] A. Vladimirescu *The SPICE* Book Wiley 1994

[3] M. Isaacs *Guide to good practices for www authors* SIMA Rep. Series, 1996

[4] JavaSoft *"Java Security API"*, `http://java.sun.com/security`

# FPGA AND DSP BASED PROGRAMMABLE SYSTEM USED FOR EDUCATIONAL PURPOSES

ANDREJ ŽEMVA ANDREJ TROST
*University of Ljubljana, Faculty of Electrical Engineering*
*Tržaška 25, 1000 Ljubljana, Slovenia*
*Phone: +386 61 1768-346 Fax: +386 61 1264-630*
*Andrej.Zemva@fe.uni-lj.si Andrej.Trost@fe.uni-lj.si*

With the computation power of today's computers and advances in CAD and EDA, the design cycle is minimized. Until the first appearance of Field Programmable Gate Arrays (FPGAs) in the mid-eighties [1], a certain gap between the circuit design phase and the circuit implementation phase existed. Implementation in a mask-programmable technology such as mask-programmable gate arrays or standard cells typically requires extensive manufacturing effort that results in a high cost and relatively long implementation period of several weeks.

User-programmability that is built into FPGAs significantly reduces the time and the cost of the prototyping realization of the design and allows fast re-design that is often imposed with the design errors in one design cycle. Another important aspect of FPGAs is that the designer can implement a circuit by himself, and thus complete the whole design process from the initial circuit specification to the physical implementation and system integration in one place. This is especially important for the educational purposes. For example, circuit design practice courses at universities are often constrained with the tight timing frames and limited financial budget which may prevent students completing large ASIC designs unless FPGAs technology is used. Through the complete design process, which ends with the working implementations, students get insight into each step of the design cycle and some specific problems that occur with the circuit implementation, hardware verification and integration of the design into the target technology.

In this paper, we present the extended and upgraded programmable system based on the prototyping system introduced in [2]. The architecture of the system is illustrated in Figure 1. The hardware programmable environment consists of an FPGA board which is connected to a DSP development board [3]. The FPGA board is composed of one Virtex XCV100 FPGA device, 128k x 24 bits of the fast static RAM and I/O connectors for the DSP development board and a CPLD interface [4].



*B. Courtois et al. (eds.), Microelectronics Education*, 189–192.

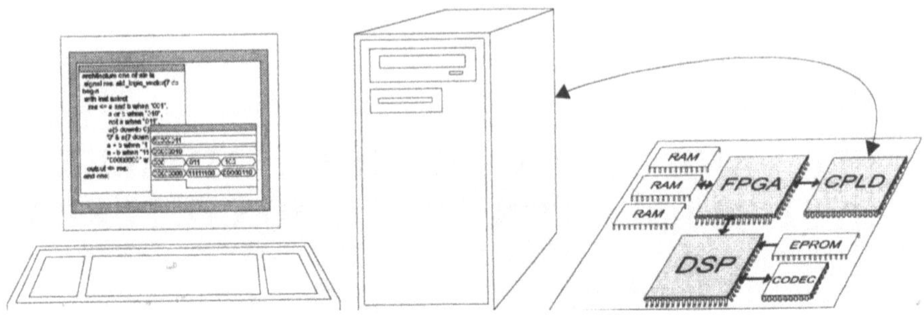


Figure 1: Architecture of the programmable system

The CPLD interface is connected to the PC and is used for downloading the design and hardware verification of the circuit implemented in the FPGA device. The XCV100 [4] has a logic capacity of up to 100.000 logic gates and can be used for hardware implementation of complete digital signal processing algorithms. The FPGA is connected to the peripheral memory port of the ADSP2181 [3] digital signal processor and the whole system can be used for implementation of complex digital systems. The DSP board includes a stereo codec which can be connected to an audio source and active speakers.

Design cycle for the programmable system is illustrated in Figure 2. Students need a deep understanding of the algorithm in order to implement it in hardware or in software, so the first design step is the study of the proposed algorithm. For this step, we use Matlab environment or C++ programming language. The next step is either a circuit design or programming phase.

The major part of the circuit design is performed in VHDL environment. We use Aldec Active-HDL [5] simulator for design and simulation of the circuit. The students are encouraged to use RTL-level VHDL for the part of the design that will be later synthesized and behavioral VHDL for testbenches and external components (e.g. external memory on the FPGA board). Synthesis and technology mapping are performed automatically with Synopsys FPGA Express and Xilinx Foundation [6] tools with no or very little user interaction. The most errors break out during the synthesis procedure due to the VHDL writing style which is not always RTL compliant. There are also some technology dependent errors, for example using both asynchronous set and reset on flip-flops that are not simultaneously available on Xilinx FPGA devices. If some major changes are required, VHDL code is modified and the high level simulation is repeated.

The programming phase consists of writing the C code for the DSP, compiling and building the code, and simulation on a DSP computer simulator. The code consists of a DSP and codec initialization routines and the algorithm usually implemented in an interrupt routine. In order to achieve better performance, we can combine C and assembler code. The simulation of the DSP program is performed on the machine code level in contrast to the circuit simulation which is usually performed on high level design stage. The low level simulation enables exact profiling and searching for critical parts of the code which can be later implemented in hardware.

The last step is hardware verification of the design on the programmable system. We have to download the design to the FPGA and the DSP code to the DSP board and evaluate the performance of the system either by listening to the processed audio signal or by measuring the response.

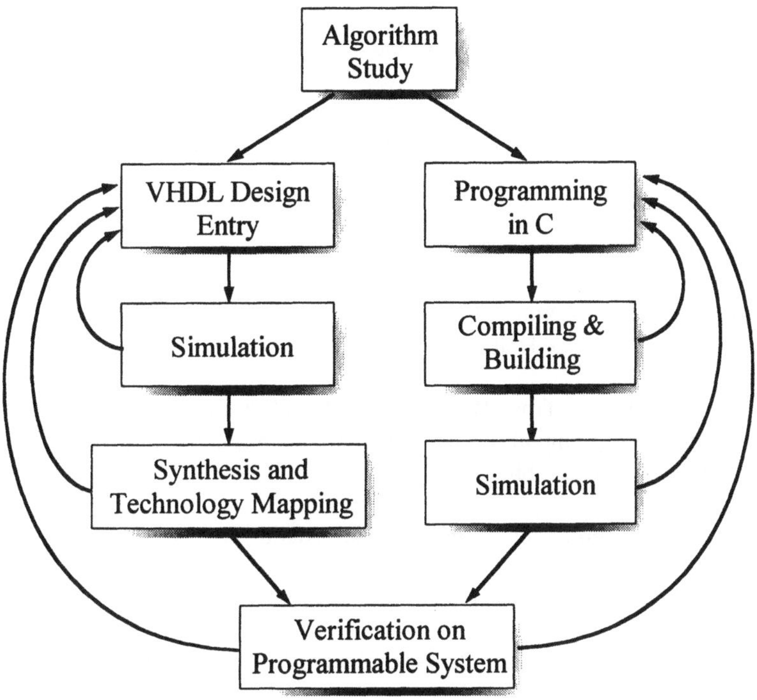


Figure 2: Design cycle for FPGA and DSP programmable system

The presented programmable system is used for laboratory practices of classes covering digital circuits (Digital Electronics, Digital Integrated Circuits) during the undergraduate electrical engineering study (4 ½ years). A typical project is the design of a digital filter [7]. Students begin first with the study and simulation of the algorithm in C programming language. After evaluating the performance of the algorithm, the digital circuit architecture is derived. The circuit blocks are then described and simulated in a high level circuit description language such as VHDL. The VHDL code is then synthesized to the netlist for the specific FPGA family. The next steps are technology implementation and downloading of the circuit to the programmable system. PC is then used for hardware verification of the circuit with the same test vectors as for the simulation. The DSP is used here only for the data transmission between the FPGA and CODEC, since the whole filtering procedure is implemented in the FPGA. Students can evaluate the performance of different hardware architectures of the digital filters (parallel implementation, serial distributed arithmetic etc.).

The next project exploits DSP and FPGA simultaneously. The signal processing algorithm is first implemented as a C program and tested on the DSP development board. The critical timing parts of the algorithm are then considered for hardware implementation following the same steps as described above. By solving this type of a project, students get familiarized with the concept of HW/SW co-design [8].

The usage of the presented FPGA and DSP programmable system for digital circuits implementation is of special interest for students working on their final project (Diploma, 6 months project). During the last 2 years, we have supervised 47 students [9] (of total of 80 students specialized in electronics) completing their electrical engineering study at our University.

## References

[1] P.K. Chan, S. Moroud, *Digital Design Using Field-Programmable Gate Arrays*, Prentice Hall, 1997.

[2] A. Žemva, A. Trost, B. Zajc, Educational Programmable System for Prototyping Digital System, *Int. Journal of Electrical Engineering Education*, Vol. 35, pp.236-244, 1998.

[3] ADSP-2100 Family User's Manual, Analog Devices, 1997.

[4] Programmable Logic Data Book, Xilinx Inc., 1996.

[5] K. Sobolewski, M. Ossysek, Active-VHDL Series Getting Started Guide, ALDEC Inc., 1998

[6] D. Van den Bout, The Practical Xilinx Designer Lab Book, Prentice Hall, 1998

[7] L. Wanhammar, DSP Integrated Circuits, Academic Press, 1999.

[8] G. de Micheli, R. K. Gupta, Hardware/Software Co-Design, Prof. IEEE, vol. 82(3), pp. 349-365, 1997.

[9] Faculty of Electrical Engineering, Laboratory for Integrated Circuits Design, http://dic3.fe.uni-lj.si

# TRAINING OF NEW EXPERTS FOR ELECTRONICS AT THE BSU

A. KOZULIN, N. YURCHUK, V. STEPANETS
*Mechanics and Mathematics Department,*
*Belarusion State University, av. F.Skoriny 4, 220050 Minsk, Belarus*

Key words: Training, Training of Expert for Electronics, Training Project, Education, Educational Plan, Mathematical Specialization, Educational Design Centre, Students' Research, Electronics, High Technologies, CAD, ASIC, Belarusian State University, Mechanics and Mathematics Department

Abstract: In this paper the experience and main distinguishing features of the new expert training for electronics have been described. The graduates of the new specialization (Mathematical Electronics) receive the degree Mathematician and Systems Designer. They gain the knowledge and skills necessary for creating ASIC chips and systems. The HP workstations and Mentor Graphics CAD have been installed at the University Scientific Centre. Its main task is the increase of students' participation in the realization of industrial hardware and software projects. Now more than 400 experts were trained.

## 1. Introduction

In Belarus, there are many research and design centres, and some of the largest microelectronics, radioelectronics, television, computer and communications enterprises. In the course of time, the creation of the electronics systems was more and more resembling a process of programming. Modern mathematical researches and improvement of CAD, CAE, CAM were conditioning its farther progress first of all. These tendencies were most brightly exhibited beginning with the 80-th years in microelectronics.

In view of this, a new specialization was opened at the Belarusian State University (BSU) in 1987. It was named as Mathematical Electronics. The



*B. Courtois et al. (eds.), Microelectronics Education,* 193–196.

Mechanics and Mathematics Department, the Belarusian National Academy of Science (BNAS) and the conglomerate Integral were the initiators of its establishment.

## 2. Characteristics of Specialization and its Training Project

The students of Mathematical Electronics are taught to become experts in Information technology of electronics and mathematics. The graduates of this specialization receive the degree Mathematician and Systems Designer and gain the theoretical knowledge and practical skills necessary for creating complicated electronic and programming systems.

They are able to choose the necessary mathematical method of data processing, to describe the electronic system being created using special high level computer language (VHDL, Verilog-HDL), to model the system, to allocate to the system software and hardware, and to design the system employing modern CAD and programming tools. The development of a special educational plan was required for the training of such highly specialized.

Practically all sections of mathematics are applied widely in the information technology field of electronics. This fact was taken into account during the realization of training project.

In the project the teaching of any part of mathematical knowledge was included in the discipline of electronics, where it received additional practical colour to raise the students' interest. On the contrary, the technical tasks were included into the base mathematical practice. As a consequence, the plan united traditional mathematical subjects with electronics and informatics (fig.1).

From the 3rd year, the students study many special disciplines: computer-aided mathematical research with Mathematica, Maple, MathCAD, MatLAB, methods of logic function minimization, automatic machine theory, VHDL, Verilog-HDL, silicon compilers, microprocessors and microcontrollers, etc.

The teaching of all subjects includes a wide application of computer practice.

Primarily the Mathematical Electronics graduates must freely use the available components of the information technology field of electronics. Also they must be able to build new mathematical models, using modern mathematical computer research tools.

This was achieved using such system of education by which the use of modern computer-aided tools penetrates all educational process from the first to the last year of tutoring. So during the first year, the students begin to study programming and use universal and special packages for mathematical practice [1,2].

The mastering of CAD methodology and software begins in the second year and then continues through all phases of tutoring. This includes an assimilation of educational programmes [3] at the initial stage. In the following years, a lot of software is studied, including Mentor Graphics tools.

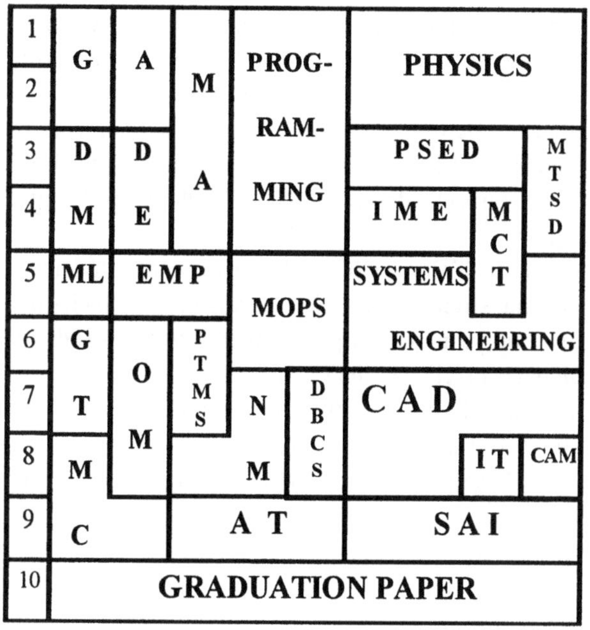


*Fig.1. Training diagram*

*1 ÷ 10 - semester numbers, G - geometry, A - algebra, MA - mathematical analysis, DM - discrete mathematics, DE - differential equations, PSED - the physics of semiconductors and electronic devices, MTSD - manufacturing technology of semiconductors and electronic devices, IME - an introduction to microelectronics, MCT - microcircuit techniques, ML - mathematical logic, EMP - equations of mathematical physics, MOPS - modern operational and programming systems, GT - graph theory, OM - optimizational methods, PTMS - probability theory and mathematical statistics, NM - numerical methods, DBCS - data base control systems, CAD - computer-aided design, MC - mathematical cybernetics, IT - information theory, CAM - computer-aided manufactory, AT - algorithm theory, SAI - systems of artificial intelligence*

This allows the future experts to find the most effective use of such means in their future scientific and industrial activities. All the students' researches (individual works, course works and graduation paper) are directed to this too.

## 3. Students' Research

The department's mathematical schools are well known and have long-standing connections with schools in Poland, Hungary, Germany,. Austria, France, Russia, China and many other countries. On this basis, new application research was begun together with the Integral, the BNAS cybernetics and mathematics institutes, and other private local and foreign companies. This is directed at the creation of new mathematical models and VLSI CAD methodology, information handling principles, silicon compilation tools, electronic devices and systems [4-6]. The results of this research have being used at the Integral and many other companies.

In the last year, this collaboration has acquired new form. The BSU High Technology Scientific Centre (HTSC) is an example of such co-operation. Modern HP workstations and VLSI CAD have been installed at the HTSC. These tools were given by the Belarusian design office NT-Laboratory Limited and the company Mentor Graphics.

The main task of the HTSC is the increase of students' participation in the realization of industrial projects. So students prepare their first ASIC chip designs during the second year. Such a project is the student's individual work during the summer computer practice at the HTSC.

The HTSC is actively collaborating with many companies of Belarus and foreign countries (Italy, Germany, USA, South Korean, Taiwan). Its engineering services contain the ASIC chips, PCB and systems design, and the development of design tools also.

## 4. Conclusion

During the existence of this specialization, more than 400 experts were trained at the BSU [7]. They successfully work at scientific and educational institutions, design centres, and enterprises, which undertake research and the creation and manufacture of various electronics products.

## References


1. Tam, P.T.: *A Physicist's Guide to Mathematics*, Academic Press, Humboldt State University, 1996.
2. Bochman, D., Steinbach, B.: *Logikentwurf mit XBOOLE*, Chemiez Technische Universitat, 1991.
3. Sicard, E., Delmas, S., Chen, X., Ferreira, A.: MICROWIND: Advanced Microelectronics Traning on PC. *Proc. of the 4th Int. Workshop MIXDES'97* (1997), Poznan, 713-718.
4. Gurski, L., Stepanets, V.: *Design of microcircuits*, Science and Technique, Minsk, 1991.
5. Stepanets, V.: The Mathematical Model for Computer-Aided Design of Integrated Circuit Layout, *Proc. of 5th Intern. Conf. MIXDES'98* (1998), Lodz, 163-166.
6. Cherniakovski, D., Kazakevich , K., Kazimirovski, I., Stepanets, V., Trofimov, S.: Computer-aided Design of Application Microcontrollers, *Proc. of 2nd Int. Conf. ECS'99* (1999), Bratislava, 173-176.
7. Kozulin, A., Yurchuk, N., Stepanets, V.: Training experience of mathematicians, *Proc. of the 2nd Int. Scientific and Practical Conf. High school science, industry, international co-operation* (1998), Minsk, 1-4.

# APPLICATION-ORIENTED MICROELECTRONICS EDUCATION

*Towards Local Industry Needs*

A. PASCHALIS
*Department of Informatics, University of Athens, Athens, Greece*
paschali@di.uoa.gr

D. GIZOPOULOS
*Department of Informatics, University of Piraeus, Piraeus, Greece*
dgizop@unipi.gr

## 1. Introduction

In this paper we present an application-oriented microelectronics education perspective to be applied in two closely collaborating university departments.

The specific application domains in which local and european industry is specialized, provide valuable information about how the university courses must be oriented so that well-educated and well-focused computer scientists and microelectronics systems designers graduate from the departments. Such qualified graduates can be easily become part of design flows in local specialized design houses.

An important aspect that we must always keep in mind in educating system designers is the target application domains where they are going to be active in their future professional steps. Usually, the local design industry that the two university departments (among others) feed with engineers is interested in designers/ scientists/engineers *with as global as possible view and knowledge* of the entire microelectronic system. This interest makes it imperative for the university professors to include as many aspects as possible in the group of courses related to microelectronics systems design. More comprehensive analysis of each aspect is naturally performed in post-graduate courses while graduate course are mainly introductory. Hands-on projects are used in most of the cases.

In Section 2 we describe the key application domains and in Section 3 we briefly discuss key aspects where special emphasis is put upon. These aspects are:



*B. Courtois et al. (eds.), Microelectronics Education,* 197–200.

- High-level synthesis and hardware description languages
- Field-programmable gate array (FPGA) technology
- System-on-Chip design and IP entities development, re-use and interchange
- Low power considerations
- Design-For-Testability considerations
- Embedded systems and hardware/software co-design

## 2. Application Domains

Microelectronic systems design in local Greek industry is particularly engaged in the following primary domains:

*Networking and telecommunications (among others)*: ATM traffic analysis and testing systems, network probing applications, ATM and ISDN interface modules, microprocessor-based multiprotocol interface modules, cardphone systems, wireless LAN applications and high-bandwidth multimedia applications.

*Digital Signal Processing (among others)*: implementation of DSP algorithms for image recognition and pattern recognition with emphasis on geographic, geological and medical images processing.

The common characteristic of the two classes of applications is that they are both *low-volume*, i.e. the total number of target systems is low. Thus in both cases, development costs are very critical and rapid time-to-market is very important.

## 3. System Design Considerations

Key contemporary topics that are covered or intended to be covered in courses are briefly described in this section.

### 3.1 High Level Synthesis, HDLs and FPGAs

The target systems for the applications mentioned in the previous section are based on the use of FPGA technology which provides flexibility for system re-configuration and rapid prototyping cycles.

Contemporary FPGA families provide very high gate counts, high performance, adequate for the specific applications and are supported by sophisticated CAD platforms available to universities.

Hardware description languages (Verilog and VHDL) are used for the design of the application specific components (FPGA-based processing elements). Well respected texts are used for teaching Verilog [1] and VHDL [2] and special effort is put on comparing the two language approaches to outline the benefits of each one.

FPGA synthesis flows such as the one provided by Xilinx are used, since they are provided in low university prices by Europractice [3] and Xilinx directly and also are widely used by the local design houses.

### 3.2 SYSTEM-ON-CHIP DESIGN & INTELLECTUAL PROPERTY RE-USE

An excellent article recently published [4] analyses the future of microelectronics education based on emerging trends like System-on-Chip (SoC) design and Intellectual Property (IP) entities re-use.

One of the key points of this article and the proposed educational flow is the close collaboration of universities, research centres and industry in developing, interchanging and integrating IP entities in an efficient way for SoC design.

IP re-use particularly when the target application domains is relatively constraint like the ones described in the previous section can resolve many problems in terms of fast development and first-time-right design.

SoC design and IP re-use in educational activities is continuously becoming a topic of interest in technical meetings worldwide [5], [6].

### 3.3 LOW POWER DESIGN

Low power optimisation design methodologies are very important for specific classes of applications like the ones mentioned in the previous section.

In the beginning of a graduate course, power modelling and power estimation are studied [7]. Then, power optimisation algorithms are studied and applied in microprocessor designs using commercial power analysis toolsets [8] provided through Europractice [3].

Low power is also discussed in parallel with design-for-testability issues and BIST methodologies described in the next subsection.

### 3.4 DESIGN FOR TESTABILITY

Design-for-Testability (DFT) is becoming a crucial issue in systems design [9], [10] since it affects the engineering trade-offs between silicon area, operating frequency and power consumption while balancing the corporate concerns of cost-of-test and time-to-market. System complexity is increasing rapidly and new types of physical failures appear when manufacturing technologies move to the deep submicron.

System-on-Chip design makes external test methodologies using Automatic Test Equipment (ATE) very difficult if at all feasible. Built-In Self-Test (BIST) methodologies for embedded cores (IP entities) is becoming a clear necessity, including memory BIST, logic BIST and scan BIST. Standardisation activities take place so that IP entities developed worldwide can be merged with common self-test mechanisms (addressed by VSI Alliance and IEEE P1500 standards).

### 3.5 EMBEDDED SYSTEMS AND HARDWARE/SOFTWARE CO-DESIGN

The design process of embedded systems has changed substantially the recent years. Nowadays, embedded systems consist of ASICs, processors, FPGAs, memories and peripherals. As a consequence, hardware/software co-design dealing with the problems of designing these systems has gained great importance [11]. A related graduate course should cover topics like: hardware/software co-synthesis algorithms, prototyping and emulation, and specification and verification. This design flow will be of great interest to the local industry the next few years.

## 4. Conclusions

We have briefly described an application-oriented educational flow for microelectronic systems. Close collaboration with local industry gives the focus for the particular applications. A comprehensive set of topics are covered or will be covered in the near future so that graduates of the departments be as ready as possible to be adopted in the design flows for the particular application fields.

## 5. References


1. P.R.Moorby, D.E.Thomas, "The Verilog Hardware Description Language", Kluwer Academic Publishers, 4th edition, 1998.
2. B. Cohen, "VHDL Coding Styles and Methodologies", Kluwer Academic Publishers, 2nd ed., 1999.
3. http://www.europractice.com.
4. H. De Man, "System-on-Chip Design: Impact on Education and Research", IEEE Design and Test of Computers, vol. 16, no. 3, July-September 1999.
5. Proceedings of the IEEE International Conference on Microelectronic Systems Education, 19-21 July 1999, Arlington, VA.
6. D.Bouldin, S.Dey, D.Gizopoulos, M.Lobetti, R.Rajsuman, Panel Session "Training IP Creators and Integrators", 3 rd International Workshop on Testing of Embedded Core-based System-chips 1999 (TECS'99), Dana Point, CA, USA.
7. W. Nebel, J.P. Mermet, "Low Power Design in Deep Submicron Electronics", Kluwer Academic Publishers, 1997.
8. Synopsys Inc., "Power Products Reference Manual", ver. 1998.08, August 1998.
9. M.Abramovici, M.A.Breuer, A.D.Friedman, "Digital Systems Testing and Testable Design", revised printing, 1998.
10. A.L. Crouch, "Design-For-Test for Digital ICs and Embedded Core Systems", Prentice-Hall, 1999.
11. J. Staunstrup, W. Wolf, "Hardware/Software Co-design: Principles and Practice", Kluwer Academic Publishers, 1997.

# INTERACTIVE LOGIC DESIGN ON THE WORLD WIDE WEB

PETER M. MAURER, MURALI VARANASI
*Dep.t of Computer Science and Engineering, University of South Florida*
*Tampa, Florida 33620, USA*

## 1. Introduction

Recently, the University of South Florida Department of Computer Science and Engineering began a program to provide web-based instruction along with web-based laboratory work for a number of our basic computer engineering courses. The philosopy of this program is to provide a series of "electronic textbooks," complete with interactive exercises. This program is actually part of a much larger program to create a comprehensive series of web-based Electrical Design Automation (EDA) tools. However, the educational portion of the program is important, and capable of standing on its own merit.

An electronic textbook consists of a series of web pages. Ideally, each web page is designed to contain the material for one lecture, along with illustrations and examples. As in traditional textbooks, a number of student exercises will be provided to further illustrate and reinforce the material. However, the electronic textbook provides much more than the traditional view of the material. Where possible, the examples will be interactive. Instead of presenting circuits as static drawings, we draw them in editor windows that can be used to simulate the circuits. The student can also modify the examples and observe changes through simulation.

In addition to providing exercises, the web pages will provide all the tools necessary to complete the exercises and submit them to the instructor. For particularly demanding exercises, student activity can be monitored, and hints can be given. For example, we have one exercise that is designed to teach students how to use our schematic capture tool. Students are instructed to draw a three-input AND gate. If students draw the wrong type of gate, a message appears identifying the error and repeating the instructions for drawing the correct type of gate.

The basis of our electronic textbooks is a set of ActiveX controls that provide the necessary functions for the interactive examples and exercises. We use a few commercial components for tool bars, menus and text editing, but most of the ActiveX controls were developed locally. The basic set of tools provides specification and



*B. Courtois et al. (eds.), Microelectronics Education,* 201–204.

simulation of text-based designs. Our two most important tools are our text-based simulator[2], and our schematic-capture and simulation tool[3]. (Before accessing any of our tools, please go to the main page for the Virtual Design Automation Laboratory[6] and read the instructions.) These tools are web-pages containing several ActiveX controls. We use commercial tools to create menus, tool bars, text editors and buttons. All other ActiveX controls were developed locally, and are distributed on the world-wide-web free of charge.

Central to our system is a standardized data structure known as the CNetList data structure. This data structure is produced by our parser control and passed to our other tools. Any software capable of producing a CNetList object can be used with any of our tools. The CNetList object is then registered with one or more simulator controls. Once a simulator control contains a registered CNetList object, it can be used for simulation. Input vectors are passed into the control in one of several different ways. When a simulation control receives an input vector, it simulates the registered circuit and produces one or more output vectors. For simulators that provide detailed timing, it is possible to request output of intermediate vectors.

One of the our most commonly used controls is the schematic capture control. This control creates netlists internally, and can be queried about the structure of the object it contains. These queries can be used to build a CNetList Structure directly, or can be used to create text-based input for our other tools. We also provide a "Looker" control that can be used to expose the components of a CNetList Structure to script-based languages. This is particularly useful in the HTML environment. For example the back-annotation that is done in our schematic capture and simulation tool, is done using the "Looker" control.

For our design exercises, we use a simple netlist language, called FHDL. This language is simple enough that it can be presented in a single lecture, or left for students to learn on their own. We provide an FHDL parser which creates CNetList objects. In addition, we provide a collection of CNetList-based simulators that provide a range of tradeoffs in simulation speed and accuracy. Because the parser is an independent component of our system, the FHDL front end could be replaced with a parser for virtually any other language We plan to provide a VHDL interface at some time in the future.

## 2. Educational Tools

We implemented our tools as a collection of ActiveX controls for several reasons. First, we believe that component-level programming is important and should be taught as a regular part of the curriculum. These tools, which are heavily used by both graduate and undergraduate students, are examples of component-level development. The design of these components is used as an example in our application development course. By creating our tools as a set of ActiveX controls, we could avoid complex user-interface issues and concentrate on the basic tools used in VLSI design. Each of our tools is provided with a rudimentary interface that can be controlled through a more sophisticated user interface, but the user-interface support in our tools is strictly minimal.

The most important reason for creating our tools as a collection of ActiveX controls, is the ability to use them in web pages. If desired, we can create conventional programs by embedding our controls in a Visual Basic program. However, this requires any user of our tools to download the installer, and explicitly install each tool on his or her system. If there are bug-fixes or other upgrades, the user must explicitly check for them and install them.

When a tool is distributed on a web page, each ActiveX control is downloaded and installed independently. Installation is automatic the first time the ActiveX control is referenced. When the version number of the control changes (signifying an upgrade) the new version is downloaded and installed automatically. Since use of the tool requires access to a web-page on the site distributing the ActiveX controls, tests for the existence of upgrades is done automatically whenever the page is referenced. The task of distributing upgrades is also simplified because the only thing that is necessary is placing the new version on the server, and updating the version number in the referring web pages. In addition to using our controls in a web page, we can also use them in PowerPoint presentations and in Microsoft Word documents. If you were reading this paper in electronic form, we could include a figure right below this paragraph that contained a running example of one of our tools. You could interact with this figure to create new circuits and simulate them.

The first educational product we produced was a set of web-pages aimed at grade-school and high-school students. This set of web pages is designed to teach students the principles of elementary logic design, and get them to the point of drawing their own elementary circuit and simulating it. There are six lessons in this set of web pages, Basic Gates, Wiring between Gates, Gate Functions, Exclusive OR and AND/OR/NOT equivalents, NAND/NOR gates, and Creation and Simulation of a simple circuit. They are currently available on the world wide web[4].

Inspired by the success of this initial effort inspired us to create a set of web pages for our Introductory Logic Design course. The first set of these is available on the world wide web[5]. These pages are based on a laboratory manual that we have used for this course for several years[1]. The lab manual contains many exercises ranging from elementary Boolean functions through sequential circuits. We have completed the first edition for the first five lessons of this course. These lessons are Elementary Boolean Functions, Properties of Boolean Functions, Theorems and Canonical Forms, Two-Level Functions, and NAND/NOR implementations. Each of these lessons includes a number of interactive examples and exercises. Each web page includes at least one editing window that the student can use to complete exercises. Depending on the exercise, the editing window can be a text editor or a schematic editor. The next edition of these pages will include features to for E-mailing the results of exercises to the course instructor. We are currently evolving these pages into a complete electronic textbook for our Logic Design course. Eventually we plan to provide electronic textbooks for all of our computer engineering courses, in particular, Computer Architecture and Design, Digital Logic Design, CMOS VLSI Design, and VLSI Design Automation.

Recent efforts have been directed toward expanding the scope of our tools to encompass physical design automation algorithms. We are currently working on tools for channel routing[7], and other standard physical design problems such as placement, grid routing, and channel routing. In addition to providing ActiveX controls, we will provide templates and sets of class definitions that can be used by students to construct their own controls.

In the next few months, we expect to complete the remainder of the logic design electronic textbook, and make some progress on that for the Computer Architecture course. Depending on the success of this project, we may also produce electronic textbooks for those software courses that have impact on the Computer Engineering program. (Analysis of Algorithms, for example.)

Although we believe that web-based education is the wave of the future, we also believe that it is not a substitute for traditional instruction. We believe that electronic textbooks will prove to be powerful tools that will assist instructors in providing a well-rounded Engineering education.

## References


1. Maurer, Peter "Course Materials," http://www.csee.usf.edu/~maurer/courses.html.
2. Maurer, Peter "FHDL Editor," http://www.csee.usf.edu/~maurer/vdal/fhdl/fhdleditor2.htm.
3. Maurer, Peter "Schematic Editor," http://www.csee.usf.edu/~maurer/vdal/schemat/webschem.htm.
4. Maurer, Peter "Logic Design Tutorial," http://www.csee.usf.edu/~maurer/vdal/tutorial/basics01.htm.
5. Maurer, Peter "Logic Design," http://www.csee.usf.edu/~maurer/vdal/logicdes/lesson01.htm.
6. Maurer, Peter "VDAL Start Page," http://www.csee.usf.edu/~maurer/vdal.
7. Maurer, Peter "Left Edge Channel Router," http://www.csee.usf.edu/~maurer/vdal/chanrout/left_edge_channel_router.htm.
8. Maurer, Peter "Dog Leg Channel Router," http://www.csee.usf.edu/~maurer/vdal/chanrout/dog_leg_channel_router.htm.

# MIXED-SIGNAL TESTING AT THE ASIC DESIGN COURSE AT TWENTE UNIVERSITY

R.J.W.T. TANGELDER, H. DE VRIES, E.A.M. KLUMPERINK,
H. SNIJDERS, H.G. KERKHOFF, J. SMIT, S.H. GEREZ, H. SPEEK
*Faculty of Electrical Engineering, University of Twente*
*P.O. Box 217, 7500 AE Enschede, The Netherlands*

*Phone number:* *+31 53 489 2645*
*Fax number:* *+31 53 489 1034*
*Email:* *R.J.W.T.Tangelder@el.utwente.nl*

**MIXED-SIGNAL TESTING AT THE ASIC DESIGN COURSE AT TWENTE UNIVERSITY**

## 1. Introduction

At our faculty the students have to follow an extensive (mixed-signal) ASIC design course in the third year of the program. In **[1]** we have presented an overview of the whole course, but in the meanwhile we have extended the mixed-signal test part of this ASIC design course considerably. In our course the students have to design and test a so-called dial-memo IC (see *Figure 1*).

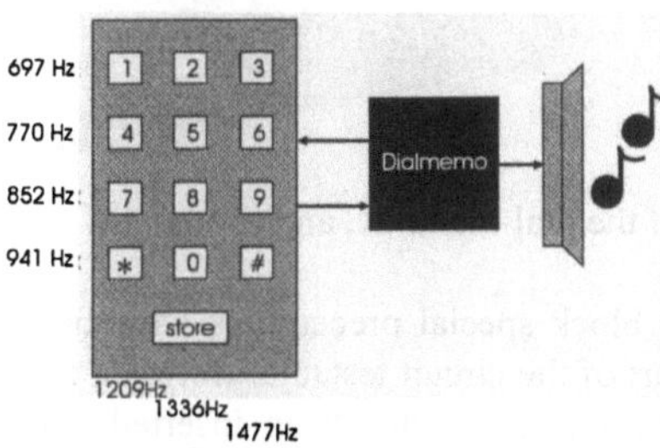


| | | |
|---|---|---|
| **Used freq.** | *L*<br>*H* | 697, 770, 852, 941 (Hz)<br>1209, 1336, 1477 (Hz) |
| **Inaccuracy freq.** | | < 1.5 % |
| **AC-signal power (effective)** | | >100 μW (± 30%) |
| **THD** | | -20 dB |
| **SNR** | | 25-35 dB |

*Figure 1:* Overview of the dial-memo IC and CCITT-specifications for telephony

The dial-memo IC is capable of storing phone numbers, and generates accordingly dual dialling sequences following the international CCITT telephony standard. While testing of ASIC designs is not present in most other ASIC design courses or takes only place at the functional level, we have integrated Design for Testability (DfT), structural digital testing and mixed-signal testing theory and practice into our course at a central part.

## 2. Testing the dial-memo IC

After the mixed-signal system and circuit design modules, the actual realisation of the dial-memo IC is tested in the last module. However, during these previous modules Design for Testability has not been neglected but been included into the design. In the test module a basic introduction to (digital) test pattern generation software and test



*B. Courtois et al. (eds.), Microelectronics Education,* 205–208.

equipment is given. The test equipment used consists of an IMS Logic Master ST System, together with a digital multimeter and a digitising oscilloscope.

The first part of the test module consists of the introduction to and the application of classical digital test pattern generators to test the digital part of the circuit for stuck-at faults. The second part, however, looks at mixed-signal test methods to see if the hardware realisation meets indeed the specifications of the CCITT standard.

**3. Testing the mixed-signal part of dial-memo IC**

The mixed-signal circuit (see *Figure 2*) in the dial-memo IC actually consists of two-tone generators (one for the lower and one for the higher frequency). In each tone generator a D/A converter (DAC) is needed. The students have to design these two generators according to the given CCITT standard, with the restriction that it should operate at a maximum clock frequency of 6.144 MHz. One other restriction is the power consumption, since the IC is intended to be used in a portable application (a key ring).

One of the main design choices for the tone generator is the number of bits needed in the DAC to meet the standard (3 bits will not be enough, but 4 bits will suffice). At the circuit module the students will again look at this block, and learn how to implement a simple DAC by means of current mirrors. Finally in the test module they have to test the whole IC, including the tone generator blocks.

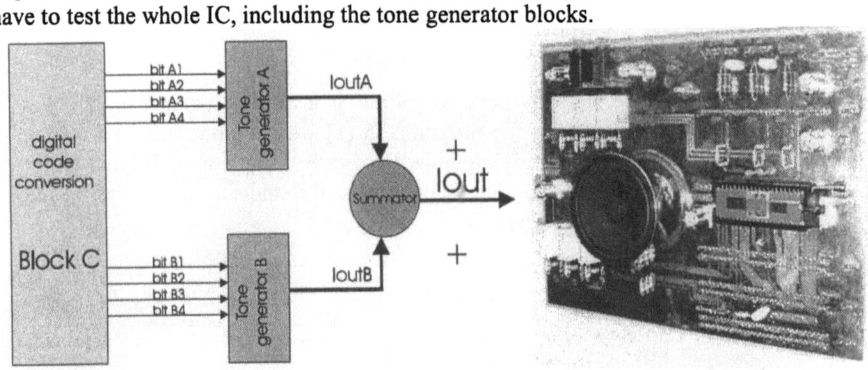


*Figure 2*: Scheme of the mixed-signal part of the dial-memo IC and test board

In order to test the tone generator block special precautions have been taken during the design phase to make also this part of the circuit testable. Between the digital parts and the tone generator block a set of flip-flops have been inserted, which are transparent during operational mode, but which are able to hold the bits (indicating the tone be generated) during test mode. Also a special test board has been made (*Figure 2*) by our technicians. This test board offers the possibility to measure the current of the whole IC, the current of the I/O pads on the IC, and the current of the IC core, to see if the power consumption meets the specs. Also it offers to measure and test the highest frequency, the lower one, or both.

We want to know and to measure if the dial-memo functions well and how well it functions. In order to know we will perform the following measurements:

- the effective AC output power,

- the output current and the current consumption to be able to derive the current-efficiency ($\eta$) (from which the power efficiency can be derived),
- the dynamic range (Signal Noise Ratio, SNR) (see *Figure 3*),
- the total harmonic distortion (THD).

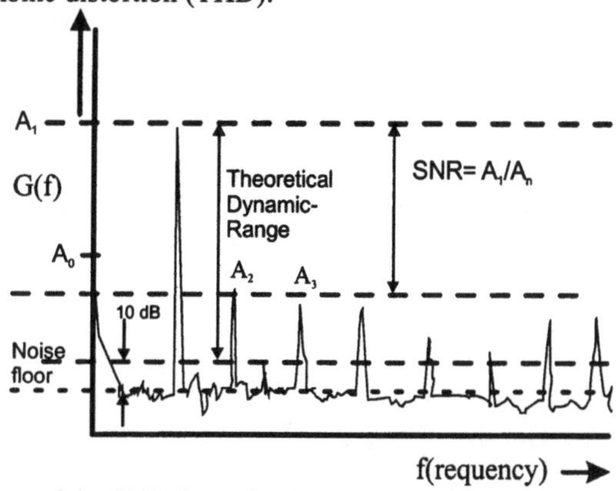


*Figure 3:* Extracting of the SNR from the DAC spectrum

Before a real test can be carried out, some investigations about the test itself must be considered. First of all a theoretical (simulated) test has to be executed to see whether unpredictable results and/or shortcomings may take place (e.g.: the cable impedance may interfere with the measurements or the amount of measurement points might be inadequate). Also with the simulated output, verification of the real measurements gives more insight how the actual test should be interpreted.

Like with digital testing an effort to detect realistic and feasible tests by means of simulation results is carried out and moreover, these results can help analysing and diagnosing the real electronic circuit under test.

The (simulation and) test–environtment consists of three commercially available CAD/CAT tools:

1. Compass with Logic Assistant (logicAsst) and Analog Assistant (AnaAsst),
2. Anacad with the Eldo circuit–simulator,
3. HP with Veetest.

As test vehicles during the simulation we use a 3-bits and 4-bits macro–model of DAC and a 4-bits component (transistor) level based model. The latter one corresponds quite well with the DAC as being realised in the dial-memo IC, which will be used afterwards for the real measurements.

The distortion is determined by:

1. the THD, which can be determined from the frequency spectrum, by performing a Fast Fourier Transformation on the output signal which is in the time-domain.
2. the residue method, in which the best-fitting carrier frequency component is subtracted from the output signal; the resulting signal (residue) is a measure for the distortion.

In *Figure 4* (left) the simulation output of the residue method on the 3-bits DAC macro and a first arbitrary guess for the carrier frequency are given in the same display. Now the students have to choose a suitable frequency, amplitude, phase and offset to match the carrier frequency and the output.

To operate the real test a virtual interface has been written in HP Vee (see *Figure 4*, right). This interface offers the students the possibility to type in a keyboard number, and to select if both tones or only the higher or lower one should be sampled. After the selection has been made, the tool will sample one or two tones, and will perform a Fourier analysis, from which both the THD and SNR can be derived, as indicated previously.

In the example *Figure 4* (right) keyboard number 1 has been pushed, and only the higher frequency is sampled. In the top left test information field the expected frequencies are given, which can be checked versus the obtain result.

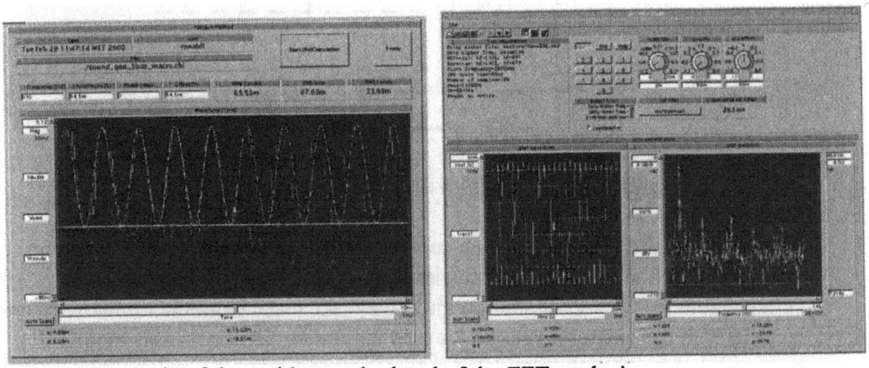

*Figure 4:* Example of the residue method and of the FFT analysis

**4. Conclusions**

In this paper we have shown in a detailed way that we have implemented mixed-signal testing as an integral part of our ASIC design course.

**5. Acknowledgements**

We would like to thank all the students, who have been trying out our ASIC design course experiment. Also we have to thank the technicians and students who implemented the test board and the HPVEE programs. Also we have to give acknowledgements to the Technical University of Delft for the distribution of their previous version of the ASIC design course.

**6. References**

**[1] The mixed-signal asic design course at Twente**
R.J.W.T. Tangelder, S.H. Gerez, H.G. Kerkhoff, E.A.M. Klumperink, J. Smit, H. Snijders, H. Speek, H. de Vries
*2nd European Workshop on Microelectronics Education, (c) Kluwer Academic Publishers*
Noordwijkerhout, The Netherlands, May 14-15 1998, pp.169-172

# MULTIDISCIPLINARY DSP-BASED CONTROL PROJECT

I. H. Al-Bahadly, and D. G. Bailey
*Institute of Information Sciences and Technology, Massey University, Palmerston North, New Zealand.*

## 1. Introduction

Traditional design and development procedures to produce a complete variable speed drive system typically involve at least, an electrical machine engineer, electronic engineer, programmer and possibly a control engineer. Communication delays between project team members and a lack of understanding of how the technologies interact, may result in a system that is delivered late, is inefficient or not easy to modify. This situation would be less likely to arise if all members of the team were educated and experienced in team and project organisation. This paper describes a group-based project laboratory for undergraduate technology students at Massey University. The project investigates the use of Digital Signal Processor (DSP) to control the speed of a DC motor. This undergraduate laboratory is an important teaching tool of exemplifying many engineering principles and introducing student to "real-life" practicalities in the area of industrial automation. It represents a typical modern industrial team project where several different technologies are involved and a variety of skills are required. The project offers the students a challenge involving many new ideas and concepts. It also gives them knowledge and new practical skills in the design and development of various technologies and how they interact to make a complete variable speed drive system.

Power electronics and DSP's together with appropriate software are extensively used for industrial automation. A particular example is their use for variable speed motor control, an area of rapid growth and importance [1]. Variable speed drive technology involves many diverse disciplines, power electronic circuits, electrical machines, microprocessors/DSP's, software and control theory. The motion control engineer today is indeed facing a challenge to maintain his/her knowledge and skills in this complex and ever-growing multidisciplinary technology. Motion control engineering as defined in Bose [2] is a specialisation of electrical machine drive engineering, and has grown at a rapid pace over the last two decades.

The project is offered to third-year undergraduate students in three different technology options at Massey University. All of the three technology programs are four-year courses. These options are Computer System Engineering (CSE), Engineering



*B. Courtois et al. (eds.), Microelectronics Education,* 209–212.

and Automation (EA), and Information Engineering (IE) [3]. Because of the different backgrounds and varied skills of the students participating in this laboratory project, it was decided to implement the design project component as a cooperative problem-based learning program [4] so that the students could work as a team to maximise their own and each others learning. The laboratory is structured as a group project and each group consists of a random selection of student from the three different options. Cooperative learning develops personal skills including conflict resolution and social skills as well as developing interdependence and individual and group accountability. Group project organisation of the laboratory has the additional advantage of reducing resource demand on the university.

## 2. Project

The project involves the use of a (TMS320C50) DSP board [5] to control a low power DC motor. The project work in each group is to be shared between two teams – one team concentrates on the hardware system, and the other team develops the software system. The students must also organise and maintain overall team management. Both teams must work together to integrate their solutions to produce a complete motor controller.

### 2.1. HARDWARE

As the DC motor being used for this project uses a permanent magnet instead of field coils to establish the required magnetic flux, speed control is achieved by switching on and off the voltage applied to the armature conductors. The mean armature current and hence the speed is controlled by varying the duty cycle. This type of controller is known as a pulse width modulator or voltage chopper. The form of control used for this project is known as a one-quadrant chopper controller as shown in the Fig. 1 [6].

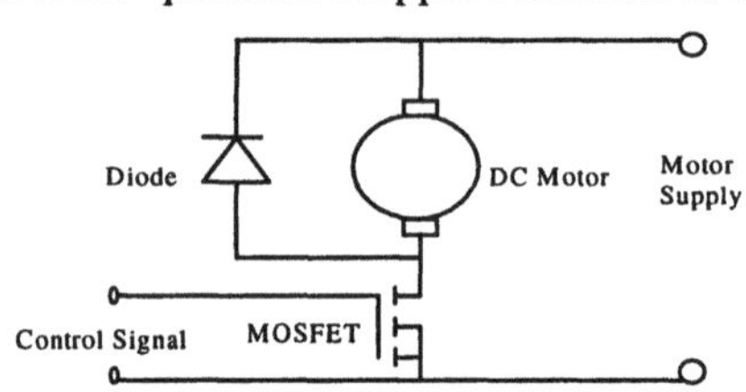


*Figure 1.* One-Quadrant chopper circuit

The hardware team tasks are to:

- Design an overcurrent protection circuit that overrides the control signal when the peak transistor current exceeds 2.5A.
- Combine the DSP controller signal and current limiting signal to switch the motor.
- Determine the frequency and duty cycle operating range of the motor the motor using a signal generator for the control signal. Fig.2 shows the complete setup of the project.

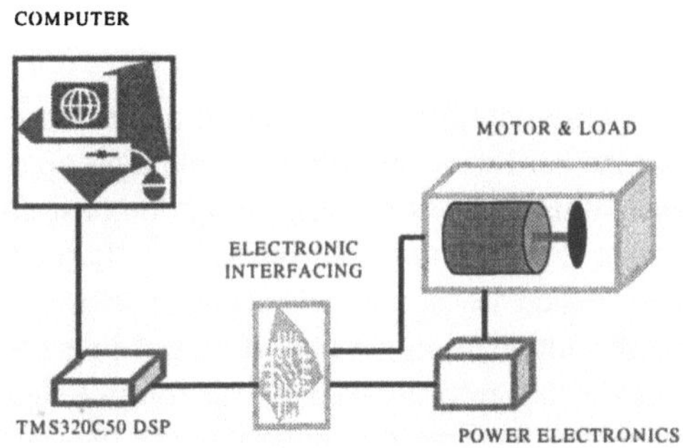


*Figure 2.* Project setup.

## 2.2. SOFTWARE

Microprocessors and Digital Signal Processors (DSP's) are being increasingly utilised for improving the performance of control systems making them more sophisticated and increasing their functionality. The students use a TMS 320 C50 as the primary control element in their project.

All group controllers were coded in the TMS assembly language and debugged using the DSK5D program supplied with the development board [7]. Additional debugging and testing was done on GO DSP Corporation's "Visual Development Environment Version 2.0 for the C5x series of DSP".

The software team tasks are to:

- Produce a program that provides PWM waveforms to control the DC motor from 0 to 2000rpm with no load as an open loop system.
- Develop a closed loop speed regulator to operate at fixed speeds (500, 1000 and 1500 rpm) as the loads change.
- Produce a fully automatic closed loop controller for full speed range irrespective of the loading conditions.

These controllers include a simple on-off controller (bang bang control), a sliding mode controller, a basic proportional controller, and a full PID controller [8]. The software team develops and integrates skills in control systems, DSP programming, and interrupt driven software design. The final controller involves a trade-off between speed of response, and stability of the controller. A block diagram representing the closed loop control system is given in Fig. 3.

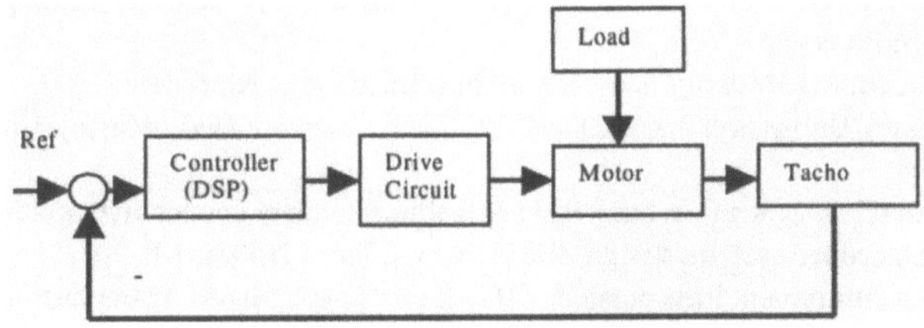


*Figure 3.* Closed loop control system

## 3. Assessment

Project assessment is based on three components: a group presentation; the standard of construction of both the hardware and software; and a group design report.

The group presentation consists of each group demonstrating their working design. This is judged against the objectives set out earlier. Part of the demonstration is an individual interview, which is used as an incentive to overcome the student's initial reluctance to cooperate and learn in groups [4]. In this interview, detailed aspects of the hardware and software design are asked of randomly selected individuals from the group. The combined results from this interview are then allocated to the group as a whole. This ensures that the group is responsible for the learning of the individual and for ensuring that each member is performing as a part of the team.

In the assessment of the standard of the construction, marks are given for tidy and neat layout of hardware components. The software should also be well-structured, documented, and easy to follow.

The formal design report documents the whole design process and should include ant theory used, their design of both hardware and software, the performance of their controller and any conclusions they have drawn from the project.

## 4. Conclusions

The students response to the project were positive and constructive. The most obvious course outcomes are a significant improvement in students' design abilities, and a development of group process skills as a result of their exposure to a "real-life" team practicalities. The students also gained an appreciation to the integration of different disciplines, and this was reflected in their willingness to learn and progress. The project offered the students a significant challenge, involving many new ideas and concepts. Yet, it also gave them knowledge and new practical skills in the design and development of various technologies and how they interact to make a complete variable speed drive system.

## 5. References


1. Bose, B. K.: *Power Electronics and Variable Frequency Drives: Technology and Applications*, IEEE Press, New York, 1997.
2. Bose, B. K.: *Modern Power Electronics*, IEEE Press, New York, 1992.
3. Massey University: *MASSEY UNIVERSITY Calendar 1999*, Massey University, Palmerston North, 1998.
4. Maskell D. L., and Grabau, P. J.: A multidisciplinary cooperative problem-based learning approach to embedded systems design, *IEEE Trans. Educ* **4** (1998), 101-103.
5. Texas Instrument Incorporated: *TMS320C5x User's Guide,* Texas Instruments, Texas, 1993.
6. Lander, C. W.: *Power Electronics*, 3rd Edition, McGraw-Hill, 1993.
7. Texas Instrument Incorporated: *TMS320C5x DSP Starter KIT User's Guide,* Texas Instruments, Texas, 1994.
8. Ogata, K.: *Modern Control Engineering*, Prentice-Hall, New Jersey, 1990.

# INTRODUCING GSM BASIC CIRCUITS PRINCIPLES

E. SICARD, S. DELMAS
*INSA/DGEI, 135 Avenue de Rangueil,*
*31077 TOULOUSE – France*
*email: etienne.sicard@insa-tlse.fr*
*web site: http://wwwdge.insa-tlse.fr/~etienne*

## 1. Introduction

The recent years have seen CMOS technology make successful inroads into radio-frequency circuits for GSM applications. The rising interest for miniaturized cellular telephone handsets has resulted in the development of new courses, tools and specific training for student engineers seeking for job opportunities in GSM IC design. Important job opportunities have attracted many students to the telecommunication world, with emphasis on mobile phones.

Although numerous books have appeared recently in this field [1,2,3], the normative approach and the complexity of the concepts have given rise to important needs for an educational approach to understand the basis of GSM.

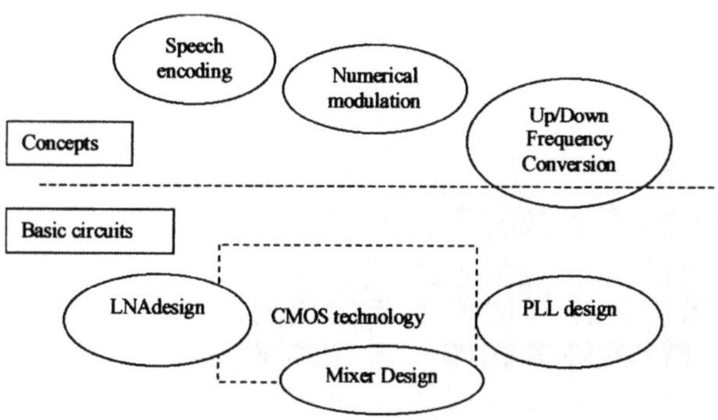


*Figure 1*: Key concepts in GSM applications and block designs addressed in this work



*B. Courtois et al. (eds.), Microelectronics Education,* 213–216.

Among the wide set of knowledge required to understand the mobile communication technology, we focus (Figure 1) on introductory concepts dealing with CMOS technology scale down, principles of GSMK modulation, speech encoding, emitters, receivers, down and up converters structures.

We developed simple interactive tools handling the illustration of these concepts, both form an abstract point of view and microelectronic implementation. These materials have been successfully used in introducing GSM concepts and design in undergraduate engineer teaching at INSA Toulouse.

## 2. Technology Scale Down

Firstly we illustrate the benefits of the technology scale down by simulating a voltage controlled oscillator thanks to a layout simulator [4] that can be configured in several technologies. In figure 2, we compare the performances of 0.5µm and 0.18µm technologies. It can be observed that although the supply voltage has decreased, the maximum operating frequency is rising to the requirements of the most recent GSM specifications (1.8 GHz for DCS 1800 and 1.9Ghz for UMTS2000).

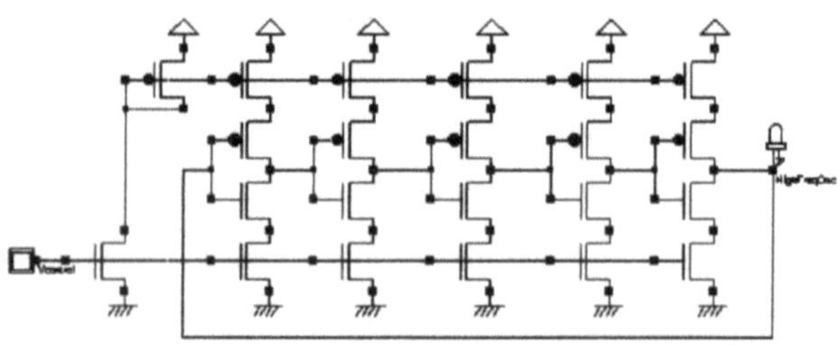

*(a)*

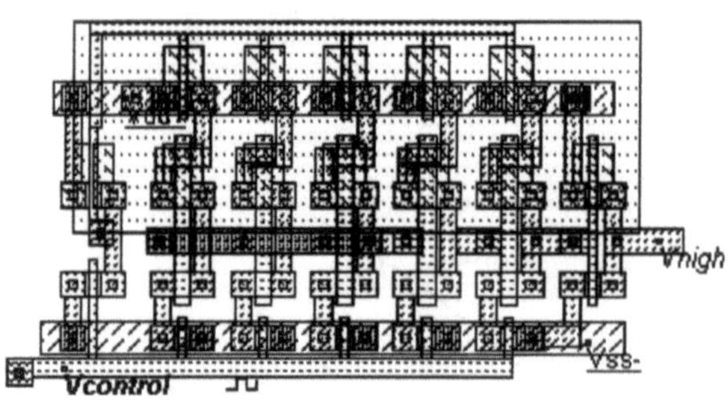


*(b)*

*Figure 2*: Schematic diagram (a) and scalable design (b) of a Voltage controlled oscillator

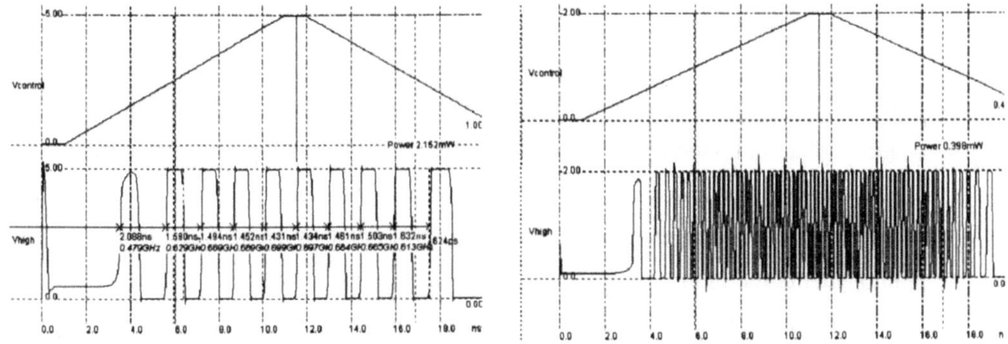


*Figure 3:* **Comparative simulation of a VCO in 0.6μm (left) and 0.18μm (right) technology**

## 3. Speech Encoding and Modulation

In GSM specifications, the speech encoding is required to compress the sampled data from 50Kb/s down to 12Kb/s [1], to enhance spectral efficiency and increase the number of simultaneous mobile communications in the same frequency channel. Therefore, the structure of the speech, the notions of fundamental harmonics, vowel must be introduced. Figure 4-a describes the spectral energy of A,I,U,E,O vowels. The compaction algorithm is not illustrated in our tool set.

The numerical modulation used in GSM, namely Gaussian Modulated Phase Shift Keying is illustrated using a combination of time domain and frequency domain tools. The time domain aspect of a series of bytes (00011011) is reported in Figure 4-b.

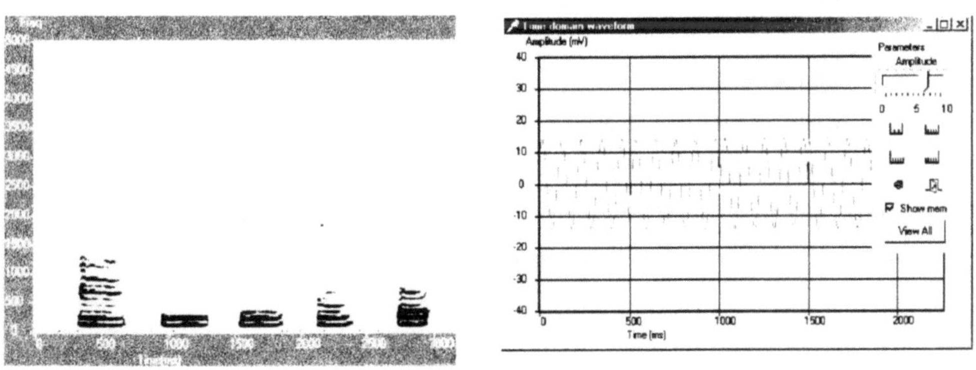


*Figure 4:* **Illustration of the vowel speech structure, and time domain aspect of the numerical modulation of speech encoding bytes**

## 4. RF Circuit Design

We illustrate the low noise amplifier (LNA) structure and function using a layout and analog simulation approach. Then the principles for up and down frequency conversion are explained both from an abstracted point of view, in time and frequency domain, and from a microelectronic point of view, with details on the implementation of mixers in CMOS technology.

A simple tool including sinusoidal wave generation, time domain representation and frequency domain calculation based on FFT has been developed for the understanding of mixer operation, with emphasis on up and down conversion operations.

The same operations are illustrated using Microwind2 at layout level, with the implementation of a CMOS mixer. The time domain simulation is performed using a SPICE like simulator linked to technology parameters and a Fourier transform illustrates the frequency shift.

## 5. Conclusion

Based on existing microelectronic layout tools, a set of tutorials has been developed to illustrate basic concepts linked to GSM principles and CMOS implementation of basic RF circuits. The course and practical training have been conducted successfully at INSA, with very positive feedback from students. Future developments will concern macro-bloc illustration of digital signal processing involved in GSM.

## 6. References

[1] Steele R. (1996) *Mobile Radio communications*, Wiley, ISBN 0-471-96419-0

[2] Znovar, Z. (1999) *GSM towards 3rd generation*, Kluwer Academic Publishers, ISBN 0-7923-8351-6

[3] Redl, S.M. (1995) *An introduction to GSM*, Artech House Publisher, ISBN 0-89006-785-6

[4] Sicard E. (1998) *Microwind2, an introduction to CMOS design on PC*, INSA Publisher

# IMPROVING STUDENT ASSESSMENT IN ELECTRONICS: OPEN EXAMS

S. DEMIDENKO
*Institute of Information Science & Technology, Massey University*
*Private bag 11222, Palmerston North, New Zealand, &*
*Department of Electronics & Communication Engineering,*
*Singapore Polytechnic, 500 Dover Road, Singapore*

## 1. Introduction

The system approach to improve quality of student learning [1] is an efficient methodology that can be applied to all the parts of the pedagogical triad: lecture-tutorial-laboratory/experimental work. The main components of the approach can be formulated as:

- to avoid overloading students with factual information and attempting to force them to learn too much too quickly;
- to limit the contents of the courses (as recommended, for example, by IEE [2]);
- to present content in a variety of ways;
- to use the released time to concentrate on the process, context, and implications of learning;
- to prepare learning materials that include an "experience" whenever possible;
- to introduce new material by linking or building on a student's previous knowledge and skills;
- to assess students on understanding rather than memorisation;
- to adopt the best practice from any source, and to be prepared to act on valid and reliable educational research.

This paper is related to two last items of list - the student assessment and the adoption of the best educational practice. Achieving both the goals can be facilitated by adoption of the *Open Exam* approach. The paper proposes to employ the approach for teaching electronic design, as well as discusses some its benefits and shortcomings.

## 2. Open Exam concept

The Open Exam concept has been proposed some years ago for teaching subject of mainly first years of the university undergraduate program [3]. The concept presumes that students sit for a timed paper under normal examination conditions. However, they have been given the *exact* text of each question with different numerical data or missed (blank) positions on the data sheets at the beginning of the course. *Students are told in advance of this arrangement.* To be successful a student needs to practice solving the



*B. Courtois et al. (eds.), Microelectronics Education,* 217–220.

question repeatedly for any arbitrary initial conditions. The goal of the approach is to combine the best features of formal examinations and coursework.
The main advantages of the open exams are:

- ensuring that students study the whole syllabus rather than guessing which particular topics would appear at the examination;
- motivating students to work steadily towards a known target by the certainty that efforts will directly lead to higher exam marks, without the influence of luck;
- better lesson attendance, higher student activity especially during tutorials and project consultations;
- substantially eliminating the stress of the examination for students;
- failing idle students, who might normally expect to earn some marks through ill-remembered definitions, descriptions and graphs;
- reducing the exam preparation and assessment workload on the lecturers.

In open exams, the students enter their answers into a simple answer script (ideally the script can done in the format ready for computer processing). At the same time, traditional answer sheet can be used either. The exam results are marked either in a traditional way, i.e., manually, or with the use of computer processing. Normally, full workings are also submitted together with the answer sheet to check how the students come to the particular results. A section scores full marks for a correct final solution, provided there is a full supporting worked solution. Usually no credit is given for partly correct solutions. In this respect, the approach is somewhat close to the multiple-choice answer question format. Obviously, the pre-programmed calculators have to be forbidden. It is possible to argue with the rule on the basis that the requirement for a fully worked solution reduces the advantage gained from the use of such devices. However, the power of today's hand-held devices should not be underestimated.

The open exam concept is obviously particularly attractive and effective for the assessment of the modules where mainly numerical and rather simple graphical solutions are possible. At the same time, it can be successfully employed in more sophisticated cases as well, including those related to the electronic design subjects. The latest is especially important, as the design is becoming a core of the engineering education. For example, the UK Engineering Council's policy document 'Standards and roles to registration' specifically requires the 'mandatory inclusion of design studies … as an expression of the practical application of theory and of accumulated experience' in all accredited engineering degree courses. The UK Design Council recommends that '...Engineering should be taught in the context of design, so that design is continuos thread running through the teaching of undergraduate engineering. Courses should expose the student to a proper mixture of analysis, synthesis, conceptual design and other wider issues' [4].

## 3. Results

The first factual report in the literature on the open exam implementation by Dr. P. Marshall and his colleagues from the University of Surrey [3] showed remarkable improvement in students' performance: an average of 77 marks compared with the usual

55 (with simpler questions). The author pointed that if the papers been marked in the usual way with some credit given to partly correct solutions the mark would have been even higher.

The results on implementation of elements of the concept in teaching design subjects in Singapore Polytechnic and Massey University also show its high efficiency. The components of the approach were tried on different design subjects taken from the teaching programs spreading from the 1st to the 4th years of study: Computer Aided Design (1st year, Electronics, Computer and Communication Engineering Diploma course, Singapore Polytechnic), Digital Electronics (2nd year, Mechanical and Manufacturing Engineering Diploma course, Singapore Polytechnic), Design on Computer & Communication Systems (3rd year, Information Engineering BTech Degree course, Massey University), Computer Systems Engineering (4th year, Computer Systems Engineering BTech Degree course, Massey University).

In all the cases, the students' performance has been improved rather significantly. For example, the results obtained in the Singapore Polytechnic while teaching Digital Electronics for dual-training students of Mechanical and Manufacturing Department (i.e., the subject was not in the mainstream of the specialization, and thus was of less interest to the students) were: the lowest score at the lab test – 67 marks, the highest – 100 marks, average – 91 (!). In comparison, the scores for this subject are usually around 10 marks lower if using a 'normal' way of teaching and assessment. The exam marks were also very satisfying indeed: the highest score was 93 marks, while the average - 72 marks. Only one student out of the group of 30 failed the sessional exam scoring 41 marks (the pass score was 50). However, he successfully passed a supplementary exam later with 63 marks.

Substantial improvement in the student performance has been registered also in a very hands-on oriented subject – Computer Aided Design. All assessments there were of an in-course type – there were two timed practical tests, the theoretical mid-semester test, and the term-long project. New course notes have been designed [5]. They included the practical tests samples of the same types and the same complexity as those of the real practical tests. Besides, several special confirmation exercises of different complexity were included into the textbook helping students to practice the design, to sort out typical design problems, and to prepare them to the assessment.

Partial implementation of the open exam concept in the Massey University for the 3rd–year subject Design of Computer and Communication Systems showed very significant improvement not only in terms of student performance (that can also be partially due to our substantial revamping the course), but also in terms of their competence in the design. This can be considered as a very important result since the design competence is becoming necessary feature of the engineering graduates. In fact, the Design Council Working Party stated that the 'Competence' was not intended to imply a necessarily high level of expertise, but rather that the level of *understanding and ability* an employer might reasonably expect of a young engineer who has fulfilled the CEng requirements. Thus, those involved could decide how best to set, achieve, and assess the required standards appropriate to the industry and field of expertise [4].

In overall, the student reaction to the open exam experiments emerges with a high interest to the subjects, good attendance at lectures (in many cases as high as 95%), hard working during class hours, active home reading beyond the set texts, low stress. In some cases, the distribution of students' performance can be somehow unusual – rather high number of students scoring full or nearly marks. Such a distribution reveals an apparent weakness in concept - a lack of discrimination for the most capable of the students. It has been right pointed [3] that comparison between the open examination marks and those in other subjects on a student by student basis shows that the greatest improvement in terms of marks is amongst middle-ranking students. This is very encouraging trend proving that the approach modifies and improves study habits of these previously middle-ranking students while almost by argument there are little room for improvement amongst the best students.

At the end of the day, the main criterion of the efficiency of the open exam approach is: do the students learn more with it? Are they more competent? The answer is very positive and encouraging. At the time of the test or examination the students know how to solve long and difficult problems covering the whole syllabus very well. Besides, the students formal (so-called SECAT – Student Evaluation of Courses, Administration and Teaching) and informal feedback supports the statement that the students have learnt more and their interest to the subject has been higher.

The second important criteria for judging the open exams approach is a staff workload. Here the savings can be very substantial. The approach leads to some sort of reusable examinations, as well as to the reduction in additional tutorial material, papers meetings and examination scripts marking. The savings in staff time could be as high as 20%-25% (depending on the program and class size). This result is especially important in the teaching environment of the technical universities and polytechnics where the academic staff is usually highly overloaded. Indeed, in the current situation in tertiary education in majority of the developed countries, with student/staff rate 15:1 and climbing, the desire for an optimal ratio of 8:1 looks just as an unrealizable hope.

The open exam concept is also very beneficial for monitoring the high quality of the course and for the degree accreditation with the professional institutions (IEE, IPENZ, IES) [2] as the course materials and contents remain of the same high standard once accredited.

Finally, the open exam concept is not something heterogeneous to the traditional educational approaches. In fact, its implementation enhances and supports the conventional teaching while substantially improving the quality of student learning.

## References


1. Button, B.L., Metcalfe, R., and Solomonides, I. P.: Improving Student Learning: Some Conceptual Dimensions - 3rd World Conference on Engineering Education, Portsmouth, 1992
2. Accreditation: Some Policy Statements. Course Content of Electrical/Electronic, Computer Related Engineering and Other Relevant Courses", IEE, London, Savoy Place, 1991
3. P. Marshall: Open Exams: An Experiment in Student Assessment - *Engineering Science and Education Journal*, **3** (1994), 15-20
4. Attaining Competence in Engineering Design – report of a joint Design Council/Engineering Council Working Party, UK design Council, 1991
5. Computer Aided Design. Workbook, 5th Edition, Singapore Polytechnic 1998

# A PROPOSAL OF THE CURRICULUM CONCERNING THE INFORMATION ENGINEERING FOR MICROELECTRONICS AND COMPUTER INDUSTRIES

YUMINOSUKE YANO
*Fujitsu Computer Technology Limited*
*2-15-16 Shin-yokohama, Kohoku-ku, Yokohama 222-0033, Japan*
*Phone: +81-45-475-8320, Fax: +81-45-475-8325*

## 1. Introduction

The microelectronics and computer industries in Japan generally hire the student graduated from the department of electrical, electronics, information engineering, mechanical engineering, physics and/or chemistry for hardware engineers and the student graduated from the department of electronics, information science and/or mathematics for software engineers. Many departments of information engineering introduce the CS Curriculum J90 based on the Proposals and Curriculum Recommendation for B.A. Degree Programs in Computer Science and Engineering made by the CS Working Group of the Curriculum Committee for B.A. Degree in Computer Science and Information Engineering, Information Processing Society of Japan in 1990. However, the CS Curriculum J90 seems to aim the computer software industry. The curriculum has to include the computer hardware and electronics studies. The microelectronics and computer industries need the student studying not only computer but also electronics. Usually the students study easy subjects for themselves, so they should study every required and/or electivesubjects for their intended courses.

## 2. Curriculum

### 2.1. GENERAL CULTURE SUBJECT

The student lacks in the study of general culture because the high school education aims at the entrance examination for the university. Table 1 shows required or elective subjects of the general culture subject, where the study time and self-study time are 15h and 30h for the lecture, 30h and 15h for the design or exercise and 45h for the experiment or practice.



*B. Courtois et al. (eds.), Microelectronics Education,* 221–224.

TABLE 1. General Culture Subject (Total studied unit: 30)

| Subjects | Unit/<br>R: Required<br>E: Elective | Study Grade and Weekly Period | | | | | | | |
|---|---|---|---|---|---|---|---|---|---|
| | | 1st. Grade | | 2nd. Grade | | 3rd. Grade | | 4th. Grade | |
| | | S1 | S2 | S1 | S2 | S1 | S2 | S1 | S2 |
| General Cultures | 16 | | | | | | | | |
| Philosophy/Ethics | 2x2/E | 1x2 | 1x2 | | | | | | |
| Psychology/ History | 2x2/E | 1x2 | 1x2 | | | | | | |
| Religion/Education | 2x2/E | 1x2 | 1x2 | | | | | | |
| Law/ Economics | 2x2/E | 1x2 | 1x2 | | | | | | |
| Literature/Logic | 2x2/E | 1x2 | 1x2 | | | | | | |
| Politics/Management | 2x2/E | 1x2 | 1x2 | | | | | | |
| Foreign Languages | 10 | | | | | | | | |
| English I/II/III | 2x3/R | 1x2 | 1x2 | 1 | 1 | | | | |
| Intensive English | 2/E | | | 1 | 1 | | | | |
| English Conversation I/II/III/IV | 2x4/E | 1 | 1 | 1 | 1 | 1 | 1 | 1 | 1 |
| German /French/Chinese/ Russian I/II | 2x8/E | 1x4 | 1x4 | 1x4 | 1x4 | | | | |
| Health and Physical Training | 4 | | | | | | | | |
| Human Body/Life and Health | 2x2/E | 1x2 | 1x2 | | | | | | |
| Sport Science | 2/E | 1 | 1 | | | | | | |
| Practical Training | 2/R | 1 | 1 | | | | | | |

## 2.2 COMMON BASIC SUBJECT

Some students do not study the physics or chemistry in the high school. Table 2 shows required or elective subjects of the common basic subject.

TABLE 2. Common Basic Subject (Total studied unit: 12)

| Subjects | Unit/<br>R: Required<br>E: Elective | Study Grade and Weekly Period | | | | | | | |
|---|---|---|---|---|---|---|---|---|---|
| | | 1st. Grade | | 2nd. Grade | | 3rd. Grade | | 4th. Grade | |
| | | S1 | S2 | S1 | S2 | S1 | S2 | S1 | S2 |
| Mathematics | 4 | | | | | | | | |
| Mathematics I/II | 2x2/R | 1 | 1 | | | | | | |
| Exercise I/II | 2x2/E | 1 | 1 | | | | | | |
| Physics | 4 | | | | | | | | |
| Physics I/II | 2x2/R | 1 | 1 | | | | | | |
| Exercise/Experiment | 2x2/E | 1 | 1 | | | | | | |
| Chemistry | 4 | | | | | | | | |
| Chemistry I/II | 2x2/R | 1 | 1 | | | | | | |
| Experiment | 2/E | 1 | 1 | | | | | | |

## 3.3. SPECIAL BASIC SUBJECT

TABLE 3shows required or elective subjects of the special basic subject.

TABLE 3. Special Basic Subject (Total studied unit/ Hardware: 30, Software: 26)

| Subjects | H: Hardware S: Software Unit/R/E | Study Grade and Weekly Period | | | | | | | |
|---|---|---|---|---|---|---|---|---|---|
| | | 1st. Grade | | 2nd. Grade | | 3rd. Grade | | 4th. Grade | |
| | | S1 | S2 | S1 | S2 | S1 | S2 | S1 | S2 |
| Basic Mathematics | 12 | | | | | | | | |
| Linear Algebra I/II | 2x2/R | | | 1 | 1 | | | | |
| Differential Equation | 2/R | | | 1 | | | | | |
| Complex Function | 2/R | | | 1 | | | | | |
| Statistics | 2/R | | | 1 | | | | | |
| Linear System Theory | 2/R | | | 1 | | | | | |
| Geometry I/II | 2x2/E | | | 1 | 1 | | | | |
| Applied Analysis | 2/E | | | | 1 | | | | |
| Basic Hardware | H:18 S:14 | | | | | | | | |
| Electro-magnetics I/II | 2x2/R | | | 1 | 1 | | | | |
| Exercises for Electromagnetics I/II | H:1x2/R S:E | | | 1 | 1 | | | | |
| Electric Circuit Theory I/II | 2x2/R | | | 1 | 1 | | | | |
| Exercises for Electric Circuits I/II | H:1x2/R S:E | | | 1 | 1 | | | | |
| Electronics Circuit Theory | 2/R | | | 1 | | | | | |
| Electronic Devices I | 2/R | | | 1 | | | | | |
| Information Theory | 2/R | | | 1 | | | | | |

## 3.4. SPECIAL SUBJECT

TABLE 4 shows required or elective subjects of the special subject.

TABLE 4. Special Subject (Total required unit/ Hardware: 68, Software: 72)

| Subjects | H: Hardware S: Software Unit/R/E | Study Grade and Weekly Period | | | | | | | |
|---|---|---|---|---|---|---|---|---|---|
| | | 1st. Grade | | 2nd. Grade | | 3rd. Grade | | 4th. Grade | |
| | | S1 | S2 | S1 | S2 | S1 | S2 | S1 | S2 |
| Extensive Mathematics | H:8 S:10 | | | | | | | | |
| Mathematical Logic | 2/R | | | | 1 | | | | |
| Numerical Methods I/II | 2x2/R | | | 1 | 1 | | | | |
| Probability Theory | 2/R | | | | 1 | | | | |
| Exercises for Applied Mathematics | S:2/R H:E | | | | 1 | | | | |
| Fundamental of Information | H:4 S:6 | | | | | | | | |
| Fundamental Information Processing | 2/R | | | 1 | | | | | |
| Code Theory | 2/R | | | 1 | | | | | |
| Language Theory | S:2/R H:E | | | | 1 | | | | |
| Hardware | H:24 S:16 | | | | | | | | |
| Electronic Devices II | H:2/R S:E | | | | 1 | | | | |
| Fundamental Measurement | 2/R | | | | 1 | | | | |
| Control Theory | H:2/R S:E | | | | | 1 | | | |
| Logic Circuits | 2/R | | | | | 1 | | | |
| Digital Circuits | :2/R | | | | | 1 | | | |
| Analogue Circuits | H:2/R S:E | | | | | 1 | | | |
| Computer Systems and Architecture | 2/R | | | | | 1 | | | |
| Experiments for Information Engineering | 4/R | | | | | 2 | 2 | | |
| Hardware Design Methodology I/II | 2x2/R | | | | | 1 | 1 | | |
| Electrical Digital Measurement | H:2/R S:E | | | | | | 1 | | |

TABLE 4. Special Subject (Cont'd)

| | | | | | | | | | |
|---|---|---|---|---|---|---|---|---|---|
| Software | H:6 S:18 | | | | | | | | |
| Algorithms I/II | 2x2/R | | | | | 1 | 1 | | |
| Exercises for Algorithms I/II | 1x2/R | | | | | 1 | 1 | | |
| Language Processing | S:2/R H:E | | | | | 1 | | | |
| Operating Systems | S:2/R H:E | | | | | 1 | | | |
| Exercises for Logic Programming | S:1/R H:E | | | | | | 1 | | |
| Exercises for System Programming | S:1/R H:E | | | | | | 1 | | |
| Software Design Methodology I/II | S:2x2/R H:E | | | | | 1 | 1 | | |
| Exercises for Applied Programming | S:2/R | | | | | | 1 | | |
| Systems | H:16 S:12 | | | | | | | | |
| Intelligent Information Processing | S:2/R H:E | | | | | | 1 | | |
| Database Theory | S:2/R H:E | | | | | | 1 | | |
| Digital Signal Processing | H:2/R S:E | | | | | 1 | | | |
| Acoustic Engineering | H:2/R S:E | | | | | | 1 | | |
| Image Science and Technology | 2/R | | | | | 1 | | | |
| Computer Network | 2/R | | | | | 1 | | | |
| Transmission Systems | H:2/R S:E | | | | | | 1 | | |
| Switching Systems | H:2/R S:E | | | | | | 1 | | |
| Human Interface | 2/R | | | | | 1 | | | |
| Reliability Engineering | 2/R | | | | | 1 | | | |
| Research Subjects | 10 | | | | | | | | |
| Seminar | 2/R | | | | | 1 | 1 | | |
| Graduate Research for B.A. Thesis | 8/R | | | | | | | * | * |
| Optional Subjects | | | | | | | | | |
| Medical Engineering | 2/E | | | | | | | 1 | |
| Computer Vision | 2/E | | | | | | | 1 | |
| Multimedia Engineering | 2/E | | | | | | | 1 | |
| Visual Communication | 2/E | | | | | | | 1 | |
| Artificial Intelligence | 2/E | | | | | | | 1 | |
| Labor Law | 2/E | | | | | | | 1 | |
| Patent Law | 2/E | | | | | | | 1 | |

## 3. Conclusion

(1) The microelectronics and computer industries need the student studying not only computer but also electronics engineering.

(2) The complete special basic subject is very effective to study the special subject.

(3) The systematic study for the special subject is necessary for the student proceeding to the hardware or software course.

## 5. References

Noguchi, S. (1991) Proposal s and Curriculum Recommendations for B.A. Degree Programs in Computer Science and Engineering, *Journal of Information Processing Society of Japan* Vol.32, No.10, 1079 – 1092

Kogakuin University (1998) *Handbook for Students*, Kogakuin University, Tokyo

# A GRADUATE PROGRAM IN INTEGRATED COMMUNICATION SYSTEMS

Ernesto Perea
EURECOM INSTITUTE
2229, route des Crêtes – BP193
06904 Sophia Antipolis
FRANCE

## 1. Introduction

Current Microelectronics technologies potentially allow complete system integration. In fact, yield is no longer the limiting factor. The new frontier is System Knowledge and Complexity Management.

Traditional Microelectronics curricula is centered on IC design based on specifications. System Architects and IC designers live independent lives. System Integration, if ever a reality, will change this state of affairs. In fact, these two independent profiles will necessarily converge in such a way that each domain will have an impact on the other through design constraints to be taken into account at the earliest stages.

EURECOM Institute, an industry-driven research establishment, is a joint initiative of ENST (France) and EPFL (Switzerland), and a consortium of European Industrial Partners. It proposes a Graduate curriculum in Integrated Communication Systems Engineering.

## 2. A new profile for engineers

The world of telecommunications is undergoing profound change. Sophisticated technologies are already available: optical fibers, integrated circuits. Digital networks can be interconnected, superimposed, globalized. The opening up of markets, deregulation and competition are the rule of the day.

Simple transmission and switching of messages is giving way to the offer of content and services. The aim is to offer the user a transparent network, accessible everywhere, that swiftly provides him with useful innovative services at a reasonable cost. The resulting communication systems are large-scale, dynamic, complex, heterogeneous systems which it is not easy to learn to operate and to design. High economic stakes are involved in such systems which, through their social impact, are laying the foundations of a new information society. To design and deploy future networks, the engineer becomes a systems architect skilled in the most sophisticated technological components, and with the ability to integrate them into coherent systems that satisfy user needs while complying with specific economic parameters and well-controlled time-frames.

This engineer-architect's profile corresponds to insistent demand from companies, which have to cope with the management of highly complex projects. For advanced education establishments, it is a challenge requiring technical excellence that cuts accross disciplinary boundaries and traditional standard teaching practices.



*B. Courtois et al. (eds.), Microelectronics Education,* 225–228.

**3. International dimension**

To respond to the globalization of networks, skills and economies, Eurécom is a truly international Institute. The engineers must understand cultural factors, conduct international technical and business negotiations, and be aware of the strategies of operators and their alliances. In this context, they are able to benefit from a job market that transcends the national borders of their diplomas. This transnational makeup is inherent in the nature of Eurécom's partners, in the origins of its students, in the recruiting pattern of its teachers.

**4. Industry**

In this sophisticated, rapidly-evolving technical area, the academic and industrial worlds must combine their efforts and transfers between them must gain pace. The trained engineers must be operational, fine-tuned to companies' needs and with a capacity for creative development. The companies in turn must become increasingly involved in the training of today's and tomorrow's executives. At Eurécom we work in partnership with our industrial members within a framework of dedicated cooperation. The Institute is located at the heart of a markedly telecommunications-oriented science park.

**5. Integration**

Integration is the art of combining the components of a system, adjusting them to their environment and ensuring they interact perfectly so that the integrated system meets user requirements and offers the expected services under acceptable economic and social conditions. It is essential for the engineer-architect to possess this capacity which gives him powers of synthetic and global vision. It is a capacity he can use at different levels (each component can be a system in itself) and which is tailored to the targeted user. The will for integration must be present right from the design stage of the system and must guide all the stages of its accomplishment.

**6. Admissions**

Several advanced European establishments have entrusted Eurécom with the responsibility of their students' final training period (three semesters) covering a communication systems engineer curriculum.
They are ENST, ENST Bretagne, INT, EPFL, EPFZ, Politecnico di Torino and, on an experimental basis, other European universities such as Linz and Helsinki.

## 7. The Faculty

The curriculum is taught by Institut Eurécom's full-time staff of professors recruited at international level, and by their assistants. Courses are also taught by professors from EPFL and ENST and by experts from the local or international scientific and industrial community.

## 8. Teaching method

Because of Eurécom's international vocation, half of the technical courses are in English. These courses cover aspects of communications systems and incorporate theory and hands-on experimentation. Eurécom draws extensively on its industrial contacts to define the courses, projects and case studies.

## 9. Curriculum

The curriculum is divided into 3 semesters beginning in March and ending in July the following year. Courses in economics, law, human sciences and modern languages complement the technical courses of the first two semesters. The third semester is devoted to an "industrial thesis" (six-month internship).

### 9.1 Semester 1

The aim of this semester is to give all the students a high level of knowledge in communication systems theory and the associated sciences and techniques.
To take into account the students' different origins, they are offered a broad menu of courses from which they can choose those ten that best complete their skills.
Typical courses include: Intelligent Systems, Distributed Advanced Programming, Network Security, Advanced Networking, Computer Networks, Real-Time Systems, Wireless LANs, Internet Application Layer Protocols, Image and Video Compression, Stochastic Processes, Digital Communication Theory, Advanced Signal Processing, Introduction to, Communications, Signal Modelling and Coding, Performance Evaluation and Cellular Radio

### 9.2 Semester 2

This semester is devoted to the option-specific courses, which provide more in-depth study of the chosen specialization. It also includes case studies guided by business experts. The students also do a 170-hour project on a subject usually proposed by Eurécom members, companies on the Sophia Antipolis site or other businesses. It can also be linked with the research activities of Institut Eurécom. Typical courses include:

9.2.1 *Corporate Communications:* Advanced Protocols and Algorithms, Network Management, Network Security, Network, Architecture and Services, Protocol Engineering and Network Technologies

9.2.2 *Mobile Communications*: Wireless Communications, Mobility Protocols, Speech Transmission, Mobile Communications Services, Management of Mobile Networks and Mobile Communication Systems

9.2.3 *Multimedia Communications:* Multimedia Networking, Multimedia Documents, Multimedia Technology, Speech & Sound Processing, Image Synthesis and Virtual Reality and Video Processing and Communications

9.3 Semester 3

This semester is reserved for the "industrial thesis", a six-month internship, usually in an industrial company, the subject of which is approved by a Eurécom professor.

This internship must have a content that is both industrial (professional) and academic (thesis). Each student is monitored by a supervisor from Eurécom and from the host company. On completion of his internship, the student submits a report and makes an oral presentation of his work before an examination board.

**10. Associated courses**

While students are studying at Institut Eurécom, they can also can follow a DEA postgraduate course (Diplôme d'Etudes Approfondies) at the University of Nice Sophia Antipolis. At present the courses offered are DEAs in Distributed Networks and Systems and the "ARAVIS" DEA (Image-Vision option). Some Eurécom courses are validated for the DEA, and additional 2 to 3-month courses count for an additional diploma.

**11. Diploma**

Successful students receive the engineering diploma of their school of origin. At the same time, Institut Eurécom awards the "Eurécom Certificate", accredited by the participating schools, according to a joint graduate system.

**12. Research and Environment**

Research subjects include radio transmission, digital cellular networks, wireless data transmission, radio local area networks, mobility, audio-visual communications, multimedia networks, communications support for dispersed groups, user centered design, man-machine interface, multimedia document indexing, network design, operation and management, network interconnection, broadband communications, network security, intelligent networks, etc.

Current projects include Software Radio Platform, Exploitation of UMTS platforms for experimenting process algorithms for mobile telecommunication systems, etc.

The industrial context in Sophia Antipolis includes a set of large companies, a number of small companies, an increasing number of start-ups, and top-quality research centers, working hand-in-hand with local organizations.

# MICROELECTRONICS

# TEACHING IN THE

# FUTURE

# CONCEPTS FOR TEACHING NEW CIRCUIT TECHNIQUES

B. VOSS, L. KABULEPA, M. SCHULZ, AND M. GLESNER
*Darmstadt University of Technology*
*Institute of Microelectronic Systems*
*D-64283 Darmstadt, Germany*
*email* voss@mes.tu-darmstadt.de

## 1. Introduction

If one introduces new circuit concepts there is always the problem of deciding on the depth of involvement in the new topic. On the one hand there could be a large potential of the new technique becoming state of the art of tomorrow, on the other hand it could prove to be not feasible and not worth the effort spent teaching students at this topic. A way out of this dilemma is to use such techniques in a project whose main aim is to teach the use of EDA tools or design flows.

In this paper we report on a project we did last year where we designed 4+4 ripple carry adders with different adiabatic logic families. The aim of the project was to teach a basic full custom design flow with the EDA tool "Design Framework II" made by Cadence Design Systems, Inc.. By the way students learned the concepts of adiabatic switching and the problems inherent in this technique.

After an introduction into the adiabatic switching principle we report about the project results and the experiences we made during the project.

## 2. Background

In conventional CMOS logic the load capacitance is charged to $V_{DD}$ via the PMOS block (pull up network) and discharged to GND via the NMOS block (pull down network). During the charging of the output, a charge $Q = C_L V_{DD}^2$ is flowing to the load capacity $C_L$. The power supply has to deliver this charge at the voltage $V_{DD}$. So the supplied energy is $QV_{DD} = C_L V_{DD}^2$. On $C_L$ only half of this energy is stored ($\frac{1}{2} C_L V_{DD}^2$). Because of the energy conservation the other half has to be dissipated by the pull up network. This amount of energy is dissipated regardless of the make-up of the network, the resistance of the pFETs or the time taken to complete the charging.

Discharging the load capacitance to ground effects the stored energy to be inevitably dissipated in the pull down network, because no energy can enter the ground rail ($Q \cdot V_{GND} = Q \cdot 0 = 0$). Because all charge is entering the



*B. Courtois et al. (eds.), Microelectronics Education,* 231–234.

circuit at the voltage $V_{DD}$ and exits it at 0V, the energy dissipation from delivery to removal of the charge is $C_L V_{DD}^2$. All of this energy is dissipated as heat.

Because the dissipated energy of a charging/discharging process is fixed at twice the signal energy in conventional CMOS, the only way to reduce the energy dissipation is to reduce the signal energy. Unfortunately, this reduction results in an increased sensitivity to background noise and thus increases the probability of malfunction.

However, if the charge transport is slowed down, energy that would otherwise be dissipated in the channel can be conserved for later reuse. This becomes clear if we consider the following. Instead of charging the output capacitance $C_L$ to $V_{DD}$, $C_L$ is first charged to $\frac{1}{2}V_{DD}$ thus dissipating an energy of $\frac{1}{2}C_L(\frac{V_{DD}}{2})^2 = \frac{1}{8}C_L V_{DD}^2$. If $C_L$ is then charged to $V_{DD}$, an energy of $\frac{1}{8}C_L V_{DD}^2$ is dissipated again, making the energy dissipation of the whole charging process $\frac{1}{4}C_L V_{DD}^2$. Making the steps infinitely small results in a charging ramp Voltage and the output capacitance is charged with a constant current. This current source delivers the charge $C_L V_{DD}$ over a period of time T. The dissipation through the channel resistance R is then:

$$
\begin{aligned}
E_{diss} &= P \cdot T = I^2 RT = \left(\frac{C_L V_{DD}}{T}\right)^2 RT \\
&= \left(\frac{RC_L}{T}\right) C_L V_{DD}^2
\end{aligned}
\tag{1}
$$

This shows that it is possible to charge and discharge a capacitance through a resistance while dissipating less than $C_L V_{DD}^2$ of energy. Making T very large, the dissipation can be reduced to an arbitrary degree. This is refered to as the principle of adiabatic charging. The term adiabatic is used to indicate that all charge transfer is done without generating heat. To the best of our knowledge, (Seitz *et al.*, 1985) were the first to formulate the relationship of equation (1) and to use the effect in practical digital circuits. The different adiabatic logics that have been developed until now can be classified as *fully adiabatic* logics (e.g. (Athas *et al.*, 1994), (Younis *et al.*, 1994)) and non-fully or *quasi-adiabatic* logics (e.g. (Athas *et al.*, 1997), (Maksimovic *et al.*, 1997)). The main difference is that the fully adiabatic logics use *reversible computing* where no information is erased which would result in dissipation[1]. Quasi-adiabatic logics just charge and discharge the capacitances slowly but do not use reversible logic.

The constant current charging needed for adiabatic switching is easily approximated using a sinusoidal power supply. To account for the then non-constant charge current, the dissipation of equation (1) must be multiplied by a constant shape factor $\xi$ (which takes the value $\frac{\pi^2}{8}$ for a sine-shaped current). Substituting $V_{DD}$ with the peak voltage of the power clock $\hat{V}_{pck}$,

[1]The theoretical limit on energy dissipation per logical step is $kTln2$ for logically irreversible operations (Landauer, 1961)

equation (1) becomes:

$$E_{diss} = 2\xi \left( \frac{RC_L}{\frac{1}{2}T} \right) C_L \hat{V}_{pck}^2$$
$$= 4\xi \left( RC_L f \right) C_L \hat{V}_{pck}^2 \qquad (2)$$

Because we consider a complete charging / discharging process, $\xi$ has to be multiplied by 2. T is the period of the sinusoidal power supply with frequency $f = \frac{1}{T}$. The sinusoidal power supply can be realized using an external inductor. Thus an LC resonant circuit with an resonance frequency of approximately $\frac{1}{\sqrt{LC}}$ is created and the energy is oscillated between the external inductor and the capacitances to be switched. Because the frequency of the power supply depends on the capacity to be switched, existing logics use the dual rail concept, where the switched capacity is nearly constant regardless of the input values.

## 3. Project task

The task given in the project was to compare different adiabatic logic families among each other and with static CMOS regarding the power consumption at given speeds. We had three groups each working on the project, each group being responsible for one logic family. This constellation had the following advantages:

- The groups had to work together to define a test vector sequence and the operating conditions.
- Each group wanted its "own logic" to be the best - so motivation was no problem.
- The students understand now the full custom design-flow *and* the principles of adiabatic logic.

At the end of the project each group had to deliver a written report and there was a discussion about the results.

## 4. Results

After the groups defined the strategy and the input patterns for the comparison, each of them chose a logic family. The constraints given were that the circuit should be easy to implement (thus each group chose a logic with a single phase or at most double phase power clock) and that a standard CMOS process is sufficient to fabricate the circuits. The following families were chosen:

- Energy Recovery Logic (ER-Logic) described in (Tzartzanis *et al.*, 1995)
- Pass-Transistor Adiabatic Logic (PAL) described in (Oklobdzija *et al.*, 1997)
- Clocked CMOS Adiabatic Logic (CAL) described in (Maksimovic *et al.*, 1997)

| Frequency | $E_{ER-Logic}$ [pJ] | $E_{PAL}$ [pJ] | $E_{CAL}$ [pJ] |
|---|---|---|---|
| 2,5 MHz | 0.80 | 4.52 | 28.90 |
| 7,5 MHz | 1.30 | 6.17 | 27.87 |
| 12,5 MHz | 1.32 | 6.28 | 31.68 |

TABLE 1. Energy needed by different logic styles for the addition of given test vectors

Each group simulated and optimised a 4 + 4 bit adder implemented with the chosen logic family and determined the power consumption of the circuit with the given input vectors. The results are summarized in table 1.
The discussions at the end of the project showed that every student who did the project understands now the full custom design-flow with its problems and advantages. Due to this project the students are now more sensitive to the problem of power dissipation and open to more possibilities to solve it.

## References


W. Athas, N. Tzartzanis, L. Svensson, L. Peterson, H. Li, X. Jiang, P. Wang, and W-C. Liu. Ac-1: A clock-powered microprocessor. In *Proc. of the International Symposium on Low-Power Electronics and Design*, Monterey, CA, August 1997.

W. Athas, L. J. Svensson, J. G. Koller, N. Tzartzanis, and E. Y.-C. Chou. Low-Power Digital Systems Based on Adiabatic-Switching Principles. *IEEE Transactions on Very Large Scale Integration*, 2(4):398 – 407, October 1994.

R. Landauer. Irreversibility and heat generation in the computing process. *IBM J. Res. Dev.*, 5:183 – 191, 1961.

D. Maksimovic, V. G. Oklobdzija, B. Nikolic, and K. W. Current. Clocked CMOS Adiabatic Logic with Integrated Single-Phase Power-Clock Supply: Experimental Results. In *International Symposium on Low Power Electronics and Design*, Monterey, CA, August 1997.

V. G. Oklobdzija, D. Maksimovic, and F. Lin. Pass-Transistor Adiabatic Logic Using Single Power-Clock Supply. *IEEE Transactions on Circuits and Systems - II*, 44(10):842 – 846, October 1997.

C. L. Seitz, A. H. Frey, S. Mattisson, S. D. Rabin, D. A. Speck, and J. L. A. van de Snepscheut. Hot-clock NMOS. In *Proc. 1985 Chapel Hill Conf. VLSI*, pages 1–17, 1985.

N. Tzartzanis and W. C. Athas. Design and Analysis of a Low-Power Energy-Recovery Adder. In *Proc. of the Fifth Great Lakes Symposium on VLSI Design*, IEEE Press, pages 66 – 69, Buffalo, NY, 1995.

S. G. Younis and T. F. Knight. Asymptotically zero energy split-level charge recovery logic. In *Int. Workshop Low Power Design*, pages 177 – 182, 1994.

# MANAGING MICROELECTRONICS EDUCATION FOR THE FUTURE AND TACKLING THE INFORMATION OVERFLOW

OLE OLESEN AND IVAN RING NIELSEN
*Centre for Integrated Electronics.*
*Technical University of Denmark, DK-2800 Kgs. Lyngby, Denmark.*
OO@it.dtu.dk, irn@technodata.dk

**Keywords:**

**NORCHIP, EUROCHIP, EUROPRACTICE, EuroTraining, Systems on Chips (SoC), embedded systems and multimedia**

## 1. Introduction

This paper is based on the progress made over the past 10 years by some of Europe's largest microelectronics training organisations. The paper commences with a short presentation on the background to and the need for an ongoing microelectronics training programme. It moves on to highlight new ideas and proposals for developing the technology transfer in the most effective way. The history of this paper is based on the NORCHIP, EUROCHIP, EUROPRACTICE and EuroTraining programmes – the key players being CMP and IMEC. Together these initiatives have formed the world's most comprehensive microelectronics training action, described in the references [1] to [4].

## 2. Motivation for continuing training:

According to Gordon Moore's law, microelectronics will continue to be an intrinsic part of the development of multilayer nano electronics and atomare electronics. On the application side, penetration of the technologies outside the electronics industry is growing faster than any other engineering area. This is due to developing information technology and the move towards nanoelectronics and optical communication; hence there is a need for extensive research and education. The most important new European initiatives are described in [5] to [12] and involve a broad spectrum of training, research and industrial initiatives.



*B. Courtois et al. (eds.), Microelectronics Education,* 235–238.

## 3. Future training

Future training will start at a considerably higher level to that of say 40 years ago. Today, six-year-old children have more computer literacy than the most advanced teachers of this time.

### 3.1 TRAINING WITH MULTIMEDIA AND DISTANCE LEARNING

Internet and distance learning methods – with slides, lecture notes, exercises and CD-ROMs – have already penetrated universities. Hannu Tenhunen [13] has presented an example of such an advanced curriculum. The methods are being enhanced by improved interactivity so that the student should be able to do most of his study using PC, CD-ROM and printer alone. If, as Tenhunen states, university education becomes more and more obsolete, there will be a growing need for continued education of a more practical nature.

The expectations placed on students now are much higher than in times when they could simply listen to lectures and read the textbook. Access to important periodicals and conference proceedings on the internet presents tremendous opportunities but also great difficulties; notably when it comes to choosing the most relevant information and building up narrow and deep competences. However, even with the evolution of multimedia study, it is expected that there will still be a need for lectures, group work and exercises. Students must be able to interact with other students and teachers for inspiration and to clarify misunderstandings as well as to gain hands-on experience. This is as true for microelectronics education: chip design, CAD tools, multimedia application and new technologies.

## 4. Leading the way in European training

Presenting EuroTraining, the latest phase of the Europe-wide microelectronics training action. EuroTraining topics focus on new technologies such as nanoelectronics, optical components, new transmission, systems on chips (SoC) and design reuse with large IP blocs and tools/methods. The paper will also cover the following areas:

### 4.1 EUROTRAINING OVERVIEW

EuroTraining is the name adopted for the new phase of the EU-supported EURO-PRACTICE Training & Best Practice Service, the one-stop shop for microelectronics training. The project maintains and develops the aims of the Training & Best Practice Service, to assess and then publicise the best and most relevant training in microsystems, sub-systems and microelectronics to engineers across Europe. The project is now in its fifth year of successful operation. The partners, comprising Technical University of Denmark, COREP (Politecnico di Torino) and Technology for Industry (UK), work alongside EUROPRACTICE Service Providers, Technology Transfer Centres and Course Providers to guarantee that courses match current needs. The current course portfolio focuses on ASICs, Packaging, Microsystems and Techno-economics.

### 4.2 THEMATIC TRAINING

As part of its current action EuroTraining is developing a series of training programmes under the heading of thematic training. Thematic training covers a selection of the most technically relevant courses either generated for a specific purpose or provided by leading European training organisations. Courses address hot topics and industry concerns and are offered at Awareness and Advanced levels. For example, EuroTraining 's microsystems programme includes scheduled and customised training for companies developing specific functionalities such as sensing, signal processing and actuating. EuroTraining is also introducing a series of advanced level techno-economic courses. These address issues such as: product development lifecycle and costing; value chain analysis; how to choose the most appropriate technology; how to choose suppliers; managing a high technology project; IPR.

### 4.3 MICROELECTRONICS COURSE DIRECTORY

EuroTraining is in the process of establishing a fully comprehensive microelectronics training reference for engineers and course providers. The Course Directory is a listing of courses with details supplied direct by European training organisations.

### 4.4 MOBILE & CUSTOMISED TRAINING

The Mobile & Customised Training Service organises structured courses tutored by experts to suit specific company needs. Courses can be in-house or at other suitable training premises in Europe.

### 4.5 QUALITY LABELLING

EuroTraining encourages course providers to apply to accredit their microelectronics training courses with the Quality Label. The Label is an EU-approved endorsement of high teaching standards and facilities, and a beneficial marker for both students and providers.

## 5. Training organisation

Realistic and effective training should be based on interaction between universities and industry as proposed by Hugo De Man [14], Ahmed Jerraya [15] and Benoit Clement [16]. The limiting factor in respect of industrial development is neither a lack of technology nor a lack of finances, but a lack of skilled personnel. This is the reason why training is given a high priority in most countries. For updated information about national key activities, please see the references [5] to [12]. They include the huge activities within Systems on Chip (SoC). Another important issue is Design reuse and the use of intellectual property (IP) as proposed by Gabrièle Saucier [17].

## 6. Conclusions

This paper presents the main technology transfer initiatives for the microelectronics industry. It examines industry and university functions and looks at a structure for future training. (Please see the references for detailed information). For the past ten years the leading partners in the initiatives have played an important role within national actions. European industry is now directly involved in projects and is influencing, for example, the university environment as regards tuition in ULSI-design and SoC using embedded systems both within hardware and software. Here, the technology transfer covers direct cooperation projects, awareness, demonstration projects as well as the delivery of the needed spectrum of qualified candidates. An important factor for future success is to make industry understand the opportunities and the industrial potential of advanced microelectronics and those areas outside traditional electronics. And technology moves fast so there is a continuous need for awareness. European coordination projects are essential to maintain momentum and ensure best use of resources. But it is essential to keep the action running smoothly both nationally and internationally; clearly some overall funding is needed. New methods for training will help the competitiveness of European microelectronics industry. The paper concludes with a look at the expanse of available information and how we can be selective.

## 7. Acknowledgement

The authors would like to thank the board members of NORCHIP, EUROCHIP EUROPRACTICE and EuroTraining for their excellent cooperation in connection with the set-up and running of the training programmes. The funding has been provided by the technology councils in various countries, the Esprit and IST programmes. DGIII's and DGXIII's interest in and support to the three EU projects has been invaluable.

## 8. References

[1] Olesen, Ole and Svensson, Christer: *"NORCHIP, A Silicon Brokers Model"*, Integration, The VLSI Journal 2, pp. 3-13, 1984.

[2] Olesen, Ole: *"Use of Fast Turn Around Chips at Nordic Universities"* G. Saucier ed. Fast Prototyping of VLSI, p. 273, North Holland 1987.

[3] Olesen, Ole: *Survey of Danish VLSI Teaching programs to Balance Engineering Education"*, Proceedings of the 3rd EUROCHIP workshop '92, October 1992.

[4] Olesen, Ole: *CAD TOOLS AND SERVICES FOR SME'S IN THE NORDIC COUNTRIES"*, CAVE'94, Sesimbra, Portugal, in the session: CAD Tools and Services for SMEs

[5] The Scottish SoC initiative http://www.cpa.ed.ac.uk/bulletin/1997-1998/04/news/01.html .

[6] IMEC http://www.imec.be/europractice/europractice.html .

[7] The Swedish Invest Initiative www.isa.se .

[8] INPG http://www.design-reuse.com/ .

[9] Berkeley Wireless Research Center, http://bwrc.eecs.berkeley.edu .

[10] IP conference in Edinburgh, 1-2 November 1999 http://www.ip99.com/ .

[11] IEEE NORCHIP http://www.technodata.dk/tc/norchip/ .

[12] ASIC personnel at DTU: http://cie.it.dtu.dk/asic/ .

[13] Tenhunen, Hannu & Isoaho, J.: *"New Computer Engineering Curriculum Development for System-on-Chip Era, in Proc of IEEE/ACM Frontiers in Education Conference"*, November 1999, Puerto Rico

[14] De Man, Hugo: *"Future Systems-on-a-Chip: Impact on Engineering Education"*, DAC 1997

[15] Moussa, Imed, Jerraya, Ahmed Amine etal.: *"Comparing RTL and Behavioral Design Methodologies in the Case of a 2M-Transistor ATM Shaper"*, DAC 1999

[16] Clement, Benoit, Pogodalla, Francois etal.: *"Fast Prototyping: a system design flow applied to a complex System-On-Chip multiprocessor design"*, DAC 1999

[17] Saucier, Gabrièle: *"IP Market Status (Summary)"*, IEEE NORCHIP '99

# A CASE STUDY IN VLSI DESIGN FOR COMMUNICATIONS: DESIGN OF A PIPELINED ADAPTIVE DFE FOR GMSK OVER INDOOR RADIO CHANNELS

L. D. KABULEPA, B. VOSS AND M. GLESNER

*Darmstadt University of Technology*
*Institute of Microlectronic Systems*
*Karlstr. 15, 64283 Darmstadt, Germany*
*email:*kabulepa@mes.tu-darmstadt.de

## 1. Introduction

Advances in microelectronics stemming from integrated circuit technology to novel system-level solutions are very promising for high speed wireless communications and mobile computing. However, a long transition time often occurs between the algorithm development and the final VLSI implementation because of the mismatch between algorithm-domain goal considerations and VLSI design requirements. While the algorithm designers focus on the performance improvement in terms of bit error rate (BER) or signal-to-noise ratio (SNR), the VLSI designers stress on implementation oriented issues such as speed, power consumption and chip area. Therefore, the skills in trading off VLSI design issues with performance constraints during the course of the VLSI implementation of digital signal processing and communications algorithms is of great interest. The education on these skills is usually not addressed in the existing courses of electronic and communications engineering.

In this paper, the design of a pipelined recursive least-squares (RLS) decision-feedback equalizer (DFE) for Gaussian Minimum Shift Keying (GMSK) modulation over indoor frequency-selective fading channels is considered. This example has been selected at Darmstadt University of Technology for a training project intended for advanced students in the area of microelectronics system design. The pedagogical objective of this project was to provide the students with a concrete application example involving a trade off between VLSI design issues and system performance.



*B. Courtois et al. (eds.), Microelectronics Education,* 239–242.

## 2. System Model

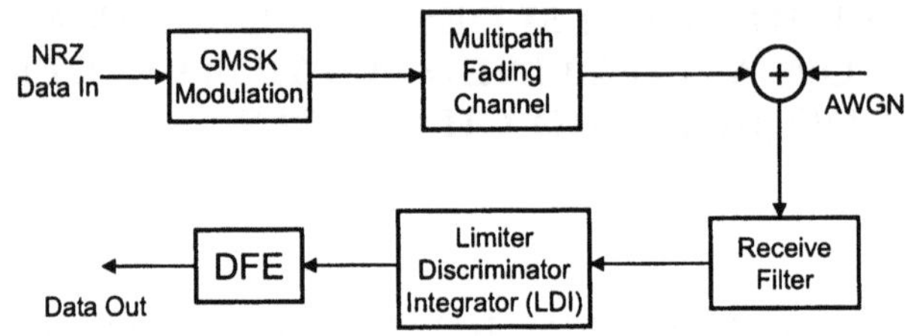


*Figure 1.* Equivalent baseband model of the data transmission system

The equivalent baseband system model used throughout our simulations in MATLAB is depicted in Figure 1. The GMSK signal at the modulator output can be represented as

$$s(t) = Re\{A \exp[j(2\pi f_c t + \phi(t))]\} \tag{1}$$

where $f_c$ is the carrier frequency, $A$ is the amplitude and $\phi(t)$ is the modulating phase given by

$$\phi(t) = 2\pi h \sum_{k=-\infty}^{\infty} a_k \int_{-\infty}^{\infty} g(\tau - kT) d\tau \tag{2}$$

In Equation (2), $h = 0.5$ is the modulation index, $a_k$ is the $k$th data symbol and $g(t)$ represents the Gaussian premodulation filter described by the product $BT$, where $B$ is the 3 dB bandwidth and $T$ is the symbol duration. The smaller $BT$ is the narrower the resulting spectrum and the greater the amount of intersymbol interference (ISI) introduced by the premodulation Gaussian filtering [2]. Assuming that no reliable line of sight (LOS) can be guaranteed, a Rayleigh fading channel model has been adopted. The frequency-selective channel can be represented as a tapped delay line [3]. The tap weighting coefficients are modeled as zero mean complex-valued stationary Gaussian random processes. For mitigating both ISI caused Gaussian filtering and multipath propagation, the DFE uses a recursive least squares (RLS) or Kalman type algorithm given by

$$k_N(t) = \frac{\lambda^{-1} R_{NN}^{-1}(t-1) U_N(t)}{1 + \lambda^{-1} U_N^T(t) R_{NN}^{-1}(t-1) U_N(t)} \tag{3}$$

$$e(t) = d(t) - W_N^T(t-1) U_N(t) \tag{4}$$

$$W_N(t) = W_N(t-1) + k_N(t) e(t) \tag{5}$$

$$R_{NN}^{-1}(t) = \lambda^{-1} R_{NN}^{-1}(t-1) - \lambda^{-1} k_N(t) U_N^T(t) R_{NN}^{-1}(t-1) \tag{6}$$

where $d(t)$ is the desired signal, $W_N(t)$ is the vector of weights, $U_N(t)$ is the vector of input signals which includes the received signals and the feedback signals in the case of DFE, $e(t)$ is the error and $R_{NN}$ is the input signals autocorrelation matrix. There are two recursive inner loop corresponding to the update of the weights vector $W_N(t)$ and the inverse of the autocorrelation matrix $R_{NN}^{-1}(t)$ in Equations 5 and 6. These loops impose a throughput bottleneck. By applying the relaxed look-ahead transformation [4], it is possible to break the inner loops and to introduce pipelining latches in them. The pipelined RLS algorithm is described by the following equations

$$k_N(t) = \frac{\lambda^{-1}R_{NN}^{-1}(t-M)U_N(t)}{\frac{1}{LA} + \lambda^{-1}U_N^T(t)R_{NN}^{-1}(t-M)U_N(t)} \quad (7)$$

$$e(t) = d(t) - W_N^T(t-M)U_N(t) \quad (8)$$

$$W_N(t) = W_N(t-M) + k_N(t)e(t) \quad (9)$$

$$R_{NN}^{-1}(t) = \lambda^{-1}R_{NN}^{-1}(t-M) - \lambda^{-1}k_N(t)U_N^T(t)R_{NN}^{-1}(t-M) \quad (10)$$

where the input signals are assumed to vary slowly over M samples and $\lambda \approx 1$. $LA$ is refered to as the look-ahead factor and is less than or equal to $M$. By adequate distribution of the introduced latches, the pipelined architecture can operate $M$ times faster. This increase in speed may be traded off for reducing either the power or the chip area by using such transformations as the folding technique [4].

## 3. Results

Since approximations have been made in the derivation of Equations 7 to 10, some performance degradation should be expected. Therefore, simulations have been conducted to verify the convergence of the pipelined RLS algorithm and to estimate its performance loss for different values of $LA$, different products $BT$ and different speedup factors in comparison to the serial RLS algorithm. The HIPERLAN/1 scenario ($BT = 0.3$, data rate = 23.5 Mb/s) has been considered for different rms delay spreads 25, 50, 75 and 100 ns. Results for DFE with 8 feedforward and 8 backwards coefficients are shown in Fig. 2. The performance loss in BER is plotted for M = 5, and LA = 3 and 5. The application of the relaxed look-ahead has resulted in the introduction of M latches or delay elements that has allowed to pipeline multiply-add operations needed for the considered algorithms.

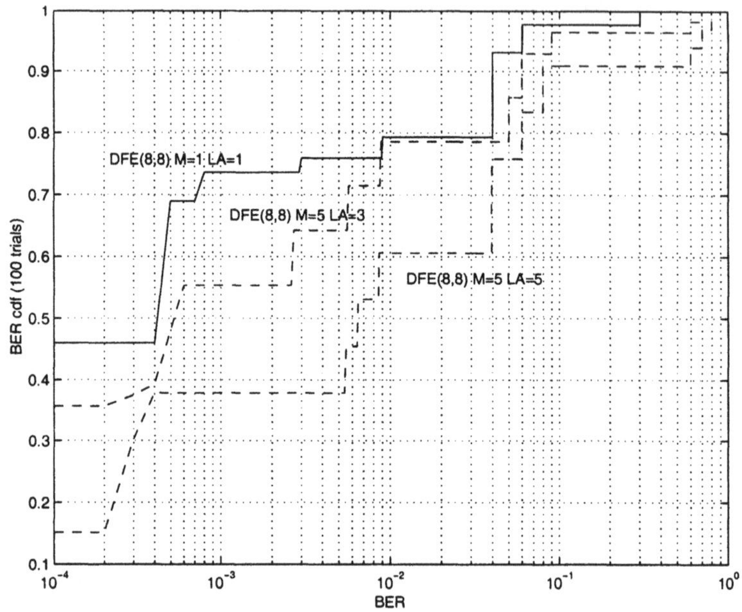


*Figure 2.* DFE(8,8) for BT = 0.3 and 100 ns rms delay spread ($E_b/N_o$ = 20 dB)

## 4. Conclusion

This project presents an illustration of an algorithmic transformation that aims at the improvement of the VLSI architecture for GMSK equalizer in an indoor radio environment. The algorithmic transformation is based upon the relaxed look-ahead technique. The students can realize how to fulfil design constraints while trading off performance loss with suitable approximation depth for a better hardware implementation of the algorithm.

## References

1. N. R. Shanbhag and K. K. Parhi, *Pipelined Adaptive Digital Filters*, Kluwer Academic Press, 1994.
2. T. S. Rappaport, *Wireless Communications*, Prentice Hall, NJ, 1996.
3. E. Biglieri, J. Proakis, S. Shamai, "Fading Channels: Information-Theoritic and Communications Aspects", *IEEE Trans. Information Theory*, vol. 44, no. 6, Oct. 1998, pp. 310-312.
4. N. R. Shanbhag and K. K. Parhi, "Relaxed Look-Ahead Pipelined Kalman Architecture", *Proc. 27th Asimolar Conference on Cir., Syst. and Comp.*, Pacific Grove, CA, 1993, pp. 1225-1229.

# Towards a Reusable SoC Design Methodology

*Experience from Project Alba*

Beth Martin , Katherine Dumont, and Stephen Winters
*Cadence Design Systems, Inc.*

Key words: System-on-a-chip, methodology, modular, reuse, documentation, Alba

Abstract: As industry generates and adopts complex methodologies for managing and designing system-on-a-chip (SoC) products, new challenges arise for effective knowledge transfer. This paper describes the specific challenges in developing the form and structure of a large information set. The information must be accessible and reusable for future development and delivery to other audiences. We present examples from our experience producing and transferring an SoC design methodology for an academic audience in a project with Scottish universities called Project Alba. We also discuss the tradeoffs and opportunities involved in a reuse approach to documenting and transferring information.

## 1. INTRODUCTION

The Design Methodology Engineering (DME) group of Cadence Design Systems authored a series of five design methodology guides with a total page count of nearly 7,400. The methodologies were developed in five sections, each building on the previous, and delivered at six month intervals. The subject matter covered includes timing driven design, block-based design, platform-based design for SoC, derivative SoC design, and deployment of design methods.

Methodologies are complex, task-based flows that are not necessarily linear. Methodologies are tool-independent, but tools and tool flows are required for practical application, and design examples are required to prove the concepts.



*B. Courtois et al. (eds.), Microelectronics Education,* 243–246.

The methodologies were created under Project Alba, with the Alba Centre as the end user. The Alba Centre grew out of an initiative by Scottish Enterprise, a government economic development body, to develop Scotland as a leading world location for system-level integration technology.

The Alba audience is comprised of masters-level students and faculty that might or might not have industrial design experience. Beyond Alba, the potential audience for methodology documentation and training includes Cadence design centers and external customers.

Because the documentation and training of the design methodologies *are* the product deliverables, they must be developed concurrently with the methodology. The development teams are design-task oriented, which lends itself to a modular approach, but which can produce many information chunks without a context.

The challenge of shaping the documentation and training grew with the size and complexity of the methodologies. The first methodology guide, *Timing Driven Design*, was about 300 pages long and contained only one project (module). The most recent deliverable, the *SoC Design Methodology Guide*, was over 3,400 pages long and contained seven modular projects.

# 2. CHALLENGES AND RESOLUTIONS

Just as management and reuse of IP becomes more important in SoC designs, management and reuse of the design knowledge also becomes crucial. The first challenge was to efficiently compile and structure a mass of "tribal" design knowledge ranging from basic design infrastructure to final verification and packaging. The second challenge was to organize the material in such a way that is was both usable to a broad audience of academics, and reusable for future development and delivery to other audiences.

Underlying those basic requirements are a number of challenges relevant to generating large knowledge, as listed below.

- Information management—how to manage complex content development?
- Information form and structure—how to combine methods, general implementation, tool flows, and design examples into one deliverable?
- Information reuse—how to deliver information that is useful to an academic audience yet reusable for industrial applications?

## 2.1 Information management

We adopted a top-down and bottom-up approach to information development. From the top-down perspective, we organized the books according to the high-level methodology flows. While form and structure was dictated at the top level, the technical content development proceeded at the module level, from the bottom up.

We then mapped those flows to the specific tool flows, keeping the methodology and implementation sections separate. We held frequent integration reviews to keep the two approaches aligned.

## 2.2 Information form and structure

We organized the information first from a top-level and defined the individual modules that would be developed. Within each module, the information is separated further into pure methodology, which is tool independent, and implementation, which contains a tool flow and detailed design examples. We used cross-references between modules as needed to reduce redundancy between modules. In the PDF versions, the cross-references are live hyperlinks.

Utmost importance was given to the only navigation aids a user has: the table of contents and the index. Each chapter within the books has it's own table of contents—generated to third-level headings—and index. A glossary for the entire guide, which is a series of books, is separate due to the very high repetition of key terms among modules. We also delivered a searchable index with the electronic versions.

## 2.3 Information reuse

The modular organization of the methodology guide lends itself to easy reuse. Each module is essentially complete as a stand-alone document, and the modules can be quickly reorganized and customized to fit the needs of future customers.

This approach to information design allows us to treat whole modules or specific parts of modules as discrete packets of information, without investing in the use of an object-oriented authoring language such as SGML. These packets will become off-the-shelf components for use in customized design methodologies and design environments.

# 3. EXAMPLES OF INFORMATION REUSE

This section describes the ways in which we reused the original information to create new documentation and training modules.

## 3.1 Creation of new information modules

Some design information spans across multiple modules. For tracking that information, we used FrameMaker's conditional text settings to tag topic and information types. For example, low-power design methods occur within the system design, hardware design, and embedded software design modules. To create a separate low power design chapter, we selected only text with the *lowpower* conditional setting from each chapter. After creating a stand-alone chapter, the file can be converted to a lecture format, in which the chapter headings become slide titles, and the first sentence of each paragraph become bullet points.

## 3.2 Creation of new information forms

The second deliverable document in Project Alba was *the Block-based Design Methodology Guide* (BBD). After publishing the BBD guide, the DME group created a BBD demonstration in a graphical flow-based tool that is capable of executing the design flow. The demo structure has the top-level design flow linked directly to design tool executables and design files. This is where the top-down and bottom-up development attained the most complete alignment. The methodology information was built into the tool as on-line help in HTML. The structure of the source material allowed for the material to be easily reused in this way.

Figure 1 shows the flow and implementation material as an executable flow diagram. Underlying methodology was linked to the appropriate boxes in the flow, as shown.

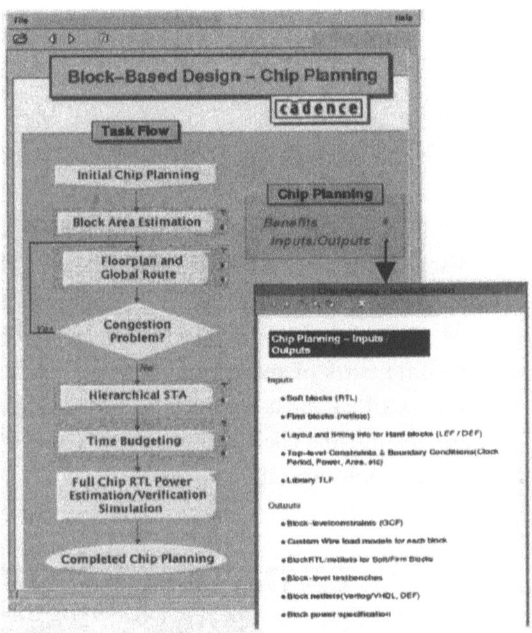


Figure 1: Design flow in executable demo form

## 4. CONCLUSION

This paper describes what we learned about structure and delivery of large knowledge sets through our engagement with Project Alba. A highly modular structure driven from a top-level organization allows for easy access to the information and lends itself to reuse opportunities. Managing the organization, consistency, and production of large books can be cumbersome. We found that keeping the concurrent top-down and bottom-up development in sync requires frequent reviews and communication between multiple project teams, as well as strong program management.

# DESIGN IN MICROELECTRONICS & COMPUTER ENGINEERING: TEACHING UNDERGRADUATES

S. DEMIDENKO, R. BROWNE, R. HODGSON
*Institute of Information Science & Technology, Massey University, Private Bag 11222, Palmerston North, New Zealand*

## 1. Introduction

Technology and knowledge along with labor and capital are intrinsic essential components of today's economies of the so-called *Knowledge-Based* type. Electronics is at the heart of innovations in the knowledge-based economy. At the forefront of development in electronics are advances in the design and implementation of circuits and systems employing full-custom, semi-custom and programmable VLSI technology. Electronic products, in particular those based on gate array technologies, have been extensively employed in the New Zealand economy for many years. A good example is field-programmable gate-array (FPGA) based implementation of the devices and systems in communications equipment. Another emerging area in New Zealand is the use of integrated microelectronic DSP devices in numerous applications. It is highly likely that such tendencies will become even more important in the future. This means that graduates in information and computer systems engineering must have knowledge of microelectronics technology, at least at the systems level.

The importance of microelectronic and computer education is related also to the fact that a substantial portion of the New Zealand graduates is traditionally employed overseas (in Australia, US, Europe). One of the recent surveys [3] shows that in terms of the likelihood of the well-educated staying in their native country, New Zealand is in the lower group of countries, between South Korea and China. This implies that the graduates have to be highly mobile and highly competitive on the international job market. This tendency provides an additional pressure on the level and contents of the engineering education in this country. To make the graduates successful and highly employable on the overseas scene the universities have to equip them with sufficient knowledge and skills in modern high technologies.

Massey University is one of the very few educational establishments in New Zealand that trains students in electronics design technology. The training is extended for both the final two years of the 4-year B.Eng. degree course in its two majors *Computer System Engineering* and *Information Engineering*. The electronics design training is mainly achieved through the following three teaching modules: *Design of Computer and Communication Systems* ($3^{rd}$ year), *Advanced Digital Technology* and *Computer Systems Engineering* (both - $4^{th}$ year). In addition, a number of other subjects that are taught during the two final years of degree study and in the postgraduate courses, also cover specific aspects of the electronics design.



*B. Courtois et al. (eds.), Microelectronics Education,* 247–250.

## 2. Curriculum and teaching methods

The heart of the Massey University educational model is the Academic Content of the degree. It has been established in accordance with Grouland's format [1] and Bloom's taxonomy [2]. The Academic Content is described by the Goals of Programs, Graduate Profile, Degree Regulations, and Practical Work Handbook, and for each option (or major) by the Option Profile, the Schedule, Subject Prescriptions and Subject Outline. The Subject Outline describes what is to be learnt by students, and this is best expressed as a set of intended learning outcomes. The following items are also presented: subject aims, syllabus, list of recommended texts, teaching methods, means of assessment, schedule, and pre- and co-requisites.

The learning outcomes of the module Advanced Digital Technology state that on the completion of the course students should be able to:

- understand technologies used in the production of VLSI devices;
- understand techniques used in the design of VLSI devices;
- be able to apply software packages to the design and simulation of VLSI devices;
- be able to incorporate VLSI devices into digital designs.

In addition to that, another final year design related subject - Computer Systems Engineering, covers hardware and software aspects of specific technologies including programmable array logic devices, field programmable gate arrays, one of the hardware description languages (HDLs), digital signal processing based systems, and several other topics.

The typical time budget for the subjects is 190 hours including: lectures – 30, tutorials – 10, project work – 120, and self-study –30. The individual and group project work supported by project centred lectures have been adopted in teaching the electronic design subjects. Intensive lecture blocks are delivered at the beginning of each major topic, thus providing the required theoretical base. The lecture blocks are reinforced though experimental and project work. This allows students to apply their knowledge in a real or simulated setting.

The lecture methods place responsibility for the development of students' knowledge primarily on a lecturer. Because the electronic design is wide ranging, the lecture materials covered in the subjects can be grouped into three main blocks:

- basics of electronic design for analog, digital and mixed-signal circuit, parameter calculations, design methods and tools.
- VLSI technologies, based principally on full custom and semi-custom CMOS devices, including a appreciation of the fabrication technology, Verilog HDL structural description, simulation, basics of logic synthesis, as well as built-in and scan based design for test.
- programmable arrays, PLD- and FPGA-based design, VHDL behavioural and structural descriptions, simulation, synthesis, CAD tools.

One of the main goals of the lecture courses is to provide the students with information sufficient for their project work. The other aim is essentially wider - to make the students aware of the spectrum of basic electronic design techniques and tools. This would enable them to efficiently communicate in their future work with the overseas third party designers (an approach that is widely used in New Zealand's electronic industry). In addition, the lecture course creates a background for independent advancement in the students' design qualifications should it become necessary in their employment. Students are required to attend the lectures in the beginning of each teaching block (several related topics grouped together), After that they are free to carry out the project work.

The project work usually comprises from 2 to 4 design assignments per subject. The complexity of the projects are steadily increased in the course of study. For example, in the

Advanced Digital Technology it progresses from creating a layout for simple logic circuits in the beginning of a semester, and up to designing of a large-scale single- or multi-chip sequential device at the semester end. The design has to be done for some pre-specified industrial grade technology (usually double-metal single-poly sub-micron CMOS). Normally, the second project has to include the following steps: obtaining the state diagram, state assignment table, initial and simplified Boolean expressions; drawing the schematic diagrams on the logic gate level and performing functional simulation in order to check the functionality and to adjust the timing characteristics; deriving structural HDL description of the circuit and performing simulation on HDL level, synthesis of the IC layout simulation on the layout level, adding input and output pads and final simulation for the completed circuit. In addition, the students have to evaluate parameters of the designed circuit (gate count, silicon size, power consumption, timing and signal characteristics), discuss the testing issues, and describe the technological processes involved in its manufacturing for the chosen technology.

For the design projects involving programmable gate array implementation, the behavioral HDL-based description is obtained first, the functional simulation and automatic synthesis are then performed, followed by the simulation on the physical level for chosen implementation (normally Xilinx FPGA 4000 series).

The following software tools are used for project work (all of them have been made available to students on the project dedicated computers): OrCAD 9.0 tools from Cadence, Xilinx Alliance and Foundation software, DSCH and Microwind packages from INSA, Evita VHDL from Aldec, and VASCO – new RTL VHDL synthesis training package from Mabex.

## 3. Assessment

In-course and end-of-course assessments are used to measure students' performance and to determine extend of the material understanding. In-course assessment includes submission of written report along with the working demonstration of the designed system and oral interview for each of 2-4 projects taken while studying a subject during a semester (60%-70% of total mark). End-of-course assessment is done by means of the semester examinations (30%-40%).

The in-course assessment is of the formative type, while the final examination is primarily summative. One of the objectives of the in-course assessment is to monitor the students' performance at mid-stream. It is done through formal written report submission, students' demonstration of the working project, and especially through the rather substantial individual project interview. The in-course assessment results are analysed during the semester, and adjustments are made if necessary.

The two hour final examination includes equal portions of theoretical and model-based problem solving questions, as well as the questions exploring student's ability to comprehend and explain modern electronic and computer design technologies and principles. Normally four or five equally weight questions are given. Out of them, the students have to answer either three or four questions. At least one of the questions is of the essay type on some specific electronic or design technology. A question asking for an explanation of the essence of built-in self-test and design for testability, as well as to characterise the most common fault models considered in functional testing, would be a good example of such an exam tasks. The other questions are related to practical problem solving in schematic or finite-state-machine-based design, in HDL-based description and synthesis, testability, physics of semiconductors. For example, a question could ask to write HDL description of a system that would be able to perform given functions, to derive and draw the schematic diagram as well as the layout circuit implementation, to find the minimum set of the test vectors required to detect given faults, and so on. Previous years examination papers are usually available to students through the library.

## 4. Feedback

The two-way feedback on the quality of teaching/learning is realised during the course. It provides students with the individualised written comments on areas of their weakness and suggestions for improvement. It is usually done as a part of the in-course assessment. In turn, the students have a chance to contribute into the teaching/learning quality audit through the student feedback scheme, called SECAT (Student Evaluation of Courses, Administration and Teaching). The scheme is an "end of process" tool as it is administered at the end of the course delivery. The student responses highlight problems in course delivery as well as any needs for participating staff to undergo personal development activities related to their teaching.

The results of the 1999 SECAT for the module Advanced Digital Technology shows strong student approval and interest to the subject. For example 75% of the surveyed students stated that the module has been well structured giving 3.75 as an average mark to the module structure (1 is the lowest mark, 5 is the highest). Almost the same was the students' opinion of the module organisation. Turning to the assessment procedure, 90% of students have found the assessment appropriate, and only 10% disagreed with it, while once again the average mark was 3.75. The differences between positive and negative evaluations were more substantial when the students presented their opinion about the workload – approximately 25% of the students disagreed with the given load, resulting in an average mark for this parameter of 3.6.

Even more interesting results were obtained from the SECAT evaluation of the other design subject – Design for Computer and Communication Systems. Besides the questions mentioned above on design and administration (the results were also very positive and rather close to those presented above), the students were asked to give their detailed comments on the module workload. Only 60% agreed that the subject required a reasonable amount of reading beyond the set text (average mark – 3.6), while 17% disagreed that the module covered the right amount of material. What was interesting was that the students voted strongly in favour of a substantial number of in-course assessments (80% of the students supported the idea, with 20% voting as indifferent), preferring however to have just one exam (95% - agreed, 5% - indifferent).

Overall, considering all of the parameters, 90-100% surveyed students rated all the subjects very high bringing the average approval mark to 4.2-4.3. Taking into account that it was the first offering for the subjects in their current form, the results are very encouraging indeed. This provided us with the incentive to bring them to this workshop for discussion – we hope the experience of the successful teaching of design in computer engineering and microelectronics in a country that does not have a strong microelectronics industry would be of interest to our colleagues in other universities.

## 5. References


1. N. E. Grouland: Measurement and Evaluation in Teaching, 5th Edition – Macmillan, NY
2. B. S. Bloom et al.: Taxonomy of Educational Objectives: Cognitive Domain – McKay, NY
3. Brain Drain, *The Economist* **352**, No 8127 (1999), 106

# MIXED-SIGNAL TEST TRAINING AT CRTC.

Y. BERTRAND, F. AZAÏS, M-L. FLOTTES and R. LORIVAL.
*Centre de Ressources de Test du CNFM (CRTC)*
*Laboratoire d'Informatique de Robotique et de Microélectronique de Montpellier (LIRMM), UMR CNRS/Université Montpellier II no 55060*
*161, rue Ada 34392 MONTPELLIER Cedex 5 France.*

## 1. Introduction

Profiting its experience in organizing test training (until now more than 280 people have been trained through 35 one-week training courses), the CRTC ("*Centre de Ressources de Test du CNFM*") proposes a new training dedicated to mixed-signal testing, in the field of both initial and continuing education. Using the mixed-signal testing capabilities of the HP83000 tester and exploiting its net accessibility, CRTC is able to deliver a complete mixed-signal test training, with remote test possibilities.
After a brief review of the running activities of CRTC, the paper gives the "philosophy" of mixed-signal testing and presents the content of the proposed training.

## 2. Digital test training at CRTC

CRTC was created by CNFM at the end of 1997 to meet industrial requirements for microelectronics engineers having the double Design & Test competency. CRTC is located in the CNFM center of Montpellier. It is hosted by LIRMM and benefits from the environment of an internationally renowned test team. CRTC is mainly equipped with a high-tech multimedia tester that allows the trainees to work in remote mode (Figure 1) [1]. This type of industrial test tool is used on production floor by several manufacturers worldwide. Its architecture and software organization makes it a good example of what an ATE is. Anyway, the basic concepts we teach during training courses are universal enough to be easily extended to any other ATE.
During the year 1998, the first task of CRTC was to ensure the education of future trainers for all the CNFM centers. So, up to 25 people have been trained at CRTC through 5 one-week training courses. As a consequence, for the past two years, several CNFM centers have organized remote training courses, thus allowing 180 students to be trained through 15 training courses.
Also, in 1998-99, we had to face a strong demand in the field of continuing education in terms of level 1 and level 2 training courses. So, about 80 technicians, engineers and service managers from circuit manufacturer industries have been trained through 15 one-



*B. Courtois et al. (eds.), Microelectronics Education*, 251–254.

week training courses. Such an activity strongly contributes in developing synergies between academic centers and microelectronics industries through the organization of partnerships, scientific projects and on-the-job training for students.

| 1998-1999 CRTC Educational Activity | # One-week training courses | # Trainees |
|---|---|---|
| Initial education | 15 | 180 |
| Continuing education (academic) | 5 | 25 |
| Continuing education (industrials) | 15 | 77 |
| **Total** | **35** | **282** |

The analysis of CRTC activity in both initial and continuing education domains demonstrates the success of this remote-learning experience. More than 280 people have been trained in 1998-99.

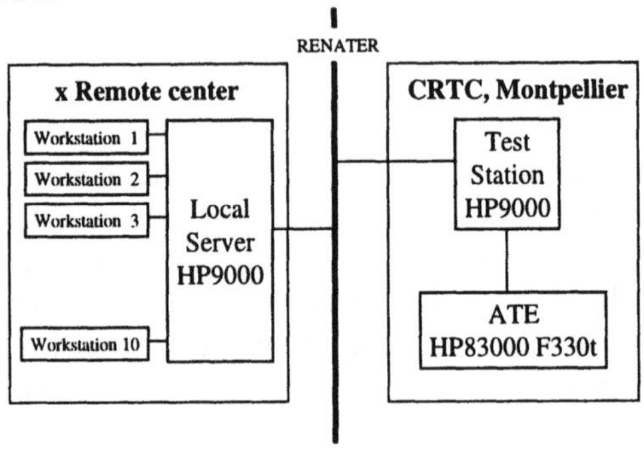


Figure 1: Network connection for remote education in IC testing

## 3. Mixed-signal test training

The sharp economic growth of the telecommunication and multimedia market makes the number of mixed-signal devices increases drastically: Converters, CODECs, RAMDACs, PLLs, audio- and video-ASICS, etc. The price of such circuits is dominated by the cost of performing production testing. The two most important factors affecting this cost are the direct cost of the test equipment and the time that a device spends on that equipment. Another factor is the indirect cost incurred when test development time becomes the critical step in the manufacture of new device [2]. Time-to-market reduction is a key issue for competitiveness in multimedia and telecom market. So, it is of prime importance for circuit manufacturers to have design engineers aware of mixed-signal testing issues and test engineers able to fully manage the test of analog and mixed-signal circuits. This explains the strong demand from industries for engineers having knowledge, skills and competencies on mixed-signal testing.

### 3.1 THE MIXED-SIGNAL TESTING PROBLEM

Of course, analog testing is very different from digital testing. Digital tests are based on logical comparisons when analog test needs both measurements and numerical comparison to a given tolerance band. So, mixed-signal testing requires instruments capable of sourcing analog and/or digital signals and analyzing analog and/or digital responses. Classically, the characterization parameters for the Device Under Test (DUT) are gain, phase, offset, non-linearity, distortion, crosstalk, etc.

In the traditional approach for automated analog test, the DUT input is fed by analog signal (DC, sine-wave, ramp, triangle-wave or any other type) and the analog response is measured and analyzed by analog instruments (voltmeters, amperemeters, phasemeter, etc.). In a production context, this approach is very time-consuming due to steady-state conditions and response settling requirements. In order to significantly reduce the test time, the nowadays approaches use the concept of DSP-based testing [3]. In this technique, the analog meters are replaced by a Digital Signal Processor (DSP) that emulates the measurement instruments. The concept of DSP-based test is given on figure 2. Arbitrary Waveform Generators (AWG) are used to generate various analog waveforms to stimulate the DUT. Waveform Digitizers (WD) are used to digitize and store the analog response from the DUT. They consists of an ADC in combination with a RAM. Of course, the test engineer needs to be familiar with the algorithms used for emulating these analog instruments.

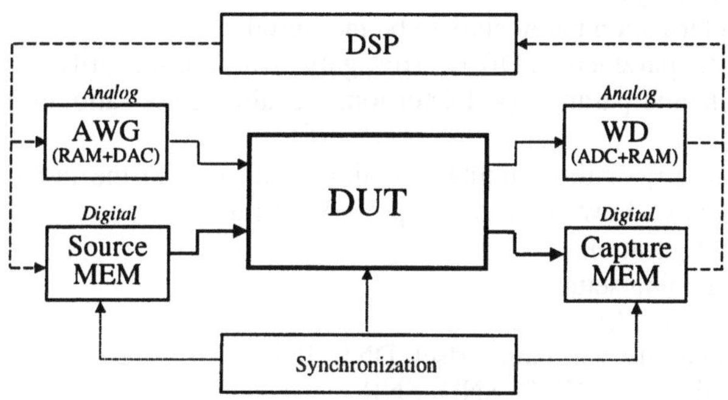


Figure 2: DSP-based testing principle

### 3.2 THE MIXED-SIGNAL TEST EQUIPMENT OF CRTC

The HP83000 F330t tester of CRTC is mainly designed for testing digital ICs in either engineering or production context. We have recently added new hardware and software resources (the so-called, SmartDSP sub-system) in order to provide it with the capability of testing mixed-signal devices. These additional functions are fully integrated into the test system and included in the overall test procedure of a device.

The SmartDSP subsystem enables to test a variety of multimedia devices, such as ADCs, DACs, CODECs, ASICS for audio and video application, and more. The

SmartDSP instruments for sourcing and analyzing analog signals, consist of boards directly plugged into the test head, namely:

- Smart Waveform Instrument (SWI) which can both sources waveforms to DUT, and analyzes the analog responses (2 channels, 1 Msamples/s)
- Smart Video Sampler (SVS) used to digitize and analyze high frequency analog signals (4 channels, 20 Msamples/s)
- Smart Capture Memory (SCM) used to capture and analyze digital responses (64K x 16-bits words at 50 Mwords/s).

### 3.3 TRAINING CONTENTS

The training course aims at giving the trainees the theoretical and practical knowledge about the fundamentals of mixed-signal testing. It consists of a tutorial followed by a set of labs.

The tutorial [4] aims at giving the basics of mixed-signal testing in order to make the trainee aware with the following points:

- Principle of the DSP-based testing
- Sampling and synchronization techniques
- Histogram-based test technique
- Fourier analysis
- Characterization parameters to be measured
    - Static parameters: Offset error, gain error, non-linearity, etc.
    - Dynamic parameters: Distortion, signal-to-noise ratio, effective number of bits, etc.

The lab part [5] of the training is mainly dedicated to the test of a 9-bits ADC (and DAC) and shows how to create setups and test functions:

- Pin configuration
- Level/Timing setups
- Mixed-signal file setup
- ADC linearity test: gain, offset, DNL, INL
- ADC Distortion: THD, SND, SNR

## 4. References


1. Y. Bertrand, F. Azaïs and R. Lorival: *Test Facilities with Distributed Remote Access for Initial and Continuing Education*, Proceedings of the SEMI Singapore 99 Conference, Singapore, 1999, pp. 65-70.
2. G.W. Roberts: *Re-Examine the Needs of the Mixed-Signal Test Community*, International Test Conference, 1995, p.298.
3. M. Mahoney: *DSP-Based Testing of Analog and Mixed-Signal Circuits*, IEEE Computer Society Press, New-York, 1995
4. HP83000 F330 System Training: *Mixed-Signal Fundamentals*, Hewlett-Packard GmbH, Böblingen Semiconductor Test Division, 1998.
5. HP83000 F330 System Training: *Mixed-Signal Testing: User Manual*, Hewlett-Packard GmbH, Böblingen Semiconductor Test Division, 1998.

# DESIGN AND

# TECHNOLOGICAL

# INNOVATIONS

# MULTIDISCIPLINARY COLLABORATIVE DESIGN COURSE FOR SYSTEM-ON-A-CHIP (*SOC*)


**R. L. Ewing**
*US Air Force Research Lab*
*Information Directorate*
*Wright-Patterson AFB*
*Dayton, Ohio 45433*

**G. B. Lamont**
*Air Force Institute of Technology*
*Graduate School of Engineering*
*Dept of Electrical & Computer Engr*
*Wright-Patterson AFB*
*Dayton, Ohio 45433*

**H. S. Abdel-Aty-Zohdy**
*Oakland University*
*Dept of Electrical & Sys Engr*
*Microelectronics Design Lab*
*Rochester, Michigan 48309*


## I. Introduction & Background

Collaborative technology is a rapidly evolving area, specifically in the arena of microelectronics. The vast developments in microelectronics and VLSIC have been changing the world's culture in many aspects since the inception of the Solid-State Transistor (1947). The following 20 years witnessed the initial IC concept as proposed by J. Kolby and R. Noyce (1956) and the resulting development of monolithic integrated circuits. Reduction in size and cost of microelectronic systems has been a consistency reoccurring phenomena. The 80's and 90's marked an era of more complex competition from the device/circuit level into effective use of design automation tools. This paper discusses contemporary integrated circuit development technologies and associated collaborative courses using a variety of methodologies including distributed education. The goal is to suggest techniques for proper identification, definition, and application of mixed-domain technologies for System-on-a-Chip (*SOC*) development involving multidisciplinary and collaborative design. These elements are crucial aspects in evolving a successful graduate class or classes on microsystems, and "smart system" integration in the new Millennium. This type of course is necessary since the new collaborative technology is creating paradigms for large complex microsystem design which requires that multidisciplinary skills be introduced and understood in academic, research and development environments.

## II. Course Objectives

The multidisciplinary topics (technologies, domains, and figures of merit) are integrated into a new collaborative class of Smart Perception Processing *SOC* products involving:

***Video Processing, Analog Design & High Speed Electronics, Digital Signal Processing, FPGAs, Opto-electronics, RF Design, Digital & Analog Controllers, Memory, Multilanguage Design Systems Microsystems, Adaptive Bus Structures & System Communication Protocols, Smart Power, and Man-machine Interaction***

Not only should students develop fundamental understanding of each topic, but also the course should motivate students to design nontrivial *SOC*. While the multi-technology topics represent significant advance in the state-of-the-art for smart



*B. Courtois et al. (eds.), Microelectronics Education,* 257–260.

perception micro-processing systems, micro-processing algorithms, to be embedded in the microelectronics, are equally significant. Standard design is generally organized as a series of discrete stages beginning with an algorithm and specification development, progressing through separate hardware and software design and prototype development phases, and finally to testing and verification. Separate paths occur for software and circuit development, and likewise, after the initial specification is approved, changes are both difficult and risky. Compounding this issue, is the fact that the engineering teams involved in the process rarely share a single, unified vision of the system under development.

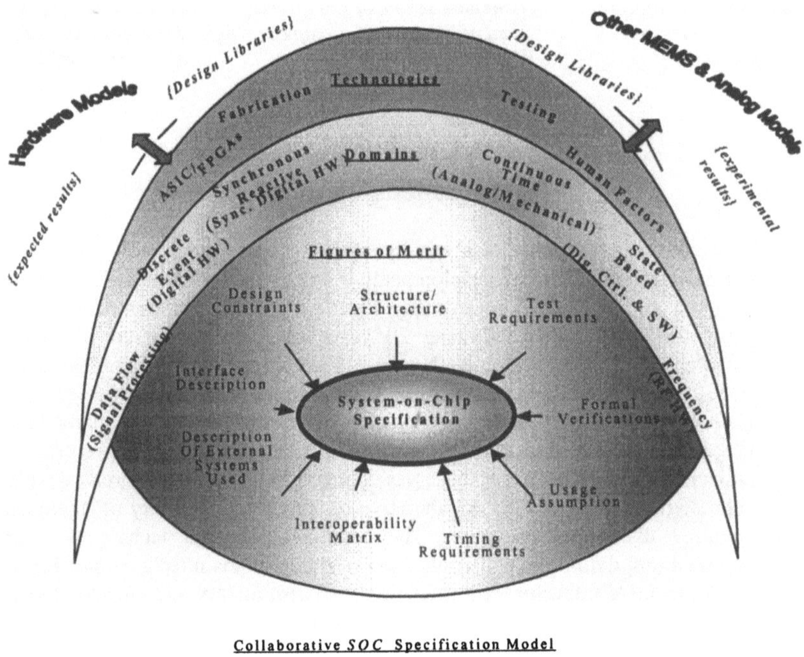


Figure 1. Collaborative Methodology for System-on-Chip Specifications

Fig. 1 attempts to unify this vision, by the use of a collaborative system-on-chip specification model. This specification model, shows three tiers; ***Technologies, Domains, and Figures of Merit***. The focus of the model is the system-on-chip specification. Necessary threads to accomplish the design ***Figures of Merit* are attained by team efforts. Within this environment, the design team creates a model of the system-on-chip** for the given ***Domains***, and using the target ***Technology***. Using this model, other design teams can collaborate through distributed multimedia tools. However, they must ensure the design compatibility and consistency in the ultimate system-on-a-chip implementation [1,2]. Also, the concept of testing the model with real world data is critical to its success.

As a practical example of collaborative smart design process, consider signal-processing mechanisms including wavelet transforms for video and audio compression as shown in Fig. 2. The key components of this collaborative system are the hardware implementation of the wavelet transforms and genetic algorithms [3,4] with embedded software. While developing this first generation of audio/video wavelets, the design team used Simulink (Mathworks, Inc.) which allowed the design team to focus on the development of the algorithms by modeling and simulation. Further collaborative tools now allow the design team to assist in rapid VHDL prototyping and hardware-in-the-loop testing of control and the wavelet transform. The wavelet transform [5], in addition to providing high-speed video/audio, also provide the ability of performing multi-resolution data mining algorithms and intelligent sensor fusion. The overall system (Fig. 2, upper right figure) consists of detector arrays integrated with CMOS circuitry to enable real-time "smart pixel" processing followed by a digital signal processor embedded with wavelet transforms and neural net based algorithms. By integrating optics, Micro-Electro-Mechanical-Systems (MEMS) based wavelength filters, image detection arrays , and specialized processing circuitry into a single compact integrated system, a powerful and robust image processing hardware, with capabilities that have not previously been demonstrated, is presented.

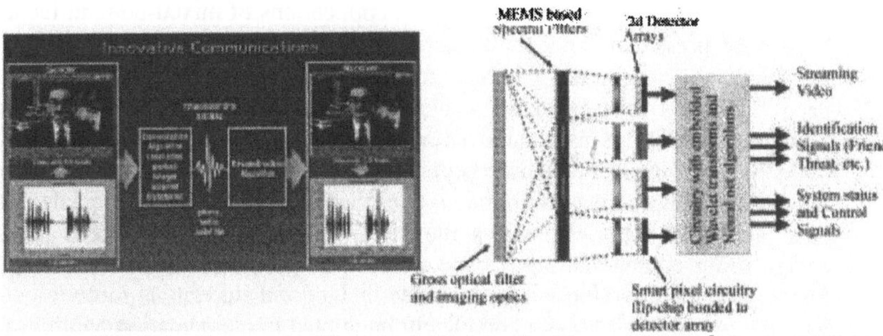


FPGA Hardware Implementation of Wavelet (Audio/Video) Algorithm {Shown on Left Side}, and Complete Multidisciplinary Smart Perception Processing {Shown on Right Side}

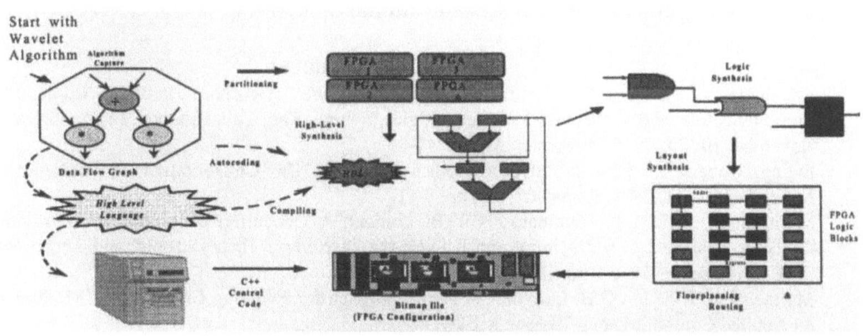


... And Finally A Programmed FPGA Wavelet Algorithm

Figure 2. Wavelet Technique and FPGA Implementation and Testing of Collaborative Design of Complete Multidisciplinary Smart Perception Processing

Other examples of collaborative and multidisciplinary SOC designs include: 1) Collaborative technologies for biomedical applications, with on-chip neural networks and genetic algorithms; 2) Environmental safety integrated SOC; 3) Complete RF transceiver system. Currently, the authors are involved in collaborative smart perception processing design and implementation of the aforementioned projects. Time and money are essential to develop the desired smart perception processing *SOC* with distributed collaborative design. It is believed that, successful collaborative SOC is possible by: (1) First, focus on a specific problem domain; (2) Second, generalize the methodology, which involves tool integration. The following summary is a guideline for collaborative interoperability tool integration, essentially addressing critical concepts and approaches for the three tiers (Fig.1): Figure of Merit , Domains, and Technology. In summary, for collaborative design flow, the *Figure of Merit* should address the signal flow performance factors of ; design capture to simulation, design capture to synthesis, synthesis to physical design, physical design to simulation. *Domains* must consider; software to compare simulated expected response, waveform w/simulated synthesized and actual waveforms, and data flow to commercial tools. *Technology* must consider; limited sets of models for given submicron technology, and design compatibility.

## III. Conclusions

Initial identifications, definitions, and applications of mixed-domain technologies have been presented for a multidisciplinary and collaborative *SOC* design course. An application example of image processing, using wavelet techniques, has also been elaborated upon. Our associated video network class involves a number of academic institutions associated with the Dayton Area Graduate Studies Institute (DAGSI) and Oakland University, Michigan. Such a facility can provide an interactive classroom environment for the *SOC* lectures. Associate hands-on pedagogical laboratories involve the implementation of the selected pedagogical design using commercial and academic software development tools as explored in Fig. 1. Collaborative interaction between faculty and students at remote locations is required in order to obtain a completely integrated implementation Multidisciplinary Smart Perception Processing *SOC* at the conclusion of the course. Expectations for the proposed courses rest on the educational objectives.

### IV. References


1. Abdel-Aty-Zohdy, Hoda S., "Neural Network and Genetic Algorithm Hardware Implementations for Communicative and Collaborative Technologies," Proc. The Collaborative Technologies Workshop, November 10 –11, 1999, Rochester, MI, pp. 40-45.
2. Ewing, Robert L., "Collaborative Microdesign," Proc. The Collaborative Technologies Workshop, November 10 –11, 1999, Rochester, MI, pp. 8-11.
3. Marmelstein, R.E., L.P. Hammack, & G.B. Lamont,"A Concurrent Approach for Evolving Compact Decision Rule Sets", Data Mining and Knowledge Discovery: Theory, Tools, and Technology, Part of SPIE's Aerosense '99, 5-9 April 1999, Orlando, Florida.
4. Marmelstein, R.E. and G.B. Lamont, "A New Approach for Evolving Clusters", ACM 1999 Symposium on Applied Computing, Feb 28-Mar 3, 1999.
5. Chengjiang Lin, Bo Zhang, and Yuan F. Zheng, "Packed integer wavelet transform constructed by lifting scheme," Proc. 1999 IEEE International Conf. on Acoustics, Speech, and Signal Processing, March 15-19, 1999, Phoenix, AZ, pp. 3177-3180.

# INTRODUCING DIGITAL CIRCUITS DESIGN AND FORMAL VERIFICATION CONCURRENTLY

Paul Amblard,
Fabienne Lagnier,
Michel Levy
Dept. of Computer Science, University of GRENOBLE, FRANCE
U.F.R.I.M.A, 60 rue de la Chimie, Dom. Univ.
F38400 ST MARTIN D'HERES
(Paul.Amblard, Fabienne.Lagnier, Michel.Levy)@imag.fr

**Abstract** This paper presents a curriculum where formal verification of digital designs is introduced concurrently with the basic techniques.

## 1. INTRODUCTION

The need of formal proofs in digital circuit design is now well established. This topic seemed to be unrealistic some years ago, but the main C.A.D. tools now contain a formal verifier. It becomes necessary to have new approaches in microelectronic education.

In the department of Computer Science of the University of Grenoble, we have introduced the use of such formal techniques. The originality of this work is the fact that this introduction is done concurrently with the introduction of design techniques. We teach at the same time digital circuits design and their proofs.

The paper first presents the context in which this work is done, then the second section describes the tools and technical background. The third section gives details about the curriculum and an example.



*B. Courtois et al. (eds.), Microelectronics Education,* 261–264.

## 2. MOTIVATION AND CONTEXT

Because of the challenge of designing complex circuits with short time-to-market constraints, we need to design correctly "at the first time". Formal tools help the designer in such a task. Many of these tools are based on Binary Decision Diagramms representation of boolean functions (B.D.D). [1] This activity must be prepared in Microelectronics Education.

At least two educational approaches must be considered :

• First, students learn the basic techniques of digital design with TTL74 or CMOS4000 family labs, in a subsequent term they begin to use VHDL and the so-called "high-level synthesis" approach. In a third step formal techniques are introduced.

• Basic design techniques (Karnaugh maps, automata synthesis) are introduced, and the circuits thus obtained are "checked" with a specification. Then high-level techniques are compared with low-level ones. The description in terms of registers, memories, adders and blocks is automatically translated in terms of gates and flip-flops and can be compared with a boolean specification.

Our work is situated in the second approach. There are many reasons :

• Our curriculum is in a department of Computer Science. Some of our students will be circuits designers, some of them will be tools designers. They must have a user knowledge of formal tools.

• Our students (in their third University year) are beginners in circuit design. For 80 per cent of them we give an actual introduction to boolean algebra, circuit synthesis techniques, etc. [3]

• In the department, introduction to programming techniques is done with a strong emphasis on proof techniques (Hoare Logic, invariants)

• The students learn both circuit design and formal logic. Some of the tools used in combinational circuit verification will also be used in propositional calculus. [5]

• We have worked very closely with the team developping the real time langage Lustre and the related environments. [4]

## 3. TOOLS AND ENVIRONMENT

The lab activities are done within the Lustre environment. [6] The language allows very pleasant description of circuits. When Lustre is used at the logic (gates, flip-flops) level, it looks like LOLA from Wirth, [7] or Model, from Solo 1400. Boolean vectors are taken into account, allowing n-bits circuits to be designed. When Lustre is used at the integer level, it looks like Verilog.

Hierarchy, modularity and regularity are welcome in the descriptions. The description can be simulated. We get timing diagrams and lists like in any C.A.D environment.

In a more interesting way the description can be *compiled.* What does mean such a compilation ?

• If the circuit is combinational, the global expression is evaluated and simplified by a B.D.D package. To compare two circuits, we introduce a comparator between the two sets of outputs. This comparator has only one output. If the two circuits are equivalent, this output is "always" true. The simplifier exhibits this property when it occurs.

• If the circuit is an automaton, the global set of states is computed, with the global transition function. Then the minimal automaton is computed. Obviously the number of states is a limitation. Here also the use of a comparator, giving only one output, allows to check for the equivalence of two automata. With only one output always true, the minimal equivalent automaton has only one state.

# 4. THE CONTENT OF THE CURRICULUM

After small exercises in Boolean algebra, the students are invited to practice different activities :

• implementing a one bit adder with TTL74 family circuits,

• verifying boolean formula with a BDD based boolean calculator,

• implementing basic function in layout (based on a library of cells and connection channels).

They so become familiar with the formal and practical manipulation of boolean formula and their corresponding gates. We have then a rather classical set of exercises (designing adders, adder/substractors, small finite-state automata...). The difference is the following : the students are invited to produce two descriptions and to verify the equivalence. One description is considered to be the specification, the other one the actual implementation.

**An example :** The students learn assembly language programming on Sparc, so we have built a simple automaton based on the actual Sparc. Our circuit, like a Sparc, has a Program Counter (PC) and a next-P-C. The (read only) memory contains two kinds of instructions : branch and no-branch. The branch instructions contain the displacement. An external condition is the input of the system. We limited the size to 3 bits for PC and nPC. It is so possible to consider ALL the (13) states of this "processor". This global set of states can be checked with a specification. This allows to deeply understand the mecanism involved in the delayed-branch mecanism. [3]

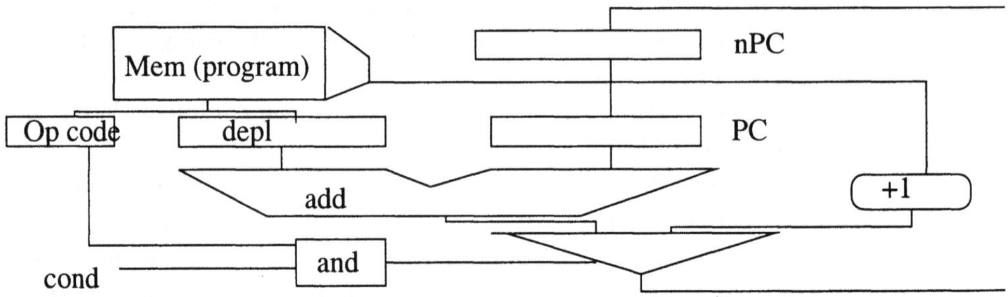


Kernel of a Sparc used as an example

# 5. CONCLUSION

Over years, we have taught digital circuits design with different C.A.D. tools based on different textual languages.

• Solo 1400 had a simple langage (Model) ; it allowed to produce silicon [2] but is no more supported in our site.

• VHDL had been used with a synthesis tool and FPGA board. The FPGA board is no more supported by the software package.

• Lustre is simple but we are far from silicon : no synthesis tool is available. But the main advantage of this new tool is that it allows to focus on formal verification.

## References

[1] S. B. Akers : Binary Decision Diagrams, IEEE Transactions on Computers, Vol C-27, June 1978.

[2] P. Amblard, How can we cook algorithms ?, 2nd Eurochip Workshop, Grenoble, Sep-oct 1991.

[3] P. Amblard, J.C. Fernandez, F. Lagnier, F. Maraninchi, P. Sicard, P. Waille : *Architectures Logicielles et Matérielles*, Dunod, 2000.

[4] N. Halbwachs, F. Lagnier and C. Ratel : Programming and verifying real-time systems by means of the synchronous data-flow programming language Lustre, IEEE Transactions on Software Engineering, Special Issue on the Specification and Analysis of Real-Time Systems, sep 1992.

[5] M. Levy, L, Trilling : A PVS-Based Approach for Teaching Constructing Correct Iterations FM'99 LNCS 1709

[6] Lustre : http://www-verimag.imag.fr/SYNCHRONE/

[7] N. Wirth : *Digital Circuit Design*, Springer-verlag, 1995.

# DESIGN OF THE HADAMARD COPROCESSOR WITH THE ALLIANCE CAD SYSTEM CARRIED BY POST-GRADUATING STUDENTS

**A. ZERROUKI, J. DUNOYER, F. WAJSBÜRT, A. DERIEUX**
*University P.et M. Curie Laboratoire LIP6, ASIM Team*
*4 Place Jussieu F75252 Paris Cedex 05 France*
*Tel : +33 1 44 27 54 15 Fax : +33 1 44 27 72 80*
*{Amal.Zerrouki, Julien.Dunoyer, Franck.Wajsburt, Anne.Derieux}@lip6.fr*

## 1. Introduction

University Pierre and Marie Curie in Paris, offers a one-year course on Advanced CAD and VLSI design for post graduate students (DEA ASIME, DESS CISAN). This course starts with a general introduction to circuit and system architecture and a presentation of ALLIANCE[1], the educational CAD system developed by the ASIM team of the university. This first part lasts 4 weeks and aims to settle the basic knowledge necessary to follow a more specific course focusing on either CAD tools design, Circuit design , or System design. In between, a 3-weeks project is proposed to allow the students practicing the concepts presented earlier. Depending on the students option, the project consists in implementing either a MIPS R3000[2] or the Hadamard coprocessor[3].
This paper presents the Hadamard project with a special focus on the methodology followed by the students from specification to layout.

## 2. Hadamard Implementation

The Hadamard coprocessor designed is actually the CMOS implementation of the Hadamard transform used for image compression. It computes the H*P*H matrix, where H is the Hadamard matrix, and P is the image matrix to be compressed. H is a square matrix and contains 8x8 bits. P is also a square matrix and contains 8x8 bytes. Computing the matrix H*P*H consists in 2 steps : computing first the intermediate matrix H*P, then the final matrix (H*P)*H.
Given the possible values of H, the product of a matrix with H is nothing more than a series of additions and subtractions. Basically, 8 operations are performed to obtain each element of a product matrix. Each element of H*P is stored in the corresponding row of a register file until a complete row of the intermediate matrix is obtained. Then, a row of the final matrix H*P*H can be calculated by multiplying the row in the register file by the corresponding columns of the H matrix. These steps are performed 8 times to complete the calculation of the 64 elements on the H*P*H product.
The circuit architecture is presented in figure 1. A *RAM* is used to fetch the matrix P corresponding to an image. *COMPUTE* is provided to compute the elements of both the intermediate and the final matrices. A *register file* is also included to store the



*B. Courtois et al. (eds.), Microelectronics Education, 265–268.*

intermediate elements. *COUNTERS* includes 3 counters C1, C2 and C3 used to handle the indexes of the different matrices in use. The H matrix values are delivered by the *MAT* bloc. A sequencer, *SEQ*, is provided to control the blocs operation.

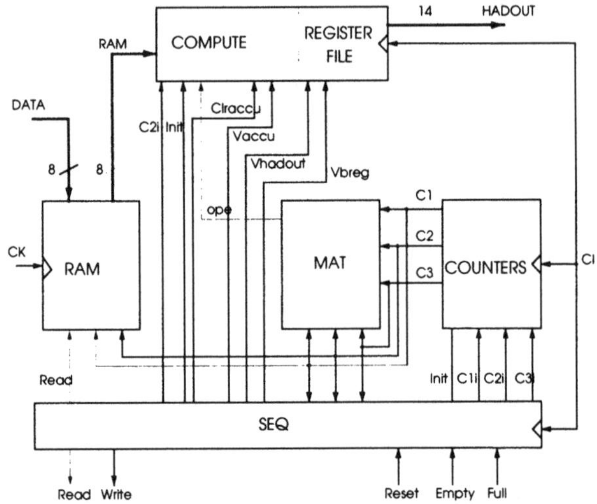


*Figure 1 : HADAMARD Architecture*

## 3. The Choice of the Hadamard Coprocessor

Although it is rarely used for image processing given its poor efficiency, this algorithm is still a very interesting example for education for various reasons. First, the algorithm is simple enough to be understood and implemented in 3 weeks. Then, the resulting architecture is complete enough to exercise the major architectural concepts (automata, register files, rams, fifo interfaces, synchronous communication, ...). Last, the VLSI implementation is simple enough to let the students get familiar with the methodology behind ALLIANCE.

This project encompasses the various aspects of VLSI design and offers the opportunity to students to measure the effects of their various implementations.

## 4. Design Methodology

### 4.1. DESIGN VIEWS

The methodology behind ALLIANCE is said *zero-default* and *top-down.* Basically, three different views are used : *the behavioral view, the structural view* and *the layout.*

The *behavioral view* is generally the starting point. It describes the functional specifications in VHDL (dataflow and finite-state-machines). The *structural view* corresponds to a netlist connecting standard cells or hierarchical blocs. It is either elaborated by the designer or obtained by synthesis including eventually an optimization loop. The *layout* of a chip is generated by Place & Route tools.

## 4.2. A HIERARCHICAL DESCRIPTION

Figure 2 presents the functional structure of Hadamard. The highest level is *Chip*, a structural view interconnecting the core to its pads. *Core* is itself a structural view. It connects the different blocs building the Hadamard function. *Seq* is a finite-state machine which controls the operation of the operators. *Ram* is a hand-written gate netlist. *Counters* is a behavioral view describing *C1*, *C2* and *C3* counters. *Mat* is the behavioral description of the Hadamard matrix. *Compute* is a behavioral description of the arithmetic unit performing calculations on matrix elements.

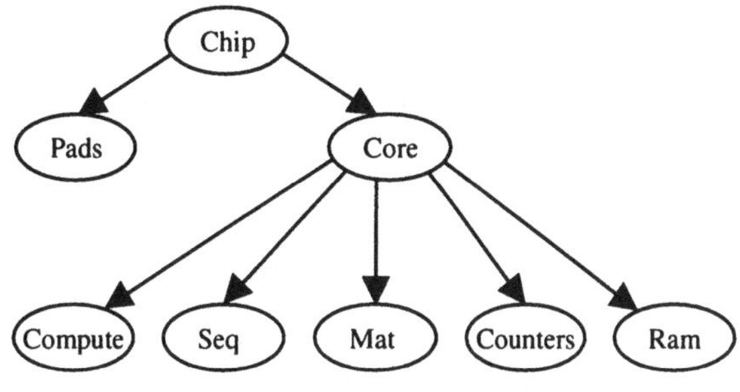


*Figure 2 : Functional hierarchy*

## 4.3. BLOC SYNTHESIS

Given the variety of input descriptions, different paths are considered to generate a gate-netlist. A finite-state-machine is first processed by SYF to generate a behavioral description. This last is then handled by the BOP optimizer to reduce its boolean expressions. The resulting behavior can finally be mapped by the tool SCMAP in standard cells using the portable cell library of ALLIANCE. This generates a gate-netlist necessary to move to the Place&Route step.

A behavioral description such as *Mat* or *Counters* is directly processed by the boolean optimizer. A new behavioral description is generated and mapped by SCMAP. The input description of *Ram* is a gate-netlist. There is no need to synthesis.

## 4.4. SCAN PATH INSERTION AND FANOUT OPTIMIZATION

In 3 weeks, a full study of chip testabilty is impossible. A scan path is therefore preferred. All registers are scanned, except those of the *Ram*. GENSCAN generates a new gate-netlist in which registers are chained. To ease the use of the scan path, students can specify the order in which the registers should be scanned.

The last step before Place&Route consists in running on each netlist a fanout optimizer to make sure that no fanout rule is violated. Such a step is not compulsary at this point of the design. If one needs to reduce the overall chip area, timing analysis will help identifying critical paths and thus those for which fanout optimization is necessary.

## 4.5. CHIP PLACE&ROUTE AND DESIGN RULE CHECKING

At this stage, all blocs building the chip are available under the form of gate-netlists. *Mat*, *Counters*, *Seq* and *Compute* are joined in an intermediate netlist named *Control*. SCR, the standard cell router is used to generate the layout of both *Ram* and *Control*. The Bi-Block Router BBR generates the *Core* layout, and the whole chip layout is generated by RING, the pad router. This last places and routes the pads in a ring around the *Core* and connects each pad to *Core*. The layout of the core and the chip are finally verified using the design rule checker DRUC to make sure that no symbolic rule has been violated.

## 5. Timing Analysis

The transistor-level netlist of the chip is extracted from layout using LYNX, the ALLIANCE extractor. The resulting netlist includes the capacitors attached to each physical segment. Using such a netlist, static timing analysis is performed with the tool TAS to identify critical paths. Depending on their analysis, students may have to return to synthesis and optimization.

## 6. Validation Environment

In ALLIANCE, each step of the design should be followed by a validation step.This is often achieved through logical simulation. In this project, system-level simulation is considered. A test bench is defined including the Hadamard core and a Stimulator bloc. This last exchanges data with Hadamard using a fifo interface. One advantage of such an approach resides in the fact that the circuit is used in its environment. Logical simulations are performed using ASIMUT, the ALLIANCE simulator. This tool accepts both behavioral and structural views. As a consequence, this unique test bench is complete enough to validate the initial behavioral view written by the students as well as all other views obtained throughout the design.

## 7. Results

The resulting Hadamard chip contains 36000 transistors for an area of around $6530*6400\lambda^2$. In a 0.35μm, this leads to 5.2 $mm^2$ area. The operating frequency determined by timing analysis varies from 45MHz to 60MHz.
The circuit has been successfully implemented in 3 weeks by groups of 3 students each. Although a real and complete circuit cannot be implemented in such a short time, this project is essential to students since it gives them the opportunity to practice circuit design from specification to layout using various CAD tools.

## 8. References


1. Greiner A., F. Pêcheux, "ALLIANCE . A Complete Set of CAD Tools for Teaching VLSI Design", *In The 3rd Eurochip Workshop on VLSI Design Training*, pp. 230-37, Grenoble, France, 1992.
2. Bazargan Sabet P., Dunoyer J., Greiner A., Rosset-Louërat M.M, A tutorial for Advanced VLSI Course : Designing the 32 bit DLX with the ALLIANCE CAD System, In *The 5th Eurochip Workshop on VLSI Design Training*, p 120-27, Dresden, Germany 1994.
3. Gonzalez R. C., Woods R.E.(1992) *Digital Image Processing*, Addision Wesley.

# MICROSYSTEM PROJECT-ORIENTED EDUCATION

M. HUSAK
*Department of Microelectronics, Faculty of Electrical Engineering,*
*Czech Technical University in Prague*
*Technicka 2, CZ-166 27 Prague 6, Czech Republic*
*Tel.: +420-2-24352267, fax: +420-2-24310792,*
*e-mail: husak@feld.cvut.cz*

## 1. Abstract

This paper deals with questions concerning teaching microelectronic courses, more precisely courses connected with the area of microsystems, at a technical university. Educational methods based on individual students' projects are described. Attention is focused on arrangement of individual courses, output form of students' projects, ways of increasing education efficiency and time utilization, material, functional, technical and instrument support of individual projects and link-up to international cooperation. Most effort is given to individual students' projects, their natural motivation in selecting courses and realization of projects.

## 2. Introduction

At present, education in microelectronic courses is a hot topic. Besides "classical" integrated electronic circuits, more attention is focused on integrated microsensors, produced in hybrid or monolithic technologies. Microsystems - new direction in interdisciplinary approach to electronic development - represent further step forward.
These problems are very interesting for students because they continuously bring new and progressive solutions and knowledge. A number of students express interest in this area in case of concrete practical tasks (i.e. solving various circuit connections based on elements available on the market). Students like to see real results that are immediately applicable in practice (measurement, control, etc.). Introducing project-oriented education has resulted in increased number of students in individual courses, namely in Sensor Systems, Microsystems, New Directions in Electronics, Microelectronics, Design of Integrated Circuits, Electronic Security Systems.



*B. Courtois et al. (eds.), Microelectronics Education,* 269–272.

## 3. Aspects of organization and contents of individual students' projects

Organization of education in individual courses. A student has a possibility of choice in selected courses. The student begins his/her decision making by choosing either "classical" seminars and laboratory work (measurement of pre-prepared laboratory tasks, prepared seminars), or individual projects – Figure 1.. The student can find a topic of the project by himself/herself. It must be related to the studied course (and usually is corrected by the teacher so that it corresponds with the aim of the course). The student can get a topic of the individual project from the teacher as well. At the beginning of the project there is an individual agreement of the teacher and student on project definition. The content of the project, including all parameters necessary for project solving, is specified. An important condition is that the project topic has direct link to the given course. Relations to other student's activities are assumed as well. In the frame of the project, the student can solve part of the Master thesis, research project, or grant, etc. When defining the project, student's profile and interest are taken into account. The choice can be enlarged in some courses by thematic division of the course (e.g. to physical, optoelectronic, or microsystem problems).

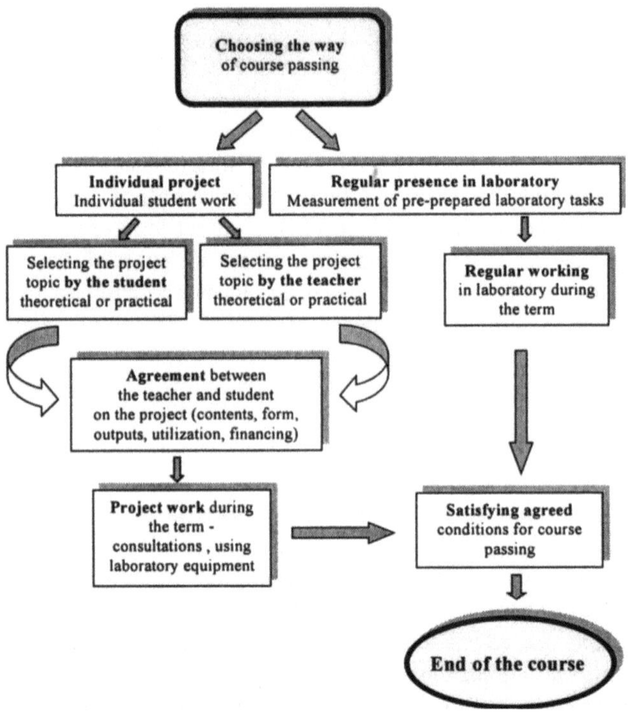


*Figure 1.* Student's decision making when selecting project activity

Outputs of individual students' projects [1]. Outputs of individual projects differ from each other depending on project character. They may be either theoretical (research reports) or practical. Let us mention as an example realised parts of more or less complex electronic equipment (e.g. programmable thermostats). It is possible to group projects for more students. In this way more complex equipment with precisely defined interface between individual parts can be realised. Then the separate parts are solved by individual students. At present there is solved a problem of complete additional equipment of a personal car that serves for its electronic security, higher comfort and monitoring proper functionality of other electronic car equipment.

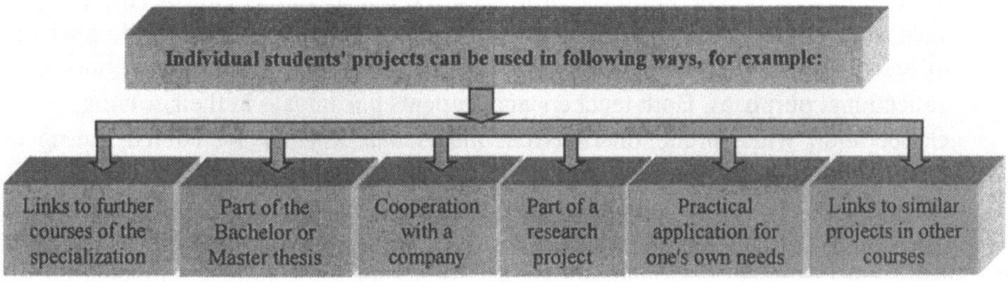


*Figure 2.* Examples of utilization of individual students' projects

Increasing effectivity of time spent on the project. The student can work individually during the whole semester, he/she can schedule the work as he/she likes. He/she can choose between work at home or in the laboratories. He/she can attend regular consultations. The project-oriented education enables to use time more effectively. It is linked to problems solved in other courses. It supports independent decision making and technical thinking.

## 4. Technical, instrument and financial support of projects

Good coordination of individual factors influencing solving students' projects is necessary for successful completion of all students' projects. The most important ones are laboratory equipment, financial support from grants and educational projects. Some of these areas are presented below.

Technical and instrument support. For individual project solving, the laboratory equipment (frequently of unique character) must be made accessible for students. Therefore individual faculty laboratories must cooperate very tightly. In this way, the students can have access not only to very specialised laboratories at the faculty or university, but also to laboratories at other universities or research institutes.

The laboratories of the Centre of Microsystems (CEMIS) and Electronic Circuit Design (ECAD) at the CTU are open for students for design of microsystems and especially integrated sensor systems and microsensors [2]. The CEMIS is focused on research, development and education in the area of microsystems design, testing and parameter measurement. Its major aim is applied research and development tightly coupled with practical applications and education of students. The CEMIS activities include hardware build-up as well as software support, setting-up measurement, diagnostics and testing

facilities determined for microsystems or microsensor and sensor systems development. The basis of the design workplace of the microsystems centre utilises the existing ECAD laboratory equipment, originally built and still used for integrated circuit design by SUN workstations with the implemented software. Predominantly, the CADENCE software system is being used now. The students acquire knowledge and skills in the area of computer-aided design using CAD tools, that correspond to the world standard.
Grant and project support [1], [2]. Financial requirements of students' projects are usually very high. Therefore the projects are often solved as part of research grants. In such a way students are directly included into research teams and learn to present results of their works in the team. Grants are awarded in public grant competition oriented to educational purposes. Both teachers and students participate in their solving.
Cooperation with foreign institutions. Individual projects are solved during stays at foreign institutions and universities as well. Projects in TEMPUS and SOCRATES programmes and other individual activities are used as a framework for them. In the frame of a TEMPUS project, the students worked at various projects in Great Britain - University of York, Bournemouth University, Leeds Metropolitan University, University of Hull, UMIST Manchester; in France - INPG Grenoble; in Belgium - Hogeschool Gent; in the Netherlands – Hogeschool West Brabant Breda. In the frame of individual projects students worked at UMIST Manchester (Great Britain), etc. [3].

## 5. Conclusion

Recent results of teaching process efficiency show that students are very much interested in this way of education both from the point of view of time efficiency and concentrated way of work. It is an example of possible way of teaching whose aim is to attract more students to the studied subject. It is necessary to realize that there is no uniform pattern and in all cases it is necessary to consider specificity of individual courses, faculty and last but not least teacher's personality. Even in project-oriented education teacher's personality is dominant.

## 6 Acknowledgment

Research and education described in this paper has been supported by the grants No. 0854/1999, No. 1040/2000 and No. 1046/2000 of the University Development Foundation and No. VS97046 of the Ministry of Education of the Czech Republic

### 7 References

[1] Husak,M. et al.: Introduction of Interdisciplinary knowledges (Sensor and microsystem ) to education. Final Report, Grant of University Development Foundation, No. 0854/1999.

[2] Husak, M. et. al.: Centre of Microsystems. Report, Grant of Ministry of Education, Youth and Sports, No. VS 97046.

[3] Husak,M.: Student Projects Integration to University Microsystem Centre. Conference MSE'99, 19.-21. July 1999, Arlington, Virginia, USA, proceedings, pp. 63-64.

# A New Trial on HDL Exercise Class for Undergraduate School in EE Department

Makoto IKEDA and Kunihiro ASADA
VLSI Design and Education Center(VDEC), University of Tokyo

**Abstract**

As the integration density increases, higher abstraction level design becomes much more important. To meet this trends, education on hardware description languages(HDL) in university becomes important as one of the topics in IC design. In 1999, we have started a new exercise class to teach HDL for undergraduate school students in the Electronics Engnieering Department at University of Tokyo. This paper describes a syllabus of this exercise class.

Table 1 describes the schedule of this HDL exercise class. We have a 3 hours classe per one week for this class and this class extended over 6 weeks. As the total time is limited, we emphasized on VerilogHDL design and FPGA mapping so that students can easily realize their designs, which can stimulate students' interests. This resulted that many students worked very hard for this exercise: 44 students have taken this class and only 1 student dropped out.

Figure 1 describes students' distributions of pre-knowledge on HDL, IC designs and programming language like C language. From fig.1(a), almost no one have heard of HDL like Verilog-HDL, except 3 out of 43 have heard the name of VerilogHDL. Prior to this exercise, we have 2 classes related to IC designs; (1) basic class for electronics circuits and (2) logics. As a result, 75% students have knowledge on logic and circuits. We have programming language exercise class prior to this class, almost all students know some programming languages. It is, however, a few students can write programs, some of the reasons for this is that students took this class were mainly belongs to electronic devices, photonic devices, electronic materials courses. From these figures, the start line of all students were almost uniform in the level of pre-knowledge.

Table 1: Schedule of the HDL exercise course.

| Day | Contents |
|---|---|
| 1st | Schooling on VerilogHDL and exercise on verilog simulation |
| 2nd | Logic synthesis for FPGA and FPGA mapping |
| 3rd | HDL design exercise targetted for FPGA (I) |
| 4th | Logic synthesis and place-and-route for ASIC design |
| 5th | HDL design exercise targetted for FPGA (II) |
| 6th | Final presentation |



*B. Courtois et al. (eds.), Microelectronics Education,* 273–276.

In this class, we use FPGA kit board that has Altera Flex10k, key inputs, LEDs, a buzzer. We have requested students to describe HDL and to make a demonstration using this FPGA board. The design flow we employed is shown in Fig. 2, where we write Verilog-HDL using text editor like Mule, carry out HDL simulation using Verilog-XL from Cadence, HDL synthesis for FPGA using FPGA Compiler II from Synopsys and FPGA mapping using MaxplusII from Altera. Instead of writing PROM, we use BitBlaster down-load cable for FPGA configuration. We taught grammer and syntax of VerilogHDL, including the relations betweeen test fixtures and the FPGA board, using sample program for this board. As for the sample program, we used "roulette" that turns on a LED in a 7-segment LED sequencially, whose description consists of frequency devider using counter and shift register. Then we requested students to write specifications for their designs and following the specifications, students have designed VerilogHDL. During HDL design, we taught logic synthesis and place-and-route for ASIC design, as shown in Fig. 3, from view points of similarities and differences between ASIC design and FPGA design.

Figure 4(a) shows distribution of time students took to complete their tasks. Most students completed tasks in about 30 hours, but some students took more than 50 hours, one of the reasons is that they have designed large scale HDL so that they have to cut their design to fit within FPGA.

Figure 4(b) shows students' expected time to master VerilogHDL language and HDL design. From this figure, many students prospect that it will still take much more time to master VerilogHDL.

Figure 5 shows distribution of number of lines students have written during the exercise class. Figure 6 shows scene of this exercise. We use Sun/HP workstations and note-PCs with FreeBSD as X-terminals, and students have logged in to server

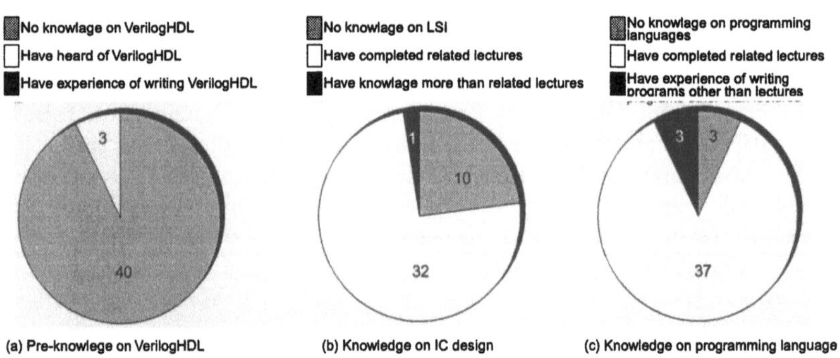


Figure 1: Enquete results: pre-knowledge of students on (a) VerilogHDL, (b) IC design including circuits and logics, and (c) programming languages.

computer (Sun Ultra Enterprise 6000), where they have executed CAD programs.

The aim of this class is to make undergraduate school students to be familier with the concept of high-level description for LSI design. The enquete results shows

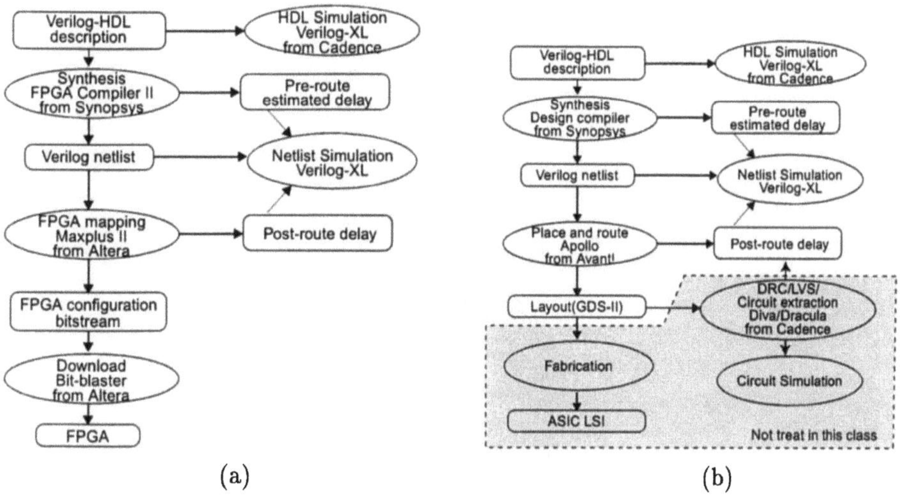


Figure 2: (a) Design flow we employed for FPGA design exercise and (b) design flow we employed for ASIC design exercise.

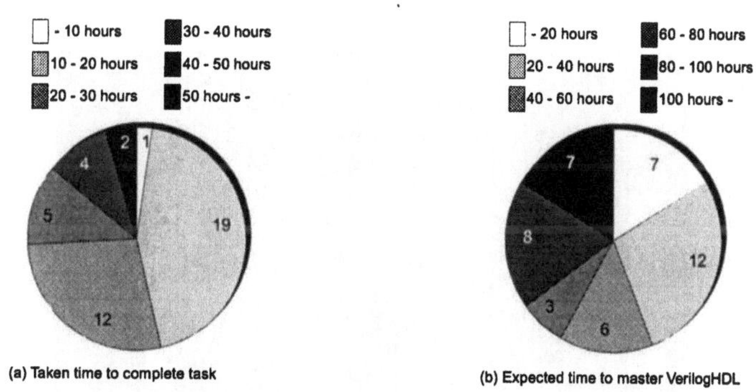


Figure 3: Enquete results: (a) taken time to complete task of this exercise including schooling time, and (b) expected time to master VerilogHDL.

that, to meet this aim, 30 hours exercise may be enough including schooling time, though, it may take much more time to master HDL and to become experienced HDL designer.

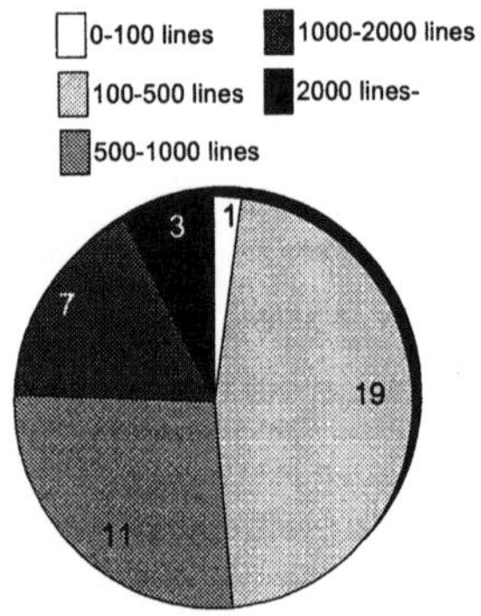


Figure 4: Distribution of number of lines students have written during the exercise class.

Figure 5: Scene of this execrcise.

# PROVIDING FLEXIBLE INTERNET INFRASTRUCTURE FOR FPGA-BASED CAD COURSES

J. BECKER, U. MAYER, M. GLESNER
Darmstadt University of Technology
Institute of Microelectronic Systems
Karlstr. 15,
D-64283 Darmstadt, Germany
*e*-mail: becker@mes.tu-darmstadt.de

L. S. INDRUSIAK, R. REIS
Univ. Federal do Rio Grande do Sul
Instituto de Informatica
Av. Bento Gonçalves,
9500 Porto Alegre, RS, Brazil
e-mail: lsi@inf.ufrgs.br

## 1. Abstract

The paper describes the large potential and a flexible internet-based infrastructure of reconfigurable computing in education of engineers and students, as well as of technicians of small and medium sized enterprises (SMEs). This paper focuses on the educational possibilities of remote FPGA programming systems. It proposes and explains the technical use of the WWW as the interface between the student and the FPGA device, introducing a flexible process for FPGA-based teaching incl. real hardware synthesis allowing some of the design steps to be done either on the client or server side of the network.

## 2. Introduction

Field Programmable Gate Arrays – FPGAs - consist of arrays of configurable logic blocks (CLBs) implementing the logical functions of gates. The logic functions of each CLB and the interconnection between all of them can be programmed by downloading bitstreams to the chip. The FPGAs programming can be repeated infinitely often, even partially and dynamically during run time. This possibility, together with its ease of use, makes FPGA devices potential tools for teaching subjects from simple digital systems design to reconfigurable architectures. Configurable Computing has demonstrated its potential of achieving high performance improvements for a wide range of applications like image processing [1], morphology [2], mobile communication [3], among many others [4]. Due to this real speed-up potential a lot of researchers have built corresponding (re)configurable prototype systems for a large number of applications. Current computer science curricula do not create awareness, that hardware has become soft, nor, that hardware, structural and sequential software are alternatives to implement problems. Principles and applications of dynamically reconfigurable devices as a basis of the new computing paradigm of *structural programming* should be included in academic and industrial main courses to increase the potential of this novel emerging technology.

The work presented in this paper aims to allow the remote use of FPGA devices through the World Wide Web. Using this feature, the student would have the opportunity to design a circuit and implement it on FPGA, even if the device is not available in his location. The concept of this approach was presented in [5], as a special feature of a reconfigurable architecture multimedia course. The present work extends that approach by allowing the distribution of the design steps either on the client or on the server side of the network. This way, the synthesis process can be used by students which already use a particular FPGA synthesis CAD suite and only need to program the device, as well as by students which don't have the CAD suite. In the later case, the design entry is done using the *Cave* Project Environment [8].



*B. Courtois et al. (eds.), Microelectronics Education,* 277–280.

## 3. Proposed Internet-based Architecture

The key aspect of the proposed FPGA courses are the practical sessions and the realistic use of reconfigurable hardware. The participants should be able to put the theory into practice by solving experimental tasks with real hardware devices.

In general, the structural code can be used to program a reconfigurable extension board [9] connected to the user via the Internet (see figure 1). The results of downloaded applications could be, for example, visible through a web camera, or be transferred back through a web-page etc.

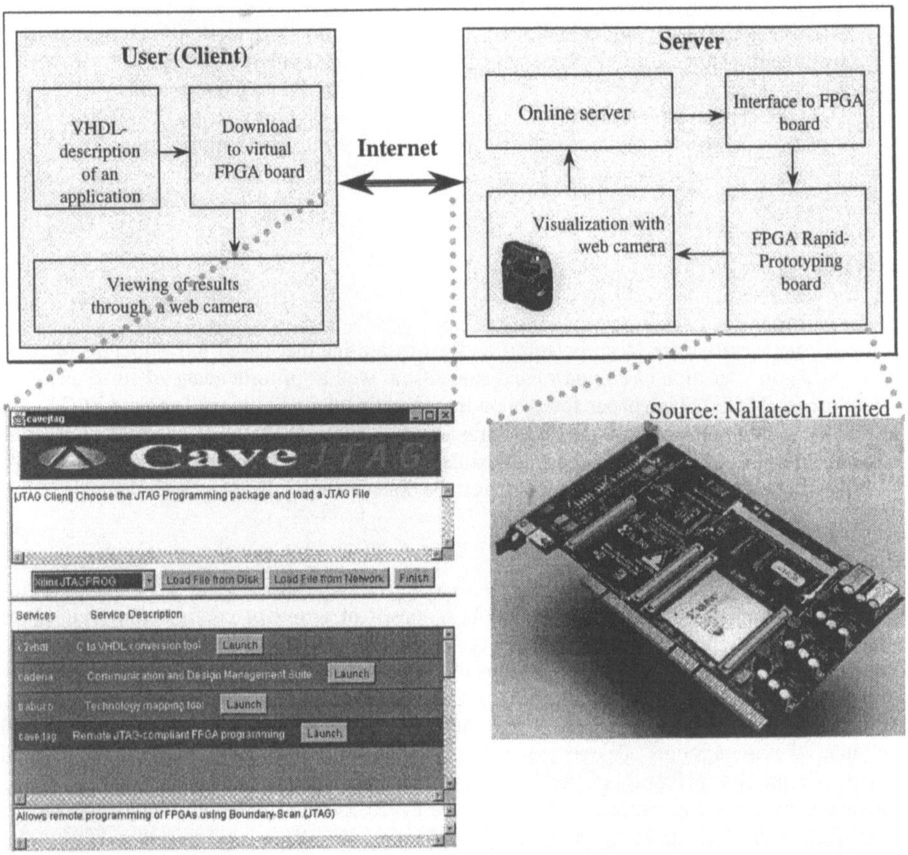


Figure 1: Client/Server Architecture and *Cave* GUI of realistic structural Programming Concept

The architecture of the system proposed and implemented here follows the *Cave* Project architecture for web-based design automation systems (see figure 1, [7]). The non-interactive design tools are stored on the server side, remotely executed through a WWW interface on the client side. However, the flexibility of access is a crucial issue in the present work, because it aims to provide infrastructure for two types of FPGA courses, as follows:

- Type 1: Courses covering a specific FPGA design tool - These courses have the focus on the design tools, exploiting the software facilities to prototype reliable circuits in short time. Once the course uses commercial software packages, the whole synthesis process is done on the client side – where the tools are installed – and the remote programming consists on uploading the bitstream for the FPGA device through the WWW.
- Type 2: Courses covering generic FPGA programming - These courses have the focus on the design itself, detailing the required steps to synthesize the FPGA circuits. So, the course can use any

software package, because the main goal is to show the student the synthesis process without tie it to a particular CAD suite. In this case, the use of the Cave Project Environment is proposed, because it provides web based design entry tools, allowing the course to be done completely through the WWW environment, from the design specification to the bitstream upload to the FPGA device. Using this approach, there is no need to install any software on the student machine but a Java-enabled web browser.

## 4. Client/Server Course Procedure Realization

The procedure for a CAD-specific FPGA course is shown in figure 1. The design entry and validation is done on the client side, where the tools are installed and the synthesis process as well as the bitstream generation is done. Once the bitstream is generated, it is uploaded to the server side using Internet

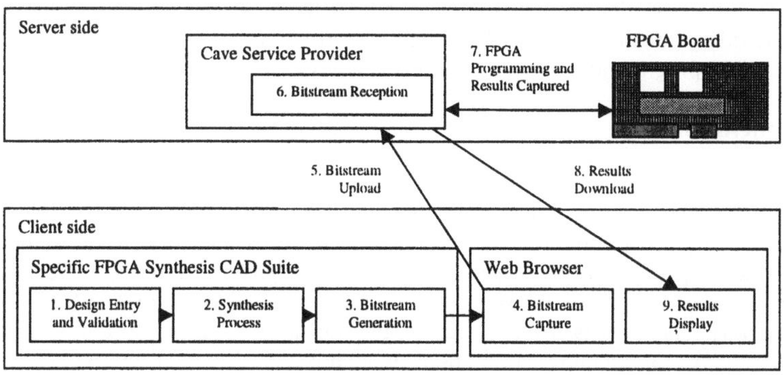


**Figure 1:** CAD-specific FPGA course procedure (Type1)

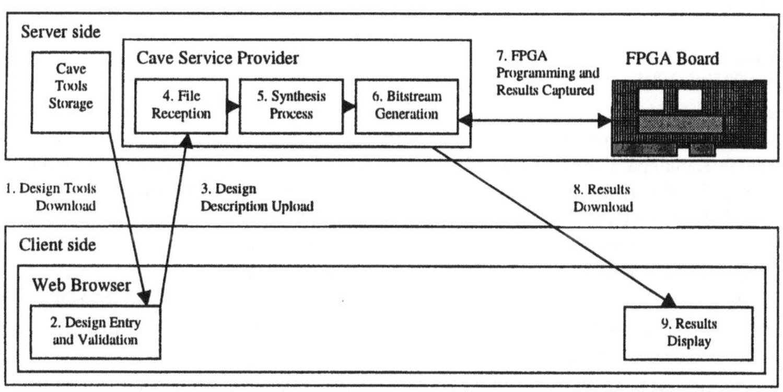


**Figure 2:** Generic FPGA synthesis course procedure (Type 2)

standard protocols. The *Cave* Service Provider receives the bitstream file, stores it and load it to the FPGA board, followed by capturing the results of the file upload, formatting them as a web page and sending the page to the client. Figure 2 shows the procedure for generic FPGA programming courses. The design entry and validation tools are downloaded from the web server – the Cave Server – and run inside the web browser. Once the design is validated, its description is uploaded to the server side using Internet standard protocols. The *Cave* Service Provider receives the design description files, stores them and invokes the synthesis process and the bitstream generation. From this point on, the process is as described for CAD-specific courses.

The input for the remote FPGA synthesis process is a netlist, created by the designer on the client side of the network. Regular synthesis tools, which are executed on the client side, process the netlist and generate a bitstream for programming the desired FPGA device. To keep with open standards, the programming must be done using Boundary-Scan [10], also known as JTAG, supported by most FPGA vendors. Once the JTAG programming bitstream is generated, it is submitted to the JTAG Service Provider module, which download the programming code to the FPGA device. In using Boundary-Scan commands one could program, re-program or verify the internal state of a FPGA device. A new Java Boundary-Scan interface can be used for developing reconfigurable hardware [11], because it allows direct manipulation of all the FPGA configurable resources at runtime - look-up tables, routing and flip-flops. The prototype system developed here uses a Virtex board [9], which is plugged to an Internet-connected PC. On the PC, the JTAG Service Provider - a Java program which listens to a TCP/IP port waiting for client requests - is running. The client module is a Java applet, used as graphic interface for FPGA device specification and programming. After the interaction with the designer, the client connects to the JTAG Service Provider and uploads the bitstream. When the upload is finished, the Service Provider uses the Boundary-Scan interface to program the FPGA, sending back to the client details about the programming process and errors, if any.

## 5. Conclusions

This paper described the educational possibilities and the technical client-server interacting structure of a remote FPGA-oriented CAD system, in addition to the perspectives of reconfigurable computing in education [6]. It presented the use of the WWW as the interface between the student and the FPGA device, introducing a flexible process for an FPGA-based synthesis allowing selected design steps to be done either on the client or on the server side of the network. Corresponding CAD exercices could be realized on different severity levels (students, technicians, engineers), so that this new and attractive way of structural programming can be used in addition to traditional programming tools, or used as alternatives for pure ASIC solutions.

## 6. References


[1] Athanas, P., Abbot, A.: Real-Time Image Processing on a Custom Computing Platform, IEEE Computer, vol. 28, no. 2, Feb. 1995.

[2] Drayer, T. H., King, W. E., Tront, J. G., Conners, R. W.: A Modular Reprogrammable Real-Time Processing Hardware, MORPH; FCCM'95, IEEE Computer Society Press, Napa, CA, April 1995.

[3] Becker, J., Kabulepa, L., Renner, F.-M., Glesner, M.: Simulation and Rapid Prototyping of Flexible Systems-on-a-Chip for Future Mobile Communication Applications; Proc. of 11th IEEE Int'l Workshop on Rapid System Prototyping (RSP'2000), Paris, France, 2000

[4] Hartenstein, R. W., Glesner, M.: 6th Int'l. Workshop on Field-Programmable Logic: Smart Applications, New Paradigms and Compilers (FPL'96); Sept. 1996, Darmstadt, Germany.

[5] Becker, J., Kirschbaum, A., Renner, F.-M., Glesner, M. Internet-based Training of Reconfigurable Technologies; In: Brazilian Symposium of Integrated Circuits Design, Proc. of SBCCI, 11., 1998, Armacao de Buzios, Los Alamitos, IEEE Computer Society Press, 1998.

[6] Becker, J., Renner, F.-M., Glesner, M: Perspectives of Reconfigurable Computing in Education; Proc. 2nd European Workshop on Microelectronics Education, Noordwijkerhout, The Netherlands, May 14-15 1998 (Kluwer Acad. Publ., Eds. Ton J. Mouthaan, Cora Salm).

[7] Indrusiak, L.S.; Reis, R.A.L.: A WWW approach for EDA tool integration. In: Proc. of X Brazilian Symposium of Integrated Circuits Design, SBCCI, 10., 1997, Gramado, Porto Alegre: CPGCC da UFRGS, 1997.

[8] Iindrusiak, L.S.; Reis, R.A.L.: A Case Study for the Cave Project. In: Proc. of Brazilian Symposium of Integrated Circuits Design, SBCCI, 11., 1998, Armacao de Buzios, Los Alamitos: IEEE Computer Society Press, 1998.

[9] Xilinx Corp.: http://www.xilinx.com/products/virtex.htm.

[10] IEEE Std 1149.1 (includes IEEE Std 1149.1a-1193) IEEE Standard Test Access Port and Boundary-Scan Architecture, ISBN 1-55937-350-4

[11] Guccione, Steven A., Levi, Delon; Sundararajan, Prasanna. JBits: A Java-based Interface for Reconfigurable Computing. 2nd Ann. Military & Aerospace Applications of Programmable Devices & Technologies Conf. (MAPLD).

# PICO — A MODEL MICROPROCESSOR FRAMEWORK FOR ARCHITECTURE EDUCATION

K.NISHIMURA, K.NUKATA and H.AMANO
*Dept. of Computer Science, Keio University*
*3-14-1, Hiyoshi, Kouhoku, Yokohama 223-8522 JAPAN*

**Abstract** — A model microprocessor is a key issue in education of both a computer architecture and an LSI design. Although such a model microprocessor was just a paper machine on a textbook, recent advanced FPGA technologies and design CAD tools enable to implement such model processors by students in a class or experimental lab. In Japan, several model microprocessors and education course and lab. have been reported.[1][2].

In most of them, an instruction set architecture is already defined and given to students. Although students can optimize its datapath structure and gate level design, architectural variety is strictly limited by the given instruction set. In order to increase design flexibility, and encourage students' motivation for their original design, we introduce an educational model processor framework PICO and education course based on it.

## 1. A course using PICO

PICO is an educational model processor framework which can be grown up from 4-bit simple accumulator machine to 16-bit RISC processor. TABLE 1. shows architecture course in Dept. of Information and Computer Science, Keio University and model processors used there. In this course, SFL/Parthenon by NTT[3] is introduced in the beginners' class first, then all model processors are incrementally designed in SFL, simulated with a functional level simulator SECONDS, and evaluated their speed, required gates and consuming power by logic synthesis tools in Parthenon. (Verilog-HDL is introduced in the final course.)

## 2. 16-bit model processor PICO-16

Here, 16-bit RISC processor PICO-16 is mainly introduced. Although students can design their own instruction sets freely, the architectural flexibil-



*B. Courtois et al. (eds.), Microelectronics Education,* 281–284.

TABLE 1. Computer Architecture Course in Dept.of ICS, Keio Univ.

| | |
|---|---|
| Computer Architecture<br>Elementary Class<br>2nd Fall Semester | 4bit Adder<br>4bit ALU<br>4bit Accumulator Machine 16bit RISC PICO-16 |
| Computer Architecture<br>Advanced Class<br>3rd Spring Semester | PICO-16 with Interrupt<br>PICO-16 with Cache<br>Pipelined PICO |
| Computer Architecture<br>Lab. 3 days<br>3rd Fall Semester | Implementation on FPGA<br>PICO-16<br>Pipelined PICO |
| VLSI design<br>4th Spring Semester<br>4th Fall Semester | Simple processor<br>using Verilog-HDL<br>for ASIC design |

ity is limited by the total amount of gates which can be implemented on the reconfigurable board in the experimental lab.

The following guideline of instruction set is given to the students.

- The number of registers is up to eight.
- Load/Store architecture with a fixed field should be used.
- Two operands are recommended for most of instructions.
- Instruction types: Type-Register, Type-Immediate and sometimes Type-Jump (shown in *Figure 1.*) should be used.
- The half of instruction set field should be used for the immediate in Type-Immediate.

*Figure 2.* shows an example of instruction set design. In the architecture course shown in *Figure 1.*, there are only three days for experimental lab. So, students design only instructions which is required to execute a given application programs: e.g. sorting, sieves, and a simple game. Although the example instruction shown in *Figure 2.* is based on a typical RISC instruction set like DLX[4], most of students introduces original instruction set especially in compare and branch instructions. For example, branches with flags or simple compare-and-branch are popularly used.

## 3. Advanced Course using the PICO$^2$

In advanced course, a pipelined structure is introduced based on PICO-16 architecture. We called this architecture PICO$^2$ (*PICO with PIpelined COntrol*). In this case, the basic pipelined structure is given to the student. Unlike conventional RISC processors, the pipelined structure with the fol-

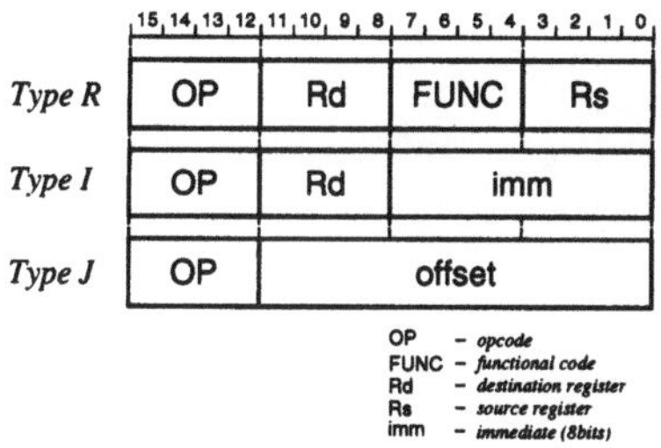


*Figure 1.* Instruction Format

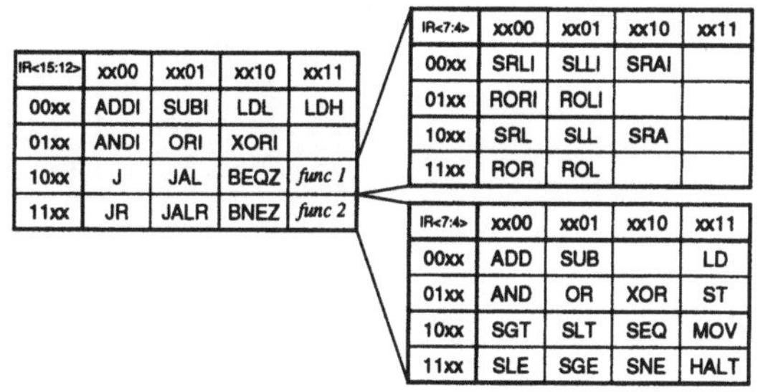


*Figure 2.* The Example of Instruction Set

lowing four stages is suitable as there are no displacement in the Load/Store instructions of PICO-16.

**IF** Instruction Fetch: Fetch an instruction from the instruction memory, and increment the program counter.
**ID** Instruction Decode and Register Fetch: Read registers from the register file.
**EX** Execution: Calculate for ALU instructions, read/write access to data memory for Load/Store instructions, and calculate target address for branch instructions.
**WB** Write Back: Write the results into the register file if needed.

For eliminating the basic structural hazard, the instruction memory is separated from data memory, and three-port memory is used for the register file.

Like the elementary course, the pipelined structure is introduced in the step-by-step approach.

**Step 1:** An elemental structure of pipeline with only simple instructions including ADD, LHI, LD, ST, and BEQZ is introduced. Simulating this structure with SECOND, students can understand the basic pipeline structure and data/structural hazard problems.
**Step 2.1:** A simple application program is implemented by inserting "nop" instruction for data/structural hazard.
**Step 2.2:** This pipelined PICO is implemented on the FPGA and the performance is evaluated and compared with non-pipelined structure.
**Step 3:** The forwarding path is added for eliminating data hazard, and branch delay slot is filled as possible by manual code scheduling. This version of PICO is implemented on the FPGA again and the performance improvement is evaluated.

*Figure 3.* The Photo of New Board

## 4. A reconfigurable board for PICO and PICO$^2$

For the experimental lab., we developed an original reconfigurable board. *Figure 3.* shows the photo of the board. A RISC microprocessor ColdFire(MCF5204) with a monitor program manages the loading of configuration data and machine code of PICO and PICO$^2$. It is also connected to common workstation through an RS232C serial interface, and controlled by a terminal emulation program. For instruction memory and data memory, dual-port RAM IDT7206 is used. They can be accessed from both MCF5204 and PICO.

## 5. Conclusion

In this paper, an educational microprocessor framework PICO is proposed. The students can implement their original instruction set according to the guideline of the framework in limited hardware and design time. It can be grown up from simple 4-bit accumulator machine to pipelined 16-bit RISC processor. Using the reconfigurable board, students can confirm their design on the real hardware.

## References


1. Nakagaki et al. Design and Implementation of the Educational Microprocessor DLX-FPGA Using VHDL. *Proc. APCHDL'94*, pages 147 – 150, Oct 1994.
2. H.Ochi. ASAver.1: An FPGA-Based Education Board for Computer Architecture System Design. *Proc. of ASP-DAC 97*, pages 157 – 165, January 1997.
3. Nakamura, Oguri, Nagoya, and Yukishita. High-level Synthesis Design at NTT Systems Labs. *IEICE Trans. Inf & Syst.*, E76-D(9):1047 – 1054, September 1993.
4. J.L.Hennessy and D.A.Patterson. *Computer Architecture - A Quantitative Approach - 2nd edition.* Morgan Kaufmann Publishers, Inc., 1996.

# A JAVA-BASED FRAMEWORK FOR SIMULATION AND TEACHING: HADES - THE HAMBURG DESIGN SYSTEM

N. HENDRICH
*University of Hamburg, Dept. of Computer Science*
*Vogt-Koelln-Str. 30, D 22 527 Hamburg*
*hendrich@informatik.uni-hamburg.de*

**Abstract.** We describe HADES (short for Hamburg Design System), a Java-based and WWW-enabled environment for simulation and visualization. As an application example, our lab-course on computer architecture is presented.

## Introduction

With increasing system complexity, a trend towards design reuse, and the use of embedded processor cores, the role of simulation in microelectronics changes. Especially, system simulation and hardware/software cosimulation become more and more important to analyse and verify complex systems. Also, understanding system interactions requires data visualization beyond simple waveforms.

## 1. A Framework for System Simulation

HADES is a Java-based and WWW-enabled environment for simulation and visualization [1]. Its simulation models are written as Java Beans, allowing for visual configuration and easy integration of new models. All signal types, events, and the simulation engine can be subclassed and extended with user-defined classes. However, so far we have used Hades mostly for digital circuit and system simulation. To this end, we provide a simulation engine based on VHDL semantics [2] with delta-cycles and signal resolution functions. Currently, about 200 simulation models are available, ranging from simple and complex gates via 74xx-series models to RT-level and system-level components, all of which use the std_logic_1164 [3] nine-



*B. Courtois et al. (eds.), Microelectronics Education,* 285–288.

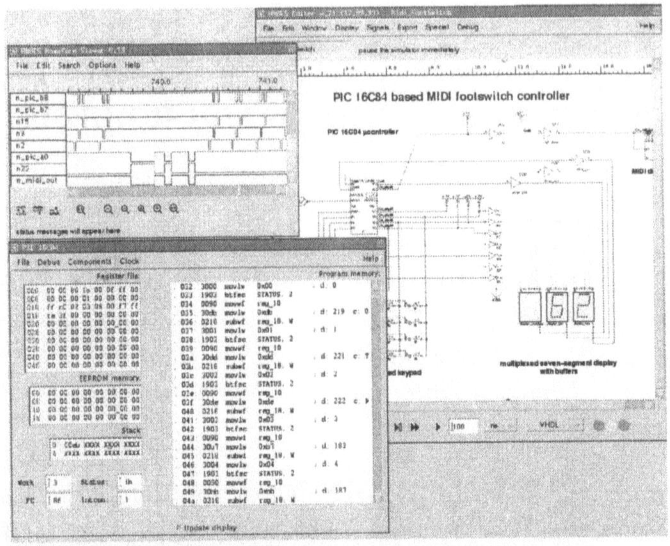


*Figure 1.* System-level simulation example: A MIDI controller based on a PIC16C84 microcontroller with multiplexed keypad and seven-segment displays. Also shown are the Hades waveform viewer and the editor for the PIC core with data registers and program memory.

valued logic model. Microcontroller models for the PIC16xx and MIPS R3000 processors allow for hardware/software cosimulation. Also, with recent advances in Java virtual machine technology [4], the performance is getting competitive with commerical simulators.

While similar simulators like DigLOG [5] are available, several features of Hades stand out in the context of education:

- free of charge and very portable (Windows, Linux, Solaris, ...)
- given a recent browser, Hades can be used as an applet
- fully interactive simulation and editing
- simulation models provide property editors with custom user-interfaces
- simulation based on VHDL semantics [2] and std_logic_1164 [3]
- microcontroller cores, memories, and I/O-models, e.g. LC-displays
- fully scriptable via JPython [6]
- visualization tools, e.g. glow-mode and context-sensitive tool tips
- real-time simulation mode to visualize gate-delays, hazards, etc.

As an example for system simulation in Hades, figure 1 shows a complete MIDI controller based on the PIC16C84 microcontroller with multiplexed keypad and display. Please note the special seven-segment components, which display time-averaged values decoded from the multiplexed inputs — a much better debugging aid than the complex waveform patterns.

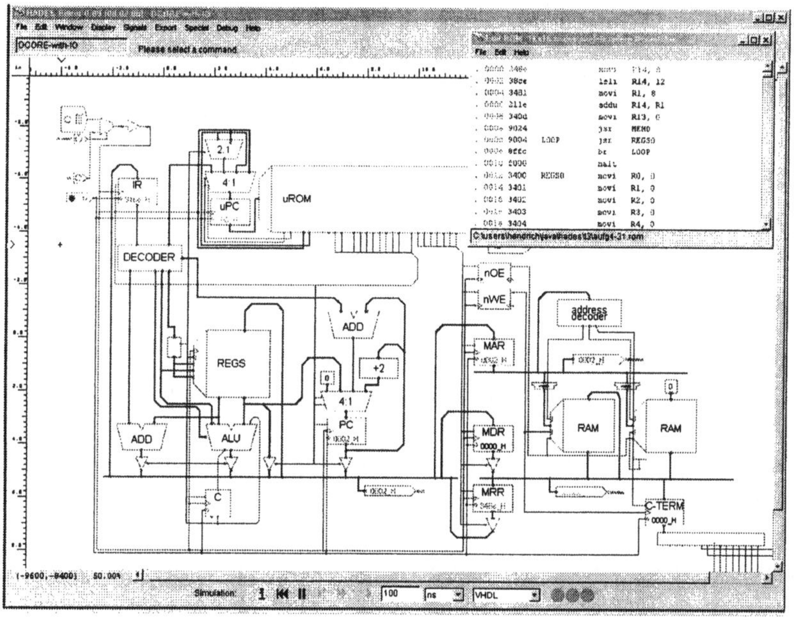


*Figure 2.* A screenshot of the RT-level processor model used in our undergraduate lab-course on computer architecture. The visualization uses both numerical values and colors for immediate visual feedback of circuit activity. All components offer custom editors with their own user interface. The figure shows the RAM/ROM editor with a disassembled program fragment.

## 2. Teaching Computer Architecture

Starting with the winter semester of 1999/2000, we introduced a new undergraduate lab-course [7] accompanying the third-semester lecture on computer architecture and operating systems, which is mandatory for all our CS students. The course relies on Hades simulations and proceeds in a bottom-up manner. Each exercise is realized as an interactive simulation using increasingly complex Hades designs. First, the students study individual processor components with exercises on datapath, microprogrammed control, and memory access. Next, they assemble these components and write their own microcode to implement a complete processor — a simplified but still assembler-compatible version of the new Motorola M•CORE architecture [8]. We chose this processor family because of the typical RISC load/store architecture and its simple programming model with 16 general purpose registers.

Because Hades offers fully interactive simulation, there is no need for lengthy "edit-compile-simulate-analysis" cycles. Therefore, the students can change both their microcode and their assembler programs while the simulation is running. Together with the animation of register and signals values for immediate visual feedback, this greatly helps understanding and allows for efficient debugging. A screenshot of the complete processor used in the lab-course is shown in figure 2.

Once the microcode for all processor instructions is ready, the students proceed to write machine and assembly language programs for their processor. The exercises cover most fundamental assembly techniques from basic machine arithmetic to array accesses and subroutine calls. As neither the architecture nor our assembler offer special support for stack manipulation, the students also have to implement (and understand) stack management. The last exercises cover interrupt processing and basic operating system principles including memory management.

Finally, the students can test their design on our FPGA-prototype board with an Altera 81500 FPGA and separate memories for microcode and main memory. Because the students only supply their microcode for an otherwise fixed hardware, there is no need to recompile the FPGA (which would be slow and might introduce subtle logic errors). Instead, we just download their microcode and their programs for instant debugging.

## 3. Summary

Hades is a Java-based simulation framework targeted at system simulation and HW/SW cosimulation. Due to the interactive simulation mode, visualization aids, component editors with custom user interface, and the portability to many platforms, the system is also ideally suited for teaching. As an example, we have described a new lab-course on computer architecture whose topics range from RT-level components up to operating systems.

## References

1. Hendrich, N., *Hades homepage*, Universität Hamburg, FB Informatik, tech-www.informatik.uni-hamburg.de/applets/hades (1998)
2. IEEE Computer Society, *IEEE Standard VHDL Language Reference Manual*, IEEE Std 1076-1993 (1993)
3. IEEE Computer Society, *IEEE Standard Multivalue Logic System for VHDL Model Interoperability (std_logic_1164)*, IEEE Std 1164-1993 (1993)
4. Sun Microsystems, *The Java Hotspot performance engine*, www.javasoft.com/products/hotspot (1999)
5. Gillespie, G., and Lazzaro, J., *DigLOG, Caltech VLSI CAD Tools*, California Institute of Technology, ftp://ftp.pcmp.caltech.edu/pub/chipmunk (1996)
6. Hugunin, J., *Python and Java: the Best of both worlds*, Proc. 6th International Python Conference, San Jose, CA (1997)
7. Hendrich, N., *Praktikum Technische Informatik T3*, Universität Hamburg, FB Informatik (1999)
8. Motorola, M•CORE reference manual, www.motorola.com/mcore/ (1997)

# DESIGN OF ANALOG INTEGRATED CIRCUITS: FREEWARE PC-BASED CAD FOR STUDENT PRACTICES

*Analog Microelectronics Design Teaching for Electronic Engineers*

FRANCESC SERRA-GRAELLS
*Institut de Microelectrònica de Barcelona. CNM*
*Departament d'Informàtica. Universitat Autònoma de Barcelona*
*08193-Bellaterra. Spain.* *pserra@cnm.es*

NÚRIA BARNIOL
*Departament d'Enginyeria Electrònica.*
*Universitat Autònoma de Barcelona.*
*08193-Bellaterra. Spain.* *nuria.barniol@uab.es*

## 1. Introduction

The aim of this work is to supply a freeware PC-based CAD environment (named WinVLSI) for the design of full-custom VLSI microelectronic circuits, from schematic capture to foundry database, for educational purposes. There are two problems related with CAD laboratories for microelectronic design education: (i) the high cost of the required hardware and the high cost of annual license maintenance for the available microelectronic CAD and (ii) the time consuming training of both the CAD software and the UNIX system. Usually students are more familiar using a PC-environment than a UNIX-environment and this is directly related with the efficiency on time used for doing the practical work. In order to overrule this problem a freeware PC-based CAD environment is being used for the design of analog integrated circuits during two courses for the electronic engineering studies. Additionally with WinVLSI, students are able to do the laboratory design work from their home, this is especially interesting for students who can not assist to established timetable. Finally the PC-based CAD environment will help to spread the microelectronic design skills for recycling engineers who will be able to use that environment from their home.

## 2. Description

WinVLSI [1] is based on two main freeware tools: LASI6 from Dr.David E.Boyce [2,4] and WinSpice3 from Mike Smith [3]. LASI 6.06 includes



*B. Courtois et al. (eds.), Microelectronics Education,* 289–292.

schematic and layout edition, Design Rule Checker (DRC), macroextraction, Electrical Rule Checker (ERC), SPICE netlister and Layout Versus Schematic (LVS), as well as final data codification for foundry submission. WinSpice3 is a Win32 port for Berkeley SPICE3F4, which can perform both electrical simulation and graphical post processing through programming scripts. Current supported version is WinSpice 3.0.7.

Current target technology only for educational purposes is the 2.5μm CMOS double-polySi double-metal process (CNM25) of the Centro Nacional de Microelectrònica (CNM) [5], for which all layer definitions, design rules, extraction information, model parameters and parasitic capacitances have been introduced in the CAD environment.

The resulting CAD system covers all design stages in a down-top approach and supplies the necessary technological information at each step for the final target VLSI process.

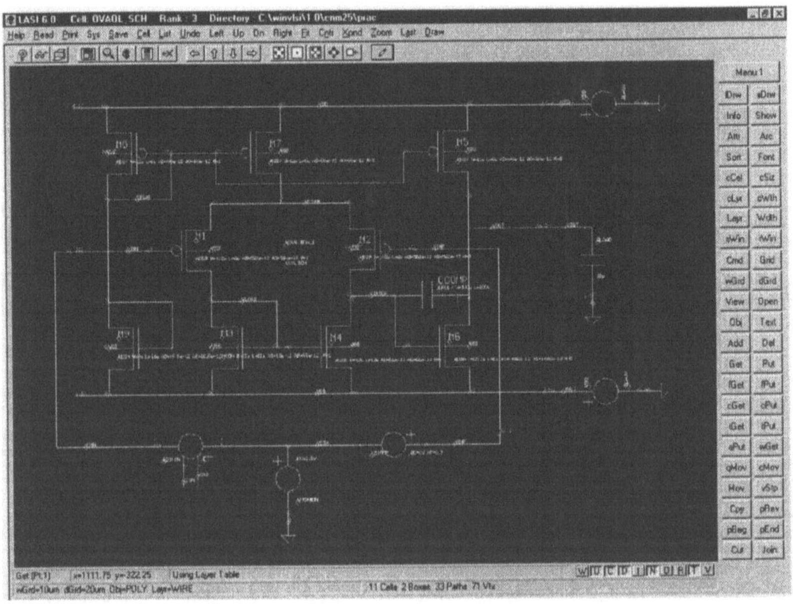

**Figure 1.** Schematics capture sample.

The main advantages of the presented tool are: free distribution, PCWindows compatible, easy of use, illustrative (in case of educational purposes) and not hardware heavy consuming. On the other hand, the following limitations can also be argued compared with the industrial CAD environments: simple DRC techniques, no full device extraction, SPICE-only modelling and SPICE3 syntax limited.

### 3. Results on the practical students work

The WinVLSI CAD is used for two courses on the design of analog integrated circuits for electronic engineers in our university. During these courses, the laboratory work done by the students is focused in two main subjects: (i) on the full-custom design of a simple CMOS operational amplifier with pre defined specifications including optimisation, and (ii) on the design of an analog-to-digital converter and a digital-to-analog converter. The available time for the student to complete the schematic, simulation, layout and post-layout simulation of the OPAMP is only 15 hours of laboratory work. Our experience over the last 4 years said that they are unable to do this work on time with industrial CAD environments such as CADENCE Design Framework. On the other hand, the same OPAMP with the same specifications has been successfully studied and designed through the new WinVLSI CAD.

An example of the work done by students is included in this presentation. In this case, the students were asked to optimise the Gain Bandwidth Product (GBW) of a simple OTA Miller. Figures 1 and 2 depict the schematic capture and the physical layout design respectively using LASI6.06 tool, while Figure 3 shows the WinSpice3 typical appearance for the electrical simulation of the analog integrated circuit designed.

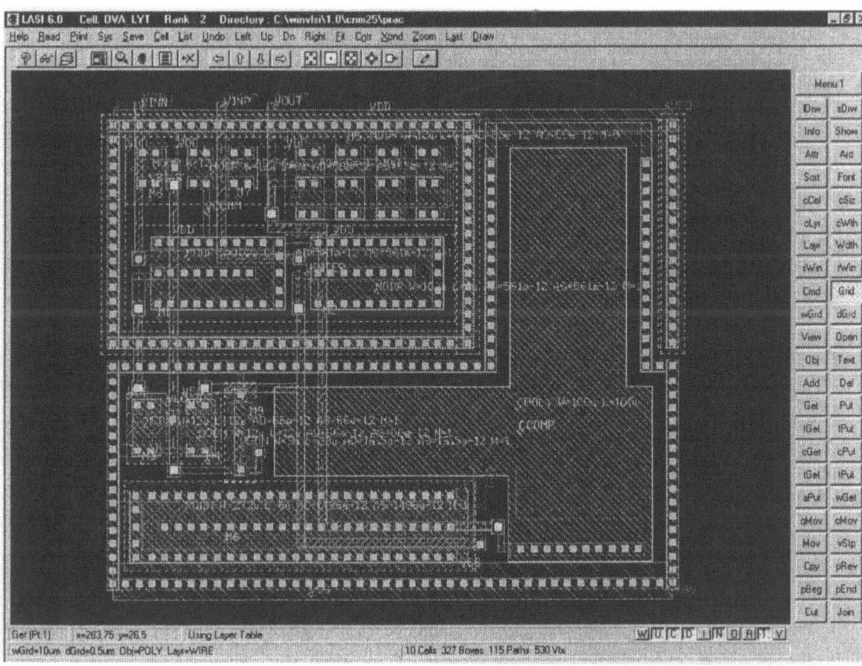

**Figure 2.** Physical layout design example.

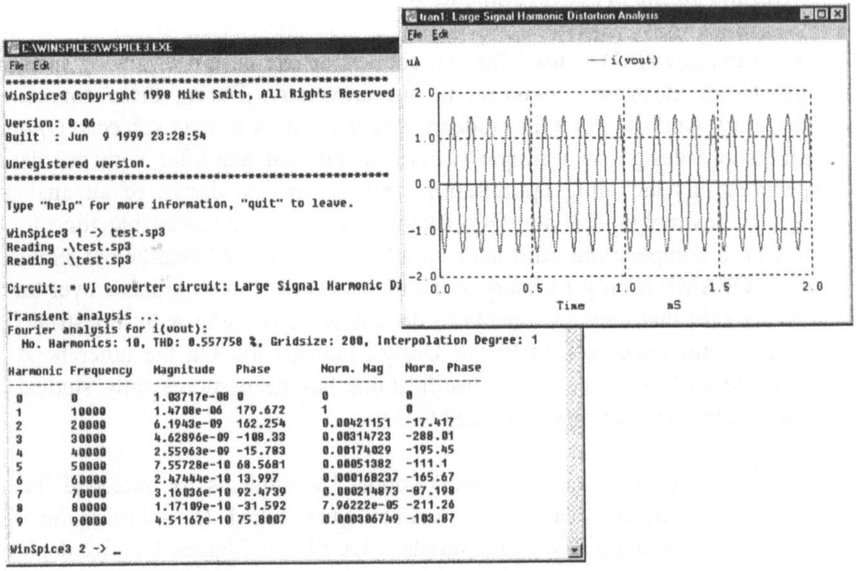


**Figure 3**. Typical pre/post layout simulation window.

## 4. Conclusions

A first experience of this freeware CAD has shown that most students are satisfied with this project since they have freedom to plan their efforts for each design task. In fact, all the qualitative steps in any full-custom industrial CAD are covered here, but simplified at the same time to increase student independence (and responsibility as well). Furthermore, this design kit facilitates both group work and teacher's evaluation through Internet tools such as e-mail.


## Acknowledgements

The authors thank Mike Smith, Dr.David E.Boyce and Xevi Formatjé for their valuable help. Also the authors appreciate the effort of all the students who have been trying out the WinVLSI CAD for the first time.



**References**

1. WinVLSI. http://www.cnm.es/~pserra/winvlsi/winvlsi.htm
2. Lasi6, from David E.Boyce. http : // members.aol.com/lasicad/index.htm
3. WinSpice3 from Mike Smith. http : //www.willingham2.freeserve.co.uk/winspice.html
4. R. J. Baker, Dr. Harry W. Li, D. Boyce. *CMOS: Circuit Design, Layout and Simulation.* IEEE Press, 1998.
5. Instituto de Microelectrónica de Barcelona. Centro Nacional de Microelectrónica. SPAIN. http://si.cnm.es/imb/

# CMP Experience On Deep Sub-Micron Design-Methodology Support and Transfer

Kholdoun TORKI, Bernard COURTOIS
CMP Service, 46 Av. Félix VIALLET, 38031 GRENOBLE, FRANCE
Kholdoun.Torki@imag.fr Bernard.Courtois@imag.fr
http://cmp.imag.fr

**Abstract:** CMP aims at providing Universities, Research Laboratories and Industries with the facility to have their integrated circuits projects fabricated for prototyping and low volume production.
During the last 3 years, CMP introduced the access to new generation advanced technologies processes which are covering a wide spectrum of microelectronics, MEMS, and MCM technologies.
In this paper we will describe and show some examples on how CMP Service succeeded transferring the knowledge of design techniques and methodology, targeting Deep Sub-Micron processes.

## 1 - Introduction

Deep Sub-Micron (DSM) processes has been introduced by CMP in 1997, with the 0.35µ CMOS DLP/4LM from AMS (Austria Mikro Systeme), and the 0.25µ CMOS SLP/6LM from STM (STMicroelectronics), and later on July 1999 with the 0.18um CMOS from STM.

DSM processes with density up to 35 kgates/mm$^2$ (the 0.25µ CMOS from STM) allows the integration of very complex designs with several millions of transistors.
Design complexity from hardware description point of view is manageable, simulation and synthesis are today accessible for such complexity when using professional EDA CAD tools. Place & Route and verifications tools algorithms has been improved by CAD vendors to offer the powerfulness necessary to generate and verify a very complex design of several million transistors. But the only difference that allows a successful working circuit is the use of an up-to-date design methodology tuned for DSM technologies.
It is well-known that DSM processes introduce new constraints on design methodology. If a conventional micronic or sub-micron methodology is used for a DSM design, the fabricated circuits will almost certainly not function as intended. Interconnection is the major actor for success or failure of a DSM design.
For DSM processes (0.25µ CMOS or 0.35µ CMOS) the interconnect delay has about the same value as gate delay. It was expected for instance in the 1997 SIA NTRS that the interconnect delay (for Alu wires) will approximately be fifteen times the gate delay.
This leads special attention to this phenomena that was ignored or less investigated in micronic and sub-micron processes. Design-flows for DSM design become more and more complex. There is a serious need for the microelectronics community to take care of those new issues, especially academic centres, since they are training the future engineers who will exclusively use those technologies in the future.

## 2 - Design flows

In an automatic CAD tool environment, the purpose of automation is to complete a fully specified design with zero errors and with no human intervention, often starting from scratch. The automation is never a complete one, the way the automation is done is controlled and tuned by the designer through a design-flow. A design flow can be viewed as a diagram depicting steps from a CAD tool to another through language interfaces and libraries, targeting circuit layout from specifications.
It can also be seen as the diagram that gives the practice of a design methodology and know-how.
Usually, the foundry and the CAD vendor are in charge of defining the design-flow to give the designer the right way using the CAD tools and libraries.
Thus, design-flows are the most important way to silicon for a designer when CAD tools and processes are available. Design-flows with a methodology give assurance to reach working silicon.

## 3 - Deep Sub-Micron Versus Micronic Design Flows

The use of a technology process from 2µ to 0.8µ CMOS needs to have "traditional" tools that are today very mature and stable allowing the circuit implementation with high accuracy and confidence on the resulted fabrication. The most important steps that a designer has to follow when designing a circuit depends on the application, it can be digital, analog or mixed analog/digital.
The digital micronic design flow is shown in figure 1. The same design-flow for DSM design is shown in figure 2.



*B. Courtois et al. (eds.), Microelectronics Education,* 293–296.

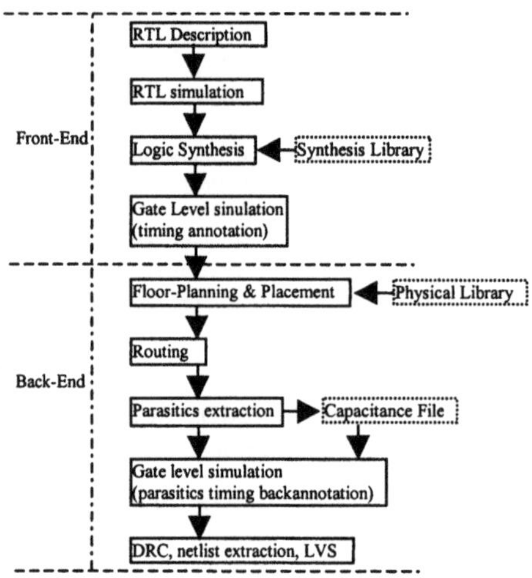


**Figure 1** : "Traditional" Micronic Design Flow

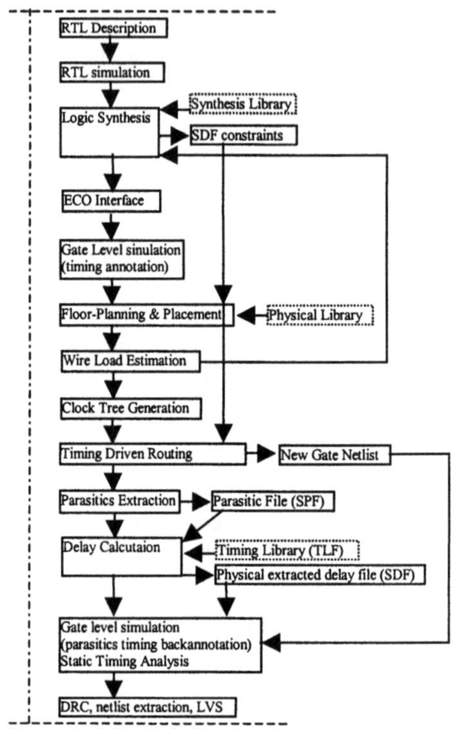


**Figure 2** : DSM Digital Design Flow

DSM processes 0.35μ and below, need more and more sophisticated CAD tools and design-flows in order to have a full functional circuit.
The main steps described in figure 1 are used, but the difference shown in figure 2 is that for a given tool, some of the steps are with additional features and more sophisticated functionalities. Additional tools and steps are also needed in order to have a powerful design in terms of speed, power consumption, signal integrity, and high density designs.
The design flow becomes more and more sophisticated, introducing new concepts :

♦ Taking into account parasitics at a very early stage of the design. At the logic synthesis step, and at the gate level simulation level, by using estimations of parasitics.

♦ Design planning will document and manage the physical architecture of the circuit taking into account the speed/power constraints and the signal distribution (busses, clocks and power supply wiring).

♦ Using place and route tools that have new features like, avoiding "antenna rules", "clock tree generation", "power stripe generation" managing routing from 4 to 6 levels of metals, "Cross-talk avoidance", and more ...
The automatic place and route engine is able to meet the timing constraints produced at the synthesis step (SDF constraint file). After the placement and routing, a report file tells if the constraints have been met. Clock skew and antenna rules checkings are also reported.

♦ Extraction of interconnect parasitics in addition to capacitances with substrate for DSM designs. They are extended to mutual capacitances between wires, and wire resistances. The accuracy of the extraction is a critical issue since the number of wires of a multi-million transistor chip can be very high, and CPU time necessary to make accurate extraction is an issue.
The use of a delay calculator will produce a delay file that annotates all the timings for the individual cell instances.

♦ The use of the extraction analysis, can be a static or a dynamic simulation, signal integrity for critical signals and buses, and power calculations.

For mixed A/D design, the analog design composed of blocks or devices can be validated together with the digital parts using schematic edition to link the two parts. Then, mixed A/D simulator can be used to validate the functionality, timing, and analog performance.
The layout can be done by using automatic place and route tools. The designer need to declare the analog parts as blocks and very often needs to make some manual interventions for having clean connections and wiring around the analog blocks as well as special powering and shielding.

The backannotation can be done by running the mixed A/D simulator engine with the digital and analog parts, both including the parasitics components.

## 4 - Deep Sub-Micron Versus Micronic Design Tools

Digital designs can be specified by an RTL language or by schematic capture. The simulation of gate level description uses the library specifications which contains the gate functions and timing information.

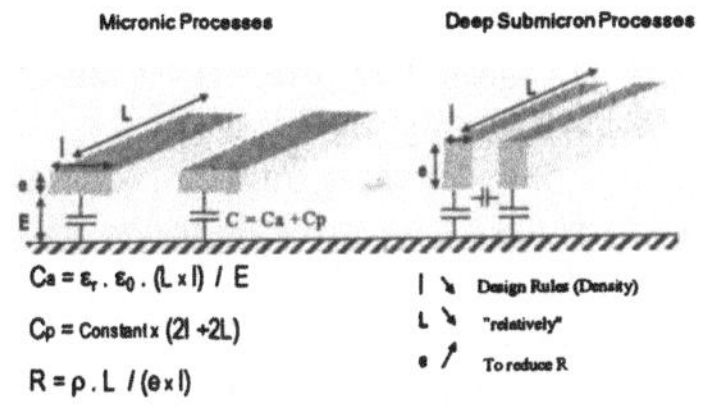


**Figure 3 :** Wire profile in Micronic and DSM processes

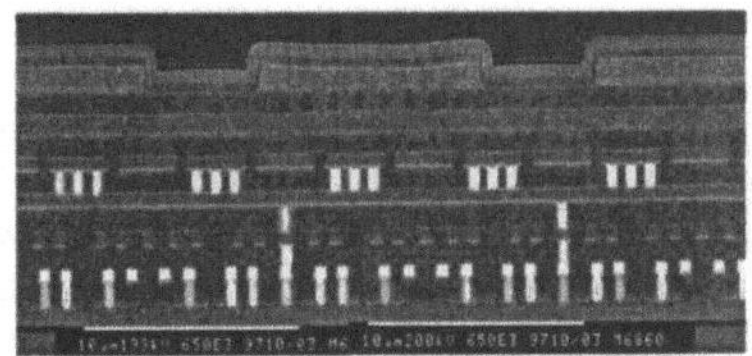

**Figure 4 :** Six levels Metal (0.25µ CMOS)
(Courtesy STMicroelectronics)

Wire load is estimated in the front-end schematic simulation. It takes the real values during the back-annotated simulation, after the layout extraction. Parasitics resistances for micronic designs are not extracted from the layout since they have very weak influence, and are negligible regarding the fanout values. This is not true for DSM designs, and the resistance extraction is becoming more and more necessary.

Layout extraction in micronic processes for capacitances is sufficient in 2D mode, as the wire width is larger than thickness (Figure 3), and wire distance will not generate wiring cross-talk for most of the signals. This is true as for those technologies very frequently the clock frequency is not exceeding some tens of MHz, and rarely reaching the 100MHz.

The 2D extraction mode is not sufficient in a DSM design, as the wire section is becoming more important than width. The clock and signal frequency is surpassing the 100MHz in most of the DSM applications, then signal integrity, cross-talk and clock-skew are becoming critical to analyze precisely for the functionality and the timing in a design.

An experienced designer may reach the design sign-off in micronic process, by making a one-shot flow without having the need to have several iterations for physical design and back-annotation.

This is never the case in DSM design. A designer needs at least 1 or 2 or more iterations between front-end and backend. The information on placement and routing is necessary in order to tune the logic synthesis and to built the design planning.

Analog designs are specified by schematic capture or Spice netlist. The analog simulation is made with Spice-like simulators, implementation of the layout is done either by manual edition, or semi-automatically using parametric cells and components. Spice level 2 or 3 for the device models used for simulation can be sufficient to specify the functionality for micronic processes. In DSM processes, the basic Spice models are not accurate. There are two major model standards that are addressing DSM devices and components :

- MM9 model developed by Philips which is widely used by the European foundries, and is being introduced by the US CAD vendors.
- BSIM model from Berkeley.

The models are enhanced models that are physics based models. Their accuracy and prediction on the MOSFET devices is recognized. Those models are continuously updated introducing more physical behavior and features as the technology is shrinking.

The know-how and experience is needed from the designer in order to secure the design and avoid cross-talk with some special layout techniques including : substrate shielding around the analog and digital parts, separating and decoupling the power supplies, and passing the signals between analog and digital, and vice versa through special interface cells and buffers.

The use of a substrate coupling analysis tool is becoming more and more necessary, as designers are mixing today high frequency digital parts with analog and RF circuitry. The substrate is collecting and transmitting noise which must be analyzed and avoided.

## 5 - CMP Service Support and Technology Transfer

CMP cares about providing microelectronics community with access to the up-to-date technology :

- Universities need to have access to technology for teaching their students. Those students will be in the industry. Therefore they have to be trained at least on the actual state of the art technology processes.
- Research Laboratories need to have high performance technologies to validate new concepts. The quality of the research results depends mostly on the quality of the technologies. Accessing to up to date technologies is a necessity.
- Industrial users also need to access to the state of the art of processes. This is vital for industry, since the development of a product is usually long (more than 1 or 2 years). It is necessary that an industrial user accesses to an up-to-date process, giving guaranty on product life.

The CMP mission is continuing providing support and know-how transfer to the microelectronics community. Here are some examples on the actions made during the 2 last years :

- CMP with Cadence Design Systems organized a 4 days seminar on DSM design in November 1998. The seminar was very informative and gave very useful information to the 15 European users who attended the seminar. A one day training on workstation has been offered showing the use of the state of the art CAD tools for making an automatic place & route in DSM technology.

- CMP is distributing DSM design-kits from the foundry for 0.35um CMOS, 0.25um CMOS and 0.18um CMOS. The use of those design-kit require an extensive experience and know-how which CMP is providing to the hundreds CMP users from Academia and Industry.

- CMP is also helping some research centers achieving their designs. An example in Figure 6, shows the layout of a circuit made by CMP for CEA-LETI-DEIN. The design has been provided as a netlist with the SDF timing constraints, together with the pad-frame diagram. CMP made the automatic place & route and sent back all necessary files and reports to CEA for making the back-annotation. The design has been placed and routed three times, and the third result was successful and fulfilled the specifications. Clock frequency of 200 MHz has been validated, and clock skew was reduced to 0.23ns with the clock tree generation. The circuit has been tested and is functional.

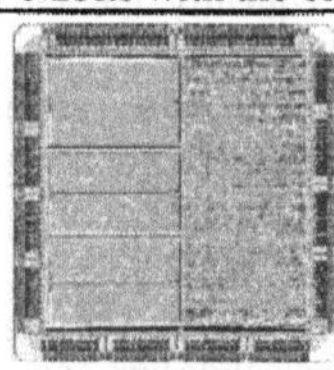

**Figure 5 :** Adaptative Flow Control Circuit For ATM Netwoks, Process : 0.6um CMOS (3 layers metal). Area : 25mm2, clock : 110 MHz, clock skew : 0.25ns.

**Figure 6** : Circuit for Fault-Tolerance Computing Process : 0.25um CMOS 6 layer metals, Area : 5.5 mm2, clock : 200 MHz, clock skew : 0.23ns.

- CMP is hosting PhD students from University of Monastir (Tunisia). They are working on the communication field, and their initial implementation is done on FPGA circuits.

The Region of Rône-Alpes together with the Gouvernorat of Monastir are funding in the frame of the TEMPRA project, the students exchange from the University of Monastir to CMP-TIMA.

There are presently 5 PhD students who are coming periodically to CMP for implementing their applications on an ASIC. Two of the projects has been finalized and are planned to be fabricated. A third one which is a circuit for image compression is planned to be finalized in August 2000.

The figure 5 represents a circuit for "Flow Control in ATM Applications". The technology used was 0.6um, and the design-flow uses advanced design techniques, since the clock frequency has been asked with more than 100MHz, and this has been achieved using the clock tree generation, for which the clock skew was 0.25ns. There are 3 RAMs embedded in the design.

**Conclusion :** CMP Service is continuing offering high performance processes and design-kits with support for the microelectronics community. CMP is maintaining a high quality support and technology transfer to the designers, and provides the up-to-date industrial tools and technologies.

**References :** CMP Annual Reports, CMP Information Bulletins. available upon request, send E-mail to cmp@ imag.fr - Web site: http://cmp.imag.fr

**H. DE MAN** "Education For The Deep Sub-micron Age: Business as Usual?", $34^{th}$ Design Automation Conference, DAC'97, Anaheim, California, USA, 1997.

**K. TORKI, S. SAULNIER** "Accurate Timing modeling and Design-Kit Development For a Sub-micron CMOS Technology", $2^{nd}$ European Workshop Microelectronics Education, EWME'98, The Netherlands, May 1998.

**K. L. SHEPARD, V. NARAYANAN** "Conquering Noise in Deep-sub-micron Digital ICs", Journal IEEE Design & Test of Computers, Jan-March 1998.

**M. ATRI, R. DJEMAL, M. ABID, K. TORKI, R. TOURKI**, "Design of an Adaptative Flow Control Algorithm for ATM Networks", Int'l Conf. On Microelectronics , ICM'98, Monastir, Tunisia, December 1998.

# AN INTRODUCTION TO MICROSYSTEMS FABRICATION

L. MICHELUTTI
*LPCS, ENSERG, Institut National Polytechnique de Grenoble,*
*23 rue des Martyrs, BP 257, 38016 Grenoble Cedex 1, FRANCE*
J.-M. TERROT
*CIME, Institut National Polytechnique de Grenoble,*
*46 avenue Félix Viallet, 38031 Grenoble Cedex 1, FRANCE*
C. SEASSAL
*LEOM, Ecole Centrale de Lyon,*
*36 Avenue Guy de Collongue, BP 163, 69131 Ecully Cedex, FRANCE*

This paper presents an introduction to microsystems fabrication mean for beginners in this field. This teaching practical experience was realised in September 1999 at the Inter-University Microelectronics Centre (CIME, France). The students were engineers, technicians and researchers in the National Scientific Research Centre (CNRS, France). Cantilevers and beams were realised, using bulk and surface micromachining techniques, and characterised.

## I. Introduction

The aim of the practical works presented here was not to realise a complete and operating microsystem. It was rather to show through the realisation of simple structures the main techniques used for microsystems fabrication. Some specific problems of microfabrication were approached such as anisotropy of KOH etching, selectivity in dry and wet etching, masking problems during deep etching, sticking, etc.

## II. Technological realisation

As bulk and surface micromachining are the most commonly used microfabrication techniques for microsystems industrial fabrication, we focused on these techniques and their particularities.



*B. Courtois et al. (eds.), Microelectronics Education,* 297–300.

## II.I. WET BULK MICROMACHINING

The wet etchant chosen was KOH, for its anisotropy [1]. Silicon nitride beams and cantilevers were realised. Two different technological processes were proposed to the students in order to show advantages and drawbacks of anisotropic etchants.

The first process uses silicon wafers oriented in the <100> direction. PECVD Silicon nitride is first deposited to a thickness of 0.8 μm. The nitride is then patterned using standard IC photolithography technology, and wet etching with a HF / $FNH_4$ solution. Anisotropic etching of the silicon is then performed in a 40% weight KOH solution at 60°C (during about 3 hours). Figure 1 shows an example of the as-obtained microstructures.

Physical measurements realised during the process allowed the determination of the etching rate and selectivity vs. nitride of the etchant.

Roughness measurements performed before and after etching show a degradation of the surface quality probably due to $H_2$ bubbles formation [1,2].

Figure 1: Released silicon nitride beam (pointed by the arrow) obtained by KOH etching of <100> silicon substrate.

The second process used <111> silicon wafers. After nitride patterning, the silicon was etched in a RIE reactor using $SF_6$ plasma in order to form a 2 μm cavity. The wafers were then subjected to a KOH etching. A cross section of the structure is presented in figure 2.

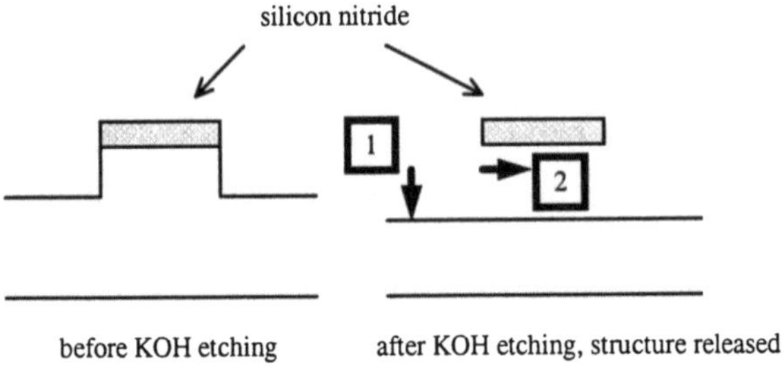


Figure 2: the release of the structure during the KOH etching is archievable thanks to a higher rate etching along lateral direction (rate 2 > rate 1).

The movable structures are in this process closer to the substrate than in the first process (gap of about 3 μm compared with tens of μm), and sticking can occur. Systems of cantilevers with varying length can show the length dependence of this phenomena [3].

## II.II. SURFACE MICROMACHINING

In surface micromachining, the used structural material was polycrystalline silicon, and the sacrificial layer was silicon oxide. A thick wet oxide layer was grown on silicon wafers (thickness of about 1 µm). A 1 µm thick film of polycrystalline silicon was then deposited by LPCVD, and phosphorus doped by thermal diffusion. The polysilicon film was patterned and the structures were finally released by removing the sacrificial layer in a HF / $FNH_4$ etchant at 30°C during approximately 5 hours. Figure 3 shows a cantilever with a platform.

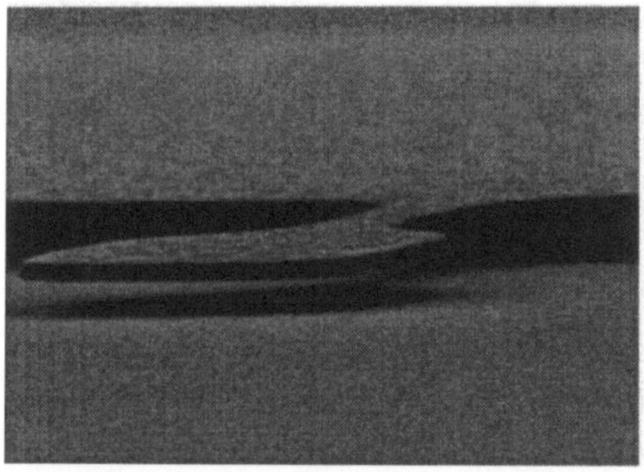

Figure 3: a short cantilever with a platform realised by surface micromachining

The doping of the polysilicon, and the short distance between suspended structures and substrate allowed an electrostatic actuation of the cantilevers (see section III).

## III. Characterisation of the structures

All the structures realised were observed in optical and Secondary electrons microscopies. Mechanical deflection of silicon nitride cantilevers was observed with a mechanical profilometer. Figure 4 shows such a deflection profile for a nitride cantilever of 20 µm wide.

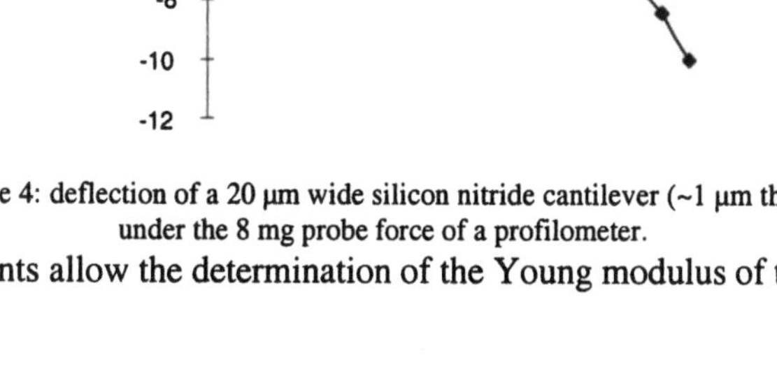


Figure 4: deflection of a 20 µm wide silicon nitride cantilever (~1 µm thick) under the 8 mg probe force of a profilometer.

These measurements allow the determination of the Young modulus of the material.

Electrostatic actuation of the cantilevers realised by surface micromachining was realised under a differential voltage (30 V) at a frequency of a few Hz. The voltage was applied using electrical probes in contact with the substrate and the cantilever embedding point, actuation was observed with optical microscopy. Figure 5 shows the experimental set-up.

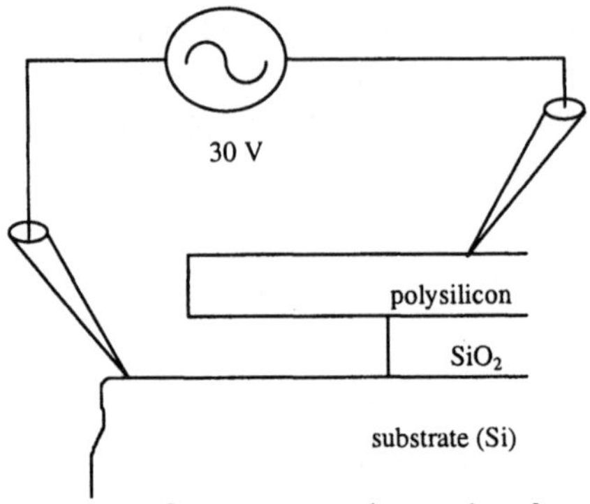


Figure 5: experimental set-up for the electrostatic actuation of suspended structures realised by surface micromachining

## IV. Conclusions

A simple practical work was presented to introduce microfabrication techniques to beginners. Both bulk and surface micromachining technologies were used. Attention was brought to the different problems one has to face when designing a microsystem.
A possible extension to that work should be the realisation of a complete (but simple) microsensor like a piezoresistive pressure sensor.

## V. Acknowledgement

The authors would like to thank all the students for their participation, and particularly T. Crozet (CRTBT, Grenoble, France) for the SEM figures elaboration.

## VI. References

[1] Madou, M. (1997) *Fundamentals of microfabrication*, CRC Press, New York.
[2] Merveille, C. and Weber, J. (1996) Surface quality of {111} side-walls in KOH-etched cavities, *Eurosensors X, Leuven, Belgium, 8-11 Sept.*, pp. 485-488
[3] Kim, C.-J., Kim, J. Y. and Sridharan B. (1998) Comparative evaluation of drying techniques for surface micromachining, *Sensors and Actuators A* 64, 17-26.

# EXPERIMENTAL AND ANALYTICAL METHODS FOR ESD PHENOMENA AND THEIR FAILURE ANALYSIS IN CMOS IC'S

V. Lange [(1)] and G. Higelin [(2)]
*[(1)] Mikrosystemtechnik, FH-Furtwangen, D-78120 Furtwangen, Germany, [(2)] School of Applied Science, Nanyang Technological University, Singapore 639798*

## 0. Introduction

Electrostatic discharge (ESD) can be generally described as an event of high current (Amperes) in a short time interval (Nanoseconds) with the result of damaged IC's. Consequently, ESD is a key problem in the reliability of IC's and as a result of downscaling from VLSI to ULSI, the sensitivity of IC's on ESD is dramatically increased. Teaching of ESD phenomena is a very challenging task. During lectures and Lab experiments of ESD problems we have to focus on theory, experimental setups, analytical methods and simulation tools. The extent of the project lab is 4 hours per week over a half semester. The flow-chart of the project lab is given in figure 1. This paper will now follow the same flow-chart.

| |
|---|
| **1. Prelab**<br>**ESD - models, ESD - protection, ESD - damage** |
| **2. Theoretical Analysis**<br>**2.1) ESD - circuit**<br>**2.2) Device simulation** |
| **3. Experiments**<br>**3.1) ESD - pùlse analysis**<br>**3.2) Chip preparation**<br>**Chip stress**<br>**Failure detection**<br>**3.3) Failure analysis** |
| **4. Postlab**<br>**Report and presentation** |

Figure 1: Flow-chart of the ESD project lab.



*B. Courtois et al. (eds.), Microelectronics Education,* 301–304.

## 1. Prelab

The intention of the prelab which is organized as a homework is to give an introduction into the physical background of ESD, models, protection circuits and damage mechanisms. For the description of ESD, several models exist, depending on the cause of the ESD event: human body model (HBM), machine model (MM), charge device model (CDM). ESD protection consists in active elements draining the ESD pulse to Vss or Vdd. These elements consist in pairs of diodes, transistors or thyristors connected between gate and source of the input transistor under consideration. ESD events result in damages, where the most common are melted metal, holes at the Si-$SiO_2$ interface, filamentation in diffused regions and $SiO_2$ breakdown. Comprehensive knowledge on ESD can be found in [1] and [2]. Methods to investigate ESD are standardized for the HBM [3], which has become industrial standard.

## 2. Theoretical background

To describe and understand ESD phenomena, the ESD model and the device under test (DUT) has to be described in terms of electronic circuit design which can be analysed by applying various methods of simulation.

### 2.1 Circuit analysis

The aim of this part is to apply standard methods for the analysis of electronic circuits. These are theoretical methods, symbolic circuit analysis tools and numerical methods. The ESD discharge model is described by the basic circuit given in figure 2.

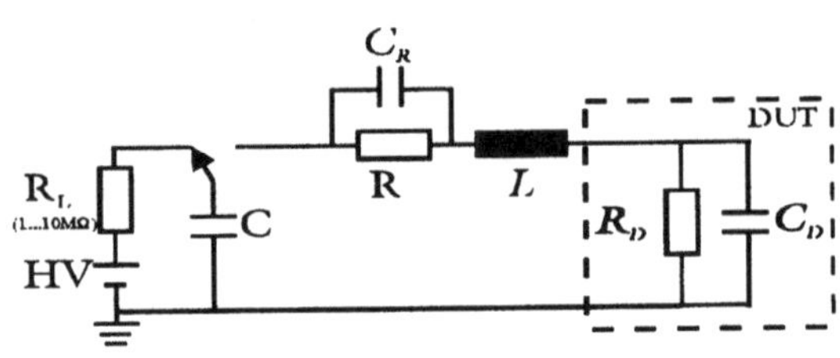


Figure 2: ESD circuit

Elementary circuit analysis with the solution of differential equations is applied on the basic RC- circuit, which is characterized by the time constant of the discharging process. The introduction of a parasitic inductance $L$ results in a pulse rise time, mainly caused by $L$. Real and imaginary solutions lead to the HBM and MM respectively. They show typical characteristics of a pulse in case of the HBM and a damped sinusoidal for the MM. In reality, further elements have to be introduced: ohmic resistance ($R_D$) of the device under test and parasitic capacitances $C_R$ and $C_D$.

The exact analytical solution of this real circuit is very complex, even with use of symbolic circuit analysis tools (Syrup package under Maple [4]). Parallel to the development of the general solution , the use of numerical circuit analysis tools like Spice [5] will be undertaken. There, the influence of parasitic elements ($L$, $C_R$ and $C_D$) on the ESD pulse shape can easily be observed.

### 2.2 Device simulation

The action of the ESD pulse on protection circuits will be examined with the physical 2D device simulator Medici [6]. Because the developement of devices under Medici requires a lot experience, only simple protection diodes are developed. The rest of the circuit is included in a Spice like circuit file. Current flow and the thermal development can well be studied with Medici. The dependance of the maximum temperature in the protection device as a function of the ESD voltage can be examined, as well as precautions to avoid extreme heating. As Medici simulations have a long runtime (hours CPU time), this lab has to be done in parallel to other jobs.

## 3. Experiments on ESD

### 3.1 ESD pulse analysis

In the first experimental task, measurements of HBM and MM pulses have to be done. Our ESD system (Verifier Systems) makes HBM and MM pulses possible. Pulse measurements will be performed with a current probe (Tektronix CT1) and a digitizing oszilloscope. Using Hspice [7], parasitic elements are determined by fitting simulated to measured data.

### 3.2 ESD stress of standard CMOS IC's

Next, commercial CMOS logic IC's (4011, 4069, ...) are revers engineered. Fuming nitric acid is used to open plastic packages without destroying the IC and keeping the full functionality [8]. The surface of the chip is now well observable. After this preparation, the chip is stressed with respect to MILSTD 883C [3]. The logic functionality, power consumption and the IV - characteristic of the protection circuit are controlled periodically until failure is observed.

### 3.3 Failure analysis

Depending on the damage of the IC, a variety of methods for the failure analysis are available. Otical microscopy makes defects on the uppermost layer visible (fig. 3).

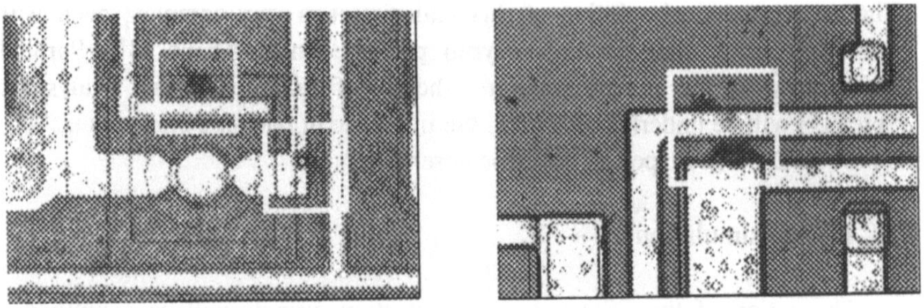

Figure 3: Damaged protection diode (left) and flashover at the gateedge (right).

The location of defects in buried layers of the chip can be performed in our lab with liquid crystal thermography [9] and with potential contrast in the scanning electron microscope (SEM) [10]. In the near future, we will introduce deprocessing methods (layer sensitive etching procedures) to make damages in buried layers visible.

## 4. Postlab

Each student has to summarize the project in a written report and to give a talk to his classmates, taking part in our project lab on semiconductor characterization.

## 5. Summary

In the above described ESD project, we introduce our students to a subject of high actuality in microelectronics engineering. The application of industrial standard tools and methods (Spice, symbolic circuit analysis, Medici, hot spot detection, SEM,...) is introduced to study a problem of high complexity.

## 6. References


[1] McAteer O.J. Electrostatic Discharge Control. McGraw-Hill (1989)

[2] Vinson J.E., and Liou J.J., Electrostatic discharge in semiconductor devices: An overview. Proceedings of the IEEE 86 (1998) 399-418

[3] Military Standard 883C method 3015.6 notice 7, (1987)

[4] Maple V R5, Waterloo Maple Software, Univ. of Waterloo (1981-1990); Syrup, Riel J., http://www.k-online.com/~joer, (1998)

[5] Spice 3 V3f3, University of California, Berkeley, (1993)

[6] Medici V4.0, Technology Associates Inc., (1997); Rollins G., TMA, Modelling ESD Phenomena, (1994)

[7] Hspice98.2, Avant! Corporation, (1998)

[8] Beck F., Präparationstechniken für die Fehleranalyse an integrierten Halbleiterschaltungen, Verlag VCH, (1988), 8-16

[9] Beck F., Use of Liquid-Crystal Thermography for Defect Location on Semiconductor Devices. Quality and Reliability Engineering International 2 (1986) 143-151

[10] Gopinath A. et al., Voltage contrast: a review. SEM 1 (1978) 375-380

# ELECTRONICS MANUFACTURING LEARNING FACTORY

HEIKKI VALMU, ARI LEHTO and KARI SALMI

*Espoo-Vantaa Institute of Technology (EVITECH)*

*Leiritie 1, FIN-01600 VANTAA*

Key words: Electronics, manufacturing, learning factory

## 1. Introduction

Espoo-Vantaa Institute of Technology (www.evitech.fi) is a polytechnical university located at the immediate vicinity of Helsinki.

The total number of students is 2500 and the annual intake of students for the electronics programmes is about 50, out of which about 20 take the so called "production oriented" programme. The students in the latter programme will extensively use the learning factory during the final year of their studies. The electronics programme at EVITECH contains for example the following courses: ASIC design, electronics manufacturing and silicon technology.

Electronics industry (including information technology) in Finland is growing very rapidly. It has already surpassed the traditionally strong paper and pulp and heavy metal industries. As a consequence there is a shortage of skilled people for the various tasks in this field. The educational institutes have increased the number of students, but the response has been limited by the number of teachers and other resources. Training the existing company personnel has been executed by the companies themselves, because a proper environment for this purpose has been lacking.

To alleviate the situation EVITECH and the city of Vantaa have undertaken the planning task for an electronics manufacturing learning factory, where both students and company personnel will be educated.

## 2. The learning factory

Special emphasis will be put on the following areas: Automated surface mount technology, microsystems and adaptive materials, precision mechanics and clean room technology.

### 2.1 SMT

The learning factory will be equipped with a professional SMD (Surface Mount Device) line, which will be used both for small scale production and training



*B. Courtois et al. (eds.), Microelectronics Education,* 305–307.

company personnel as well as new manufacturing engineers. The reason for the professional SMD-line at the learning factory is that the real production lines operate in two to four shifts and it is very difficult to use these lines for training without interfering with the production.

## 2.2 MICROSYSTEMS AND ADAPTIVE MATERIALS

There is already a relatively strong silicon cluster in the area including companies and research institutes. Okmetic Oy is manufacturer of silicon wafers especially for microsystems technologies, MAS Oy is mass producer of integrated circuits, VTI Hamlin Oy is world leader in low-g capacitive silicon accelerometers for the car industry and others, Vaisala Oyj is a well known manufacturer of sophisticated equipment for environmental and atmospheric measurements. Suunto Oyj is known for excellent diving and orienteering equipment. Microsystems and sensor technologies research and development is carried out both at the companies and VTT Electronics and the Helsinki University of Technology. VTT is the state technical research centre with about 3000 employees having excellent facilities for the research and development in the field microsystems technologies. Helsinki University of Technology is carrying out basic research in this field at their Center of Microelectronics.

Adaptive materials change shape under the influence of external excitation. Known phenomena include piezoelectricity and magneto-striction. The new phenomenon is the so called magnetic shape memory (MSM) material, which changes shape under the influence of magnetic field. The change can be very large as compared to the previously known materials and phenomena. AdaptaMat Oy is a new enterprise specialising in the development and production of magnetic shape memory materials. The electronics manufacturing learning factory will help find and develop applications for these materials in co-operation with AdaptaMat Oy.

## 2.3 PRECISION MECHANICS

Precision mechanics is an integral part of microsystems technologies as concerns the packaging and assembly of microcomponents. The learning factory will focus in on automated microassembly and packaging.

## 2.4 CLEAN ROOM TECHNOLOGY

Controlled environments for manufacturing, research and development are needed in many industries, not only in semiconductor industry. It has therefore deemed expedient to build a clean room with several cleanliness categories in the factory for educational and production purposes. Education will include clean room culture (i.e. how to be dressed and how to behave in the clean room etc.), clean room measurements (air and water), pure water technology and clean air technology.

## 2.5 CENTER OF EXCELLENCE

Adaptive materials and microsystems Center of Excellence will be located in the premises of the learning factory.

## 2.6 SMALL SCALE PRODUCTION

As the equipment in the learning factory will be professional, it is well suited for low volume production and making prototypes. The equipment will be leased to the companies for the period of time necessary.

# 3. Co-operation

International co-operation in the field of microsystems technologies and adaptive materials has already been established with the Fachhochschule Wiener Neustadt (Austria), Fachhochschule Kaiserslautern (Germany), Hogeschool van Utrecht (The Netherlands) and University of Nancy (CNRS, France). National co-operation network includes Helsinki University of Technology, Tampere University of Technology, Centre for Technical Training and vocational schools.

The research and development group will also actively offer services to CERN and EU-projects.

Several companies will establish new R&D activities in the learning factory.

Joint courses and utilisation of the factory with the upper secondary technological education is an important part of the co-operation. Upper secondary education includes vocational schools and scientifically oriented high schools.

# PHYSICS OF SEMICONDUCTOR DEVICES : A 2D SIMULATION TRAINING COURSE

C. MANEUX, P. CAZENAVE, P. FOUILLAT, N. SAYSSET,
V. POUGET, H. LAPUYADE, A. TOUBOUL
*IXL, UMR 5818, ENSERB, University Bordeaux 1*
*354, Cours de la liberation, 33400 Talence, France*

Key words: Physics of semiconductor device, 2D physical simulation, work station

Abstract: An example of device 2D simulation : the BJT forward operation.

## 1. Introduction

Bordeaux 1 University which sets, in 1978, the first engineering degree in Microelectronics named "DESS microélectronique", performs student formation in physics of semiconductor devices. Unlike the other electronic matters, physics of semiconductor devices is hardly a subject for conventional training. The exploitation of the C(V) measurements from a MOS or Schottky structure, of the resistivity characterisation, of the Hall effect or of the Haynes-Shockley experiment are the only realistic practical approaches that can be performed. To fill in this gap, a set of 2D physical simulation training is implemented on work station since 4 years.

## 2. 2D Simulation Tool and Pedagogic Objective

The 2D device physical simulation is practised by students from "licence EEA" to "DESS microélectronique" equivalent to the master degree. The main objective is to analyse numerical simulation results in order to verify that the simplified hypothesis made during lecture and leading to 1D analytical formulations can be justified. A 2D simulation of device requires two types of input information



*B. Courtois et al. (eds.), Microelectronics Education*, 309–312.

- a text file that contains commands to solve transport mechanism equations.
- a structure file defining the device description and the mesh definition.

The device simulator produces two types of outputs:

- The electrical output information (current, bias, frequency).
- The solution files store two-dimensional data related to the equation root (electrical field, carriers concentration, energy band diagram) within the device.

**A 3-Hour Training Course** - The training courses based on 2 D simulation proposed to "licence EEA" students are restricted to the solution file analysis. Among the five three-hour sessions of practical work of the training set, the BJT study is one of the most significant due to the large number of transport mechanisms involved in its operation. This example presents the repartition of physical variables within the BJT biased in forward mode. Each 4 main steps required 30-minutes results observation during which students focus on the analysis of a major physical effect (figure 1, 2, 3, 4). A pre-print report is given to the students, with blank fields to fill, according to the topics previously presented. The two first hours are open for questions. During the last hour, the report has to be fulfilled and an individual evaluation is discussed at the beginning of the next session.

| | |
|---|---|
| **Step 1**<br>The device structure | - Evaluation of device dimensions and justification of half structure,<br>- Location of ohmic contact,<br>- Evaluation of type and profile of doping concentration. |
| **Step 2**<br>Energy band diagram | - Analysis of the influence of forward bias,<br>- Evaluation of potential barrier height,<br>- Comparison with: $V_\Phi=U_T\ln(N_AN_D/ni^2)$ |
| **Step 3**<br>Electrical field variation | - Evaluation of space charge region widths and of space charge region maximum field value.<br>- Comparison with: $W=(2\varepsilon_{Si}V\phi/q\ N_{A,D})^{1/2}$ and $E_{max}= V_\Phi/2W$ |
| **Step 4**<br>Electrical curves | - Analysis of output characteristics.<br>- Conclusion : Impact of access region and Early effect. |

## 3. Conclusion

For several years, Bordeaux 1 University teaches physics of semiconductor device to students from the bachelor level to the engineering degree. Today, 2D simulation has demonstrated the benefit of computer-assisted learning for this education. The tool is complementary to the lecture since it allows to:

- display electrical variable evolution within a device for several bias conditions,
- quantify the impact of electronic properties on the device electrical curves.

Since 4 years, an increasing interest for this approach is clearly observed. Although the number of student in microelectronics is decreasing over the world, the decrease in Bordeaux 1 is still lower when compared with other academic institute.

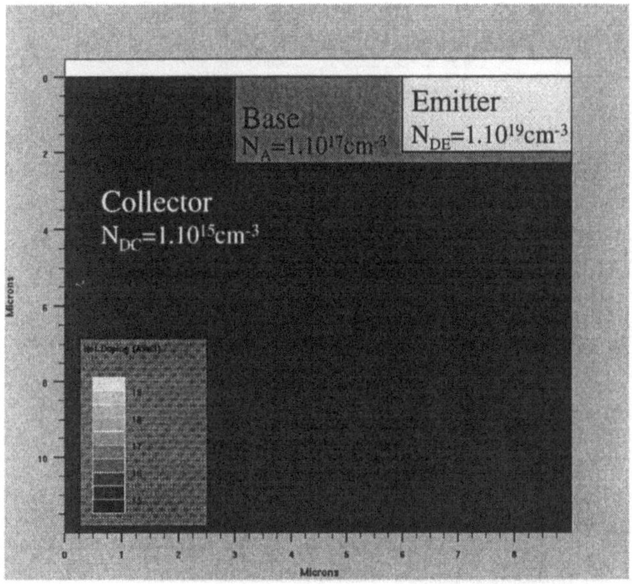


*Figure 1.* The device structure, the third dimension is assume to be equivalent to 1μm. This simplest structure can be complete with a buried layer for example.

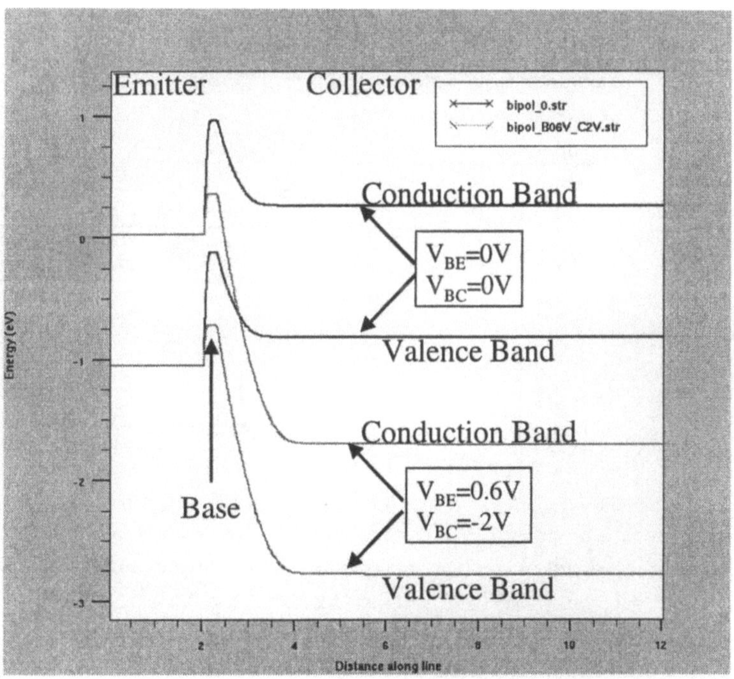


*Figure 2.* Energy band diagram in intrinsic transistor where the energy band gap is presented

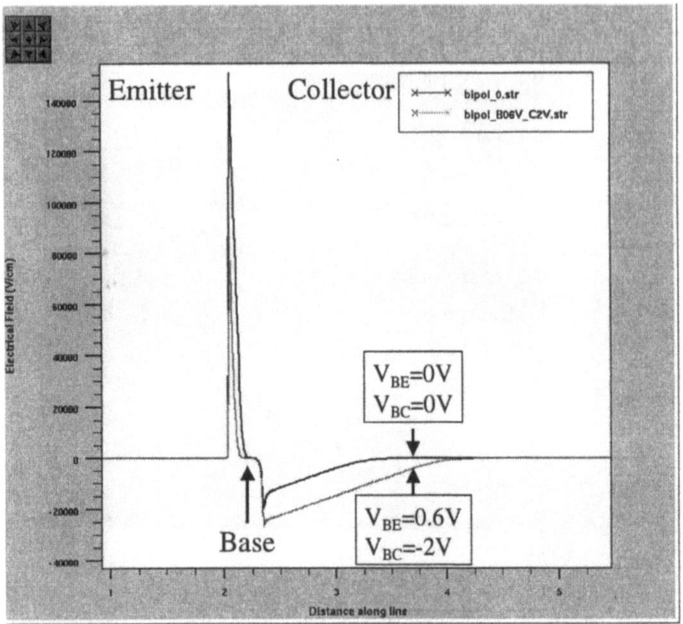


*Figure 3.* Electrical field variation confirming the validity of the depleted region hypothesis.

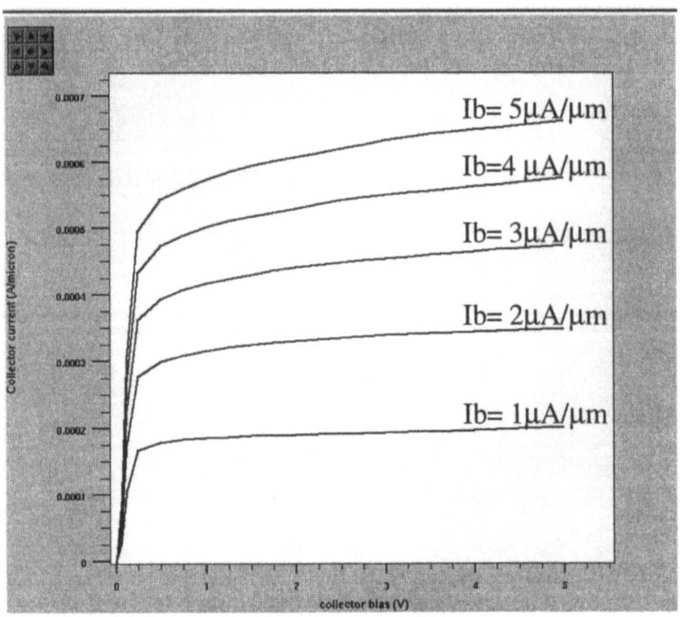


*Figure 4.* Electrical characteristics showing the output curves collector current versus emitter-to-collector bias as a function of base current injection level.

# Index

MIX
Papier aus verantwortungsvollen Quellen
Paper from responsible sources
FSC® C105338

If you have any concerns about our products, you can contact us on
**ProductSafety@springernature.com**

In case Publisher is established outside the EU, the EU authorized representative is:
**Springer Nature Customer Service Center GmbH**
**Europaplatz 3, 69115 Heidelberg, Germany**

Printed by Libri Plureos GmbH
in Hamburg, Germany